AF676341

Integrated Energy Systems towards Carbon Neutrality

Integrated Energy Systems towards Carbon Neutrality

Editors

Pei Liu

Ming Liu

Xiao Wu

MDPI • Basel • Beijing • Wuhan • Barcelona • Belgrade • Manchester • Tokyo • Cluj • Tian

Editors

Pei Liu
Department of Energy and Power Engineering
Tsinghua University
Beijing
China

Ming Liu
Energy and Power Engineering
Xi'an Jiaotong University
Xi'an
China

Xiao Wu
School of Energy and Environment
Southeast University
Nanjing
China

Editorial Office
MDPI
St. Alban-Anlage 66
4052 Basel, Switzerland

This is a reprint of articles from the Special Issue published online in the open access journal *Processes* (ISSN 2227-9717) (available at: www.mdpi.com/journal/processes/special_issues/Integrated_Energy_System).

ion purposes, cite each article independently as indicated on the article page online and as
below:

.; LastName, B.B.; LastName, C.C. Article Title. *Journal Name* **Year**, *Volume Number*,

Hbk)
F)

in

ok are Open Access and distributed under the Creative
ch allows users to download, copy and build upon
lisher are properly credited, which ensures maximum
ions.
under the terms and conditions of the Creative Commons

Contents

About the Editors

Pei Liu

Dr. Pei Liu is an Associate Professor at Department of Energy and Power Engineering. He obtained his PhD in Process Systems Engineering from Imperial College London in 2009. His research interest lies in energy systems engineering, where he uses methodologies of process systems engineering to model, simulate, and optimize the design and operation of complex energy systems. He has developed a series of methods and tools for optimal design of polygeneration energy systems, distributed energy systems, energy systems in buildings, hydrogen infrastructure planning, optimal design of pipelines network for carbon dioxide transport at supercritical state, power sector planning with constraints of carbon dioxide emissions, and coal supply chain planning with constraints of water resources. He has authored/co-authored more than 200 papers in international peer-reviewed journals and conferences, with more than 3000 citations. He has co-authored two books on energy systems engineering published by Springer, and more than ten book chapter contributions published by Springer, Elsevier, and others. He is a subject editor of the journal *Chemical Engineering Research and Design*, and an editorial board member of the journals *Computers and Chemical Engineering*, *Processes*, *Frontiers in Chemical Engineering*, and others. He leads projects funded by the National Science Foundation of China, the Ministry of Science and Technology, the European Commission, Energy Foundation, and several projects from the industry, including BP, Siemens, and Toyota.

Ming Liu

Prof. Ming Liu received his PhD degree in Power Engineering and Engineering Thermo-Physics from Xi'an Jiaotong University in 2013. His research interests are the optimization and control of energy power systems, including thermal system optimization, load cycling and frequency regulation control strategy of thermal power plants, energy saving, and heat and power decoupling of combined heat and power plants. He has authored over 130 peer-reviewed papers and 70 patents, and some technologies have been applied.

Xiao Wu

Xiao Wu is an associate professor and doctoral supervisor at School of Energy and Environment, Southeast University. He received his B.S. degree and Ph.D. degree in Energy information and Automation in Southeast University, Nanjing, China, in 2008 and 2014. Since 2019, he has been a visiting professor in Department of Chemical and Biological Engineering in the University of Sheffield, UK.

Prof. Wu has been an investigator in over 20 research grants with funding from European Union, Royal Society, NSFC, MOST and Industry. He has published over 100 peer-reviewed papers including Applied Energy, Energy, Fuel, IEEE Trans. Energy Conversion. He has been awarded by the Jiangsu Outstanding Youth Foundation, Royal Society-Sino-British Trust International Fellow, Excellent young scholar of Jiangsu Engineering Thermophysics Society, Young Science and Technology Talents of Jiangsu Province in the last three years. His main research area is in Process Systems Engineering for Energy and Environment, including Process Modelling, Simulation, Control and Optimization, Big Data and Artificial Intelligence (AI), Power PlantsCarbon Capture, Utilization and Storage (CCUS) and Integrated Energy System (IES).

Preface to "Integrated Energy Systems towards Carbon Neutrality"

Energy systems have played an essential role in the history of human civilization. As our civilization evolves, energy systems are expected to adapt to the environment and desire of people for more sustainable development whilst meeting the ever-increasing energy demand of society. To address global warming and its threats to sustainable development to multiple ends, major economies around the world have announced low-carbon, carbon-neutral, or negative-carbon development targets. To meet these goals, the energy systems as we know them today need to undergo substantial structural changes in terms of the way primary energy is extracted from nature, converted to secondary energy, transmitted from conversion sites to end use, and shifted between time slots to coordinate supply and demand. The share of renewable and fossil energy in the overall energy portfolio could experience unprecedented structural change of a kind not witnessed since industrialization. To cope with this harsh transition, energy systems should be planned, designed, retrofitted, and operated in a revolutionary manner.

This reprint aims to present the most recent advances in energy systems analysis towards low/zero/negative carbon emission targets via integration amongst different primary energy supplies, between multiple energy supplies and demands, across geographically separated regions, and over different time scales from seconds to seasons. We cover topics including:

- The theoretical development of integrated energy systems and/or applications;

- Integrated energy systems modeling, simulation, optimization, and case studies;

- Dynamic behavior studies of integrated energy systems and dynamic modeling, simulation, and control;

- Novel energy extraction, conversion, transmission, storage, onsite generation/dispensing technologies that enable integrated energy systems.

These studies present the most recent theoretical and technical approaches in low-carbon energy systems research, with potential applications in various circumstances. We hope these works collected in this reprint can contribute to a transition from our current energy systems to low-carbon ones in the future in a more energy-efficient, cost-effective, and realistic way.

We would like to thank all the authors who contributed to this reprint, the Editor-in-Chief of *Processes*, Michael A. Henson, for their enthusiastic support of the reprint, and the editorial staff of *Processes* for their efforts.

Pei Liu, Ming Liu, and Xiao Wu
Editors

processes

MDPI

Editorial

Special Issue on "Integrated Energy Systems towards Carbon Neutrality"

Pei Liu [1,*], Ming Liu [2] and Xiao Wu [3]

1 State Key Lab of Power Systems, Department of Energy and Power Engineering, Tsinghua University, Beijing 100084, China
2 State Key Laboratory of Multiphase Flow in Power Engineering, School of Energy and Power Engineering, Xi'an Jiaotong University, Xi'an 710049, China
3 National Engineering Research Center of Power Generation Control and Safety, School of Energy and Environment, Southeast University, Nanjing 210096, China
* Correspondence: liu_pei@tsinghua.edu.cn

Energy systems have played an essential role in the history of human civilization. As our civilization evolves, energy systems are expected to adapt to the environment and fulfill people's desire for more sustainable development whilst meeting the ever-increasing energy demands of society. To address global warming and its threats to sustainable development on multiple fronts, major economies around the world have announced low-carbon, carbon-neutral, or negative-carbon development targets. To meet these goals, energy systems as we know them today need to undergo substantial structural changes in terms of the way primary energy is extracted from nature, converted to secondary energy, transmitted from conversion sites to end use, and shifted between time slots to coordinate supply and demand. The share of renewable and fossil energy in the overall energy portfolio could experience unprecedented structural change of a kind not witnessed since industrialization. To cope with this harsh transition, energy systems should be planned, designed, retrofitted, and operated in a revolutionary manner.

This Special Issue, "Integrated Energy Systems toward Carbon Neutrality" (https://www.mdpi.com/journal/processes/special_issues/Integrated_Energy_System, accessed on 1 November 2022), aims to present the most recent advances in energy systems analysis towards low-, zero-, or negative-carbon emission targets via integration amongst different primary energy supplies, between multiple energy supplies and demands, across geographically separated regions, and over different time scales from seconds to seasons. We cover topics including:

- The theoretical development of integrated energy systems and/or their applications;
- Integrated energy systems modeling, simulation, optimization, and case studies;
- Dynamic behavior studies of integrated energy systems and dynamic modeling, simulation, and control;
- Novel energy extraction, conversion, transmission, storage, and onsite generation/dispensing technologies that enable integrated energy systems

Typical topics and specific applications of energy systems engineering in various circumstances are summarized as follows.

The energy systems in the future may exhibit rather different structural styles. Shen et al. [1] present methods for improving the thermoeconomic energy efficiency of integrated energy systems. Feng, Wang, and colleagues [2] present an optimization approach to pipeline networks in an integrated energy system with multiple heat sources. Wang et al. [3] analyzed the dynamic behavior of yet another promising type of energy system—the supercritical carbon dioxide power cycle. Dai and colleagues [4] further analyzed the control strategy of a supercritical carbon dioxide power cycle. Fuel cells are expected to play an important role due to their theoretical high energy conversion ratio and potential applications closer to

Citation: Liu, P.; Liu, M.; Wu, X. Special Issue on "Integrated Energy Systems towards Carbon Neutrality". *Processes* **2023**, *11*, 439. https://doi.org/10.3390/pr11020439

Received: 15 November 2022
Accepted: 20 December 2022
Published: 1 February 2023

Copyright: © 2023 by the authors. Licensee MDPI, Basel, Switzerland. This article is an open access article distributed under the terms and conditions of the Creative Commons Attribution (CC BY) license (https://creativecommons.org/licenses/by/4.0/).

end consumers. Ramadhani et al. [5] reviewed development of solid oxide fuel cell-based polygeneration systems in residential applications from the asepects of technology, planning, and design. Lee and Chen [6] investigated energy saving potential of heat recovery networks in industrial processes.

In a low-carbon world, renewable energy is expected to be used extensively, and gas turbines, due to their fast load-changing properties, can compensate the intermittency renewable energy might bring to a power grid. Ren, Li, and colleagues [7] studied heat transfer issues in gas turbine vanes via data-driven analysis. Qian et al. [8] present regional short-term load forecasting methods for the purpose of better matching power supply to consumption.

Carbon mitigation in the industrial and consumer sectors are of special importance for a deeply decarbonized economy. Liu, Li, and colleagues [9] delivered a low-carbon transition pathway for the transport sector of China using multi-scale temporal and special modelling and optimization methods. Zhao and colleagues [10] present a two-step intelligent control strategy for the energy supply of an electric vehicle station. Peng, Ou, and colleagues [11] present a life-cycle analysis of greenhouse gas emissions in the primary and recycled aluminum sector in China. Wang and colleagues [12] discuss potential of cascaded cooling in the polysilicon industry.

These studies present the most recent theoretical and technical approach in low-carbon energy systems research, with potential applications in various circumstances. We hope the works collected in this Special Issue can contribute to the transition from our current energy systems to low-carbon ones in the future in a more energy-efficient, cost-effective, and realistic way.

Funding: This research received no external funding.

Acknowledgments: We would like to thank all the authors who contributed to this Special Issue, the Editor-in-Chief for their enthusiastic support of the Special Issue, as well as the editorial staff of *Processes* for their efforts, Pei Liu, Ming Liu, Xiao Wu, *Guest Editors*.

Conflicts of Interest: The authors declare no conflict of interest.

References

1. Liu, S.; Shen, J. Improved Thermoeconomic Energy Efficiency Analysis for Integrated Energy Systems. *Processes* **2022**, *10*, 137. [CrossRef]
2. Cui, Z.; Lin, H.; Wu, Y.; Wang, Y.; Feng, X. Optimization of Pipeline Network Layout for Multiple Heat Sources Distributed Energy Systems Considering Reliability Evaluation. *Processes* **2021**, *9*, 1308. [CrossRef]
3. Song, P.; Zhao, Z.; Chen, L.; Dai, C.; Huang, C.; Liao, M.; Lao, X.; Lin, Y.; Wang, W. Research on Dynamic Modeling of the Supercritical Carbon Dioxide Power Cycle. *Processes* **2021**, *9*, 1946. [CrossRef]
4. Dai, C.; Song, P.; Ma, C.; Zhang, K.; Zheng, W.; Chen, L.; Guo, X.; Lin, Y.; Qiu, Z. Research on Response Characteristics and Control Strategy of the Supercritical Carbon Dioxide Power Cycle. *Processes* **2021**, *9*, 1943. [CrossRef]
5. Ramadhani, F.; Hussain, M.; Mokhlis, H.; Erixno, O. Solid Oxide Fuel Cell-Based Polygeneration Systems in Residential Applications: A Review of Technology, Energy Planning and Guidelines for Optimizing the Design. *Processes* **2022**, *10*, 2126. [CrossRef]
6. Lee, J.-Y.; Chen, P.-Y. Optimization of Heat Recovery Networks for Energy Savings in Industrial Processes. *Processes* **2023**, *11*, 321. [CrossRef]
7. Cui, H.; Wang, L.; Li, X.; Ren, J. Data-Driven Conjugate Heat Transfer Analysis of a Gas Turbine Vane. *Processes* **2022**, *10*, 2335. [CrossRef]
8. Qian, K.; Wang, X.; Yuan, Y. Research on Regional Short-Term Power Load Forecasting Model and Case Analysis. *Processes* **2021**, *9*, 1617. [CrossRef]
9. Li, C.; Liu, P.; Li, Z. A Long-Term Decarbonisation Modelling and Optimisation Approach for Transport Sector Planning Considering Modal Shift and Infrastructure Construction: A Case Study of China. *Processes* **2022**, *10*, 1371. [CrossRef]
10. Shi, S.; Fang, C.; Wang, H.; Li, J.; Li, Y.; Peng, D.; Zhao, H. Two-Step Intelligent Control for a Green Flexible EV Energy Supply Station Oriented to Dual Carbon Targets. *Processes* **2021**, *9*, 1918. [CrossRef]

11. Peng, T.; Ren, L.; Du, E.; Ou, X.; Yan, X. Life Cycle Energy Consumption and Greenhouse Gas Emissions Analysis of Primary and Recycled Aluminum in China. *Processes* **2022**, *10*, 2299. [CrossRef]
12. Yang, S.; Wang, Y.; Wang, Y. Optimization of Cascade Cooling System Based on Lithium Bromide Refrigeration in the Polysilicon Industry. *Processes* **2021**, *9*, 1681. [CrossRef]

Disclaimer/Publisher's Note: The statements, opinions and data contained in all publications are solely those of the individual author(s) and contributor(s) and not of MDPI and/or the editor(s). MDPI and/or the editor(s) disclaim responsibility for any injury to people or property resulting from any ideas, methods, instructions or products referred to in the content.

processes

Article

Improved Thermoeconomic Energy Efficiency Analysis for Integrated Energy Systems

Sha Liu [1,2] and Jiong Shen [2,*]

1 School of Mechanical and Electrical Engineering, Jingling Institute of Technology, Nanjing 211169, China; liusha@jit.edu.cn
2 Key Laboratory of Energy Thermal Conversion and Control of Ministry of Education, School of Energy and Environment, Southeast University, Nanjing 210096, China
* Correspondence: shenj@seu.edu.cn

Abstract: The structure of an integrated energy system is complex. Thermoeconomics can play a significant role in the analysis of IES because it makes up for the deficiency of traditional thermodynamic analysis and provides new information on the cost and energy conversion efficiency. When using thermoeconomics to analyze the energy efficiency of an IES, one key issue that needs to be solved is how to transfer irreversible loss across thermal cycles, so that the mechanism of system performance degradation can be fully revealed. To this end, an irreversible cost and exergy cost integrated analysis method based on improved thermoeconomics is proposed, in which the cumulative and transmission impact of irreversible loss across thermal cycles is evaluated using linear transformation of <KP> matrix. A case study on a 389MW combined cooling, heating, and power IES demonstrates the effectiveness of the proposed approach. The proposed approach can reveal the key links impairing the overall energy efficiency and transfer of irreversible loss across thermal cycles. The approach can be extended to various types of IES to provide directions for the assessment and optimization of the system.

Keywords: integrated energy system; performance analysis; thermoeconomics; exergy cost modeling; irreversible loss

Citation: Liu, S.; Shen, J. Improved Thermoeconomic Energy Efficiency Analysis for Integrated Energy Systems. *Processes* **2022**, *10*, 137. https://doi.org/10.3390/pr10010137

Academic Editor: Pei Liu

Received: 28 November 2021
Accepted: 5 January 2022
Published: 10 January 2022

Publisher's Note: MDPI stays neutral with regard to jurisdictional claims in published maps and institutional affiliations.

Copyright: © 2022 by the authors. Licensee MDPI, Basel, Switzerland. This article is an open access article distributed under the terms and conditions of the Creative Commons Attribution (CC BY) license (https://creativecommons.org/licenses/by/4.0/).

1. Introduction

An integrated energy system (IES) is a new energy system based on advanced technology and management modes to achieve a complementary and cascading utilization of multiple types of energy [1–4]. It can integrate coal, oil, natural gas, renewable power and other resources in the region to meet diversified energy demand for electricity, heat and cooling through coordinated planning, operation, and management. The IES consists of various thermal cycles and energy conversion equipment, contributing to a variety of system topology structures. This feature makes the IES more complicated than the conventional single thermal cycle system. Energy efficiency analysis of the IES is vital to improve the energy conversion efficiency of the system and the synergistic performance of multi-energy cogeneration [5–8].

To date, three major methods have been used for IES performance analysis, i.e., energy analysis, exergy analysis, and exergy economic (thermoeconomics) analysis [9]. Energy analysis reveals the energy loss of the whole system from the perspective of energy "quantity", but it cannot represent the energy quality deterioration during conversion and dissipation processes [10–12].

Exergy analysis focuses on the change in the "quality" of energy. Although the amount of energy remains unchanged in the process of energy conversion, the amount of available energy (exergy) gradually decreases and even eventually disappears with the increase in irreversible loss. Exergy analysis can be applied to analyze the degree of irreversible loss in the production process and identify the location of irreversible loss in the system [13–15].

However, the irreversibility generated from different types of equipment within the entire system has the property of unequal valence, that is, the technical non-equivalence principle of local irreversibility. Taking a thermal power unit as an example, the same amount of irreversible loss occurring in a boiler and generator will lead to different amounts of external fuel consumption. Valero proposed in [16] that "The local savings of exergy which can be achieved in the different units or processes of an installation are not equivalent. The same decrease in the local irreversibility of two different plant components leads, in general, to different variations of the total plant energy consumption".

Exergy economics is a new analytical field that considers the technical non-equivalence principle of local irreversibility. The concept of exergy cost is applied in the analysis system, which makes the production performance of components comparable within the entire system. It thereby provides a new idea for the evaluation and optimization of IESs. Exergy cost theory connects thermodynamics and economics. It assumes that the cost formation process of products runs parallel to the continuous and inevitable energy degradation process of resources. The result is a rigorous program to calculate the cost of all process flows in the system according to the cost matrix, which defines the system interaction from the perspective of thermoeconomics [17].

In recent years, the application of exergy economic structural theory to the performance analysis and optimization of IESs has become a hot topic. Haydargil et al. [18] conducted an exergy economic analysis of a cogeneration system and calculated its overall exergy economic cost. Marques et al. [19] carried out an exergoeconomic analysis on a compact electricity–cooling cogeneration unit. This study applies the specific exergy costing (SPECO) methodology for the exergoeconomic assessment of a compact electricity–cooling cogeneration system, give the cost rate of chilled water. Boyaghchi et al. [20] performed the thermodynamic, economic and environmental analyses of a solar-geothermal driven combined cooling, heating and power (CCHP) cycle integrated with flat plate collectors containing water/copper oxide (CuO) nanofluid as the absorbing medium. Twelve main parameters were selected as the decision variables of the desired system while the daily exergy efficiency, total product cost rate and total product environmental impact associated with the exergy rate were chosen as the three main objective functions. NSGA-II (Non-dominated Sort Genetic Algorithm-II) was individually applied to obtain the final optimal solutions for the multi-objective optimization of the desired system for four working fluids from the exergy, exergoeconomic and exergoenviromental points of view. Ghaffarpour et al. [21] carried out an exergy analysis and exergy economic analysis of a new type of power generation system in which biomass is combusted with coal and calculated the impact of different operating parameters on the exergy efficiency and exergy cost of the system; the results show that the fuel mass flow rate and current density are the dominant factors affecting the variation in energy and exergy efficiencies as well as unit production cost. Eder et al. [22] conducted an exergy economic analysis of a thermal power plant with an air Brayton cycle and analyzed the exergy economic cost of the system under different operating parameters. Mohammad et al. [23] analyzed an integrated solar combined cycle (ISCC) power plant and evaluated the performance of the plant. Thermodynamic and thermoeconomic analyses were carried out for the ISCC, and zones with high exergy destruction were identified by their study. Wang et al. [24,25] carried out an exergy economic analysis for a solar energy system, and the overall exergy economic cost of the system was calculated.

The above studies all used the exergy economic cost expressed in the monetary dimension to analyze the cost allocation of polygeneration systems and did not focus on the exergy cost expressed in the energy dimension to analyze the change process of irreversible loss of the system. The change process involving irreversible loss is still assessed using the exergy analysis method (local energy efficiency index: exergy efficiency). As mentioned above, irreversible loss is characterized by the principle of technical inequivalence of local irreversibility, and the transmission process of irreversible loss in a multiple thermal cycle coupling system cannot be analyzed with the exergy analysis method. Therefore, the existing literature does not pay attention to the transmission of irreversible loss in IESs, and it is difficult to quantify the formation process of irreversible loss between thermal cycles.

For an IES, understanding the mechanism of increasing system exergy costs can effectively improve the performance of the entire system, but it is challenging to reveal the cumulative transfer process of irreversible losses across thermal cycles. Therefore, it is necessary to establish an evaluation and analysis mechanism that considers the relationship between process irreversible loss transfer and system exergy cost.

Therefore, based on exergy economic structure theory and exergy cost theory, this paper will present a production structure diagram construction method and exergy cost modeling method for IESs, carry out linear transformation on the <KP> matrix, and reveal the transmission characteristics of irreversible loss in an IES and the formation process of multiple product exergy costs in the system. This paper will unify the overall and local energy efficiency evaluation indexes of the system, so as to fully reveal the performance degradation mechanism of the IES, reveal the key links affecting the overall energy efficiency of the IES, provide an important direction for the overall performance improvement of the IES, and lay the foundation for system optimization.

The novel contributions of this paper are as follows:

(1) The exergy economic analysis method is applied to analyze the performance degradation process of an IES;
(2) For an IES, an integrated model for calculation of exergy costs and irreversible loss costs based on the classification of exergy flow attributes is proposed;
(3) Linear transformation of the <KP> matrix is carried out to obtain the relationship between process irreversible losses and exergy costs for an IES. The transfer of irreversible losses in the thermal cycle and the influence of irreversible loss accumulation on the formation of exergy costs are revealed.

The background and problem formulation are given in Section 1. Section 2 presents an integrated model for calculation of exergy costs and irreversible loss costs based on the classification of exergy flow attributes. Section 3 presents the linear relationship between exergy costs and irreversible costs. Simulation studies are carried out in Section 4. The last section concludes this paper.

2. A Integrated Calculation Model of Exergy Cost and Irreversible Loss Cost Based on Exergy Flow Attribute Classification

Exergy economics usually constructs the exergy cost balance equation based on the production structure diagram. The exergy economics evaluation index can show the production function and irreversibility of each component in a system. Literature [26] holds that each component is defined by its production function, and puts forward the concept of branch point. It holds that the point where system products are allocated to multiple components is a branch point, which is a fictitious unit. In this paper, problems in the generation rules for the production structure diagram were as follows: firstly, there is no clear distinction and definition between branch component and collection component; secondly, the setting rules for the number of virtual components and the input and output composition rules for virtual components are not given.

The production structure diagram is the graphical representation of the exergy economics model, in which each stream can be the real fuel flow and product flow; it can also be a virtual negative entropy flow. The resource allocation of the whole system can be clearly represented by the production structure diagram. There are complex energy interactions in the integrated energy system, and the traditional empirical drawing method is not applicable. Based on the exergy flow attribute classification method, this paper puts forward a principle for constructing virtual components and determining the connection relationship between various virtual components and related thermal components, thus providing a basis for standardizing the preparation of production structure diagrams.

2.1. The Logical Topology Structure Construction Based on the Exergy Flow Attribute Classification

Based on the exergy flow attribute classification, the logical topology structure construction mainly consists of the following five parts: (1) system fuel input, including coal, natural gas, photovoltaic, wind power, and the external grid power supply, represented by a solid line; (2) energy conversion components, including steam turbines, gas turbines, coal-fired boilers, gas boilers, electric refrigerators, absorption chillers, and heat exchangers, and energy conversion components essential to complete a certain thermal cycle, represented by rectangles; (3) thermal cycle components, including gas turbine unit cycles, Rankine cycles, organic Rankine cycles, gas turbine engine cycles, refrigeration cycles, etc., and thermal cycles representing a complete energy conversion (production) closed thermal process, represented by ellipses; (4) energy collection and distribution, based on collection and branch components represented by diamonds and circles, respectively; (5) energy transmission. The energy transfer process is represented by exergy flow, indicated by a dotted line. The process of energy transmission, transformation, and cascade utilization can be displayed intuitively and clearly through the construction of logical topology.

For IESs, the modular construction idea is adopted. The component is taken as the analysis object to analyze the exergy flow classification in the thermal cycle. On the other hand, the thermal cycle is applied as the analysis object, and the energy cascade utilization relationship is used as the basis to analyze and complete the exergy flow classification in the system. The construction of the overall logical topology of the IES can lay a foundation for the graphical modeling of exergy costs. The modular construction process is as follows

2.1.1. Exergy Flow Classification in the Thermal Cycle

The energy conversion equipment in the thermal cycle is defined as an energy conversion component. Its input and output exergy flow attribute classification is analyzed. According to the function of energy conversion components, the exergy flow attribute classification is conducted, and the exergy flow collection module of the same attribute is established. The number of collection components and branch components is determined by the number of exergy flow attribute categories. Moreover, the collection component is taken as the core to determine the connection relationship between each component within the thermodynamic cycle. The energy conversion component is connected with the corresponding collection component according to its output exergy flow attribute, and the output exergy flow of the collection component under the same attribute is the input exergy flow of the branch component under this attribute.

Figure 1 is a schematic diagram of the exergy flow classification connection in a thermal cycle system. If the input and output exergy attributes of energy conversion components are divided into three categories, i.e., thermal exergy, mechanical exergy, and electrical exergy, the number of collection components and branch components is 3, namely, thermal exergy collection component J1, mechanical exergy collection component J2, and electrical exergy collection component J3. Corresponding branch components are set with the collection components. The corresponding branch components are denoted as thermal exergy branch component B1, mechanical exergy branch component B2, and electrical exergy branch component B3, responsible for thermal exergy, mechanical exergy, and electrical exergy, respectively.

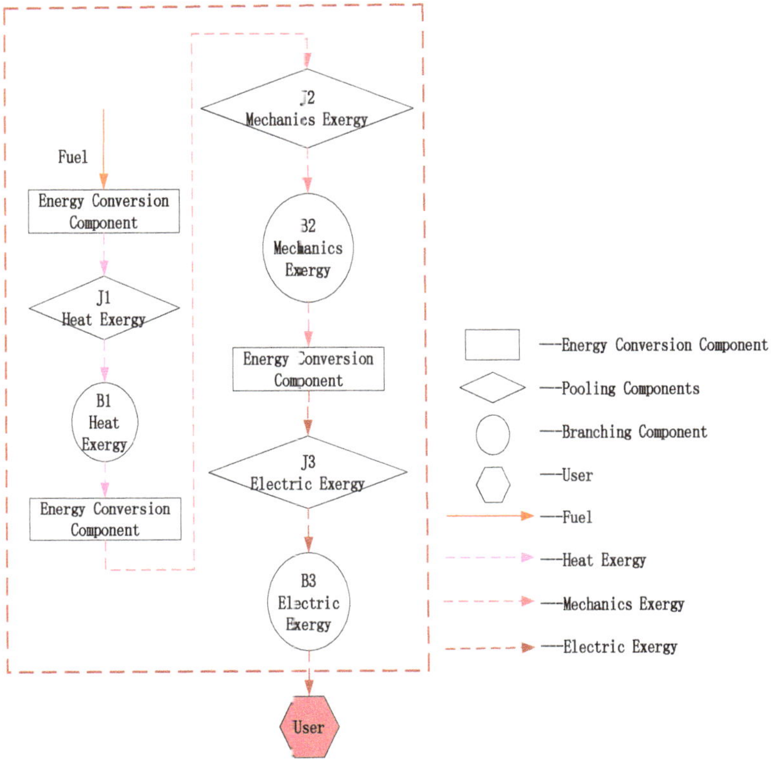

Figure 1. Classification of exergy flow in a thermal cycle.

For an energy conversion component with thermal exergy output, the thermal exergy it produces is the input of J1. For an energy conversion component with mechanical exergy output, the mechanical exergy it produces is the input of J2. For an energy conversion component with electrical exergy output, the electrical exergy it produces is the input of J3. In addition, the output of J1 is the only input of B1. The output of J2 is the only input of B2. The output of J3 is the only input of B3.

2.1.2. Exergy Flow Classification in System

In an IES, a variety of prime movers and energy conversion equipment are used to form different thermal cycle processes; various types of energy products can be output to users and realize energy cascade utilization. With the increase in system coupling links, exergy classification of coupling links should be considered in the system.

Firstly, the relationship of energy cascade utilization among thermal cycles in the system is analyzed. The number of coupling nodes is determined by the number of waste heat utilization processes. The exergy flow attribute of each coupling node indicates different levels of thermal exergy. Secondly, each thermal cycle process is defined as a thermal cycle component. According to the function of thermal cycle components, the exergy flow attribute classification is completed, and the exergy flow collection module of the same attribute is established. The number of collection components and branch components is determined by the number of exergy flow attribute categories. Finally, taking the thermal cycle component as the core, the connection relationship of each component in the system is determined. The thermal cycle component is connected with the corresponding collection component according to its output exergy flow attribute, and the output exergy flow of the collection component under the same attribute is the input exergy flow of the branch component under this attribute.

Figure 2 is a schematic diagram of the exergy flow classification connection in the system. The number of waste heat utilization processes and the number of coupling nodes are both 2. In addition to BS1, there are also coupling node branch components BS11 and BS12 corresponding to thermal exergy collection component JS1 in the system. When the input and output exergy attributes of thermal cycle components are divided into three categories, i.e., thermal exergy, electrical exergy, and cold exergy, the number of collection components and branch components is 3, namely thermal exergy collection component JS1, electrical exergy collection component JS2, and cold exergy collection component JS3. Corresponding branch components are set with the collection components. The corresponding branch components are called thermal exergy branch component BS1, electrical exergy branch component BS2, and cold exergy branch component BS3, respectively responsible for thermal exergy, electrical exergy, and cold exergy.

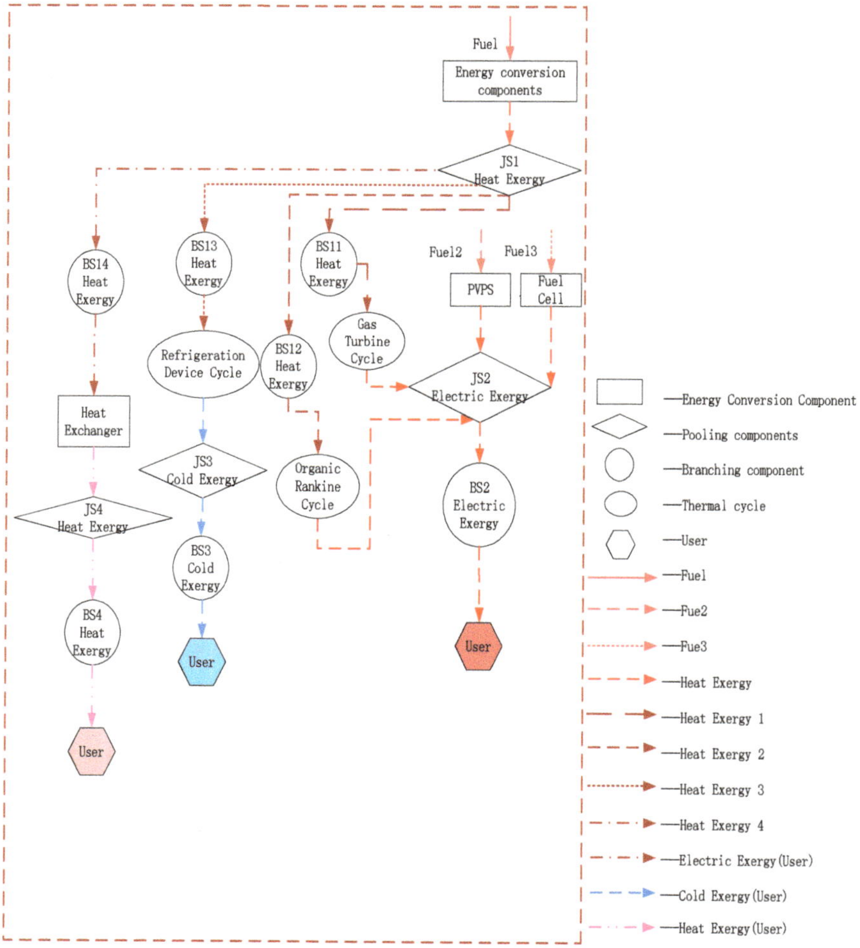

Figure 2. Classification of exergy flow in the system.

For a thermal cycle component with thermal exergy output, the thermal exergy it produces is the input of JS1. For a thermal cycle component with electrical exergy output, the electrical exergy it produces is the input of JS2. For a thermal cycle component with cold exergy output, the cold exergy it produces is the input of JS3. In addition, the output of JS1 is the only input of BS1. The output of JS2 is the only input of BS2. The output of JS3 is the only input of BS3.

2.2. Exergy Cost Modeling Method Based on Exergy Flow Attribute Classification

According to the logical topology of the system, the exergy cost modeling criteria of input exergy flow, output exergy flow, and coupling link are defined respectively.

2.2.1. Exergy Cost Modeling Criteria of Input Exergy Flow

(1) For a component with external resources, the exergy cost of input exergy flow is 1, as shown in Equation (1):

$$k^*_{FBn} = 1 \tag{1}$$

where k^*_{FBn} is the input exergy flow exergy cost of component n, kW/kW.

(2) For an energy conversion component with different exergy attributes, the exergy cost of input exergy flow is equal to the output exergy cost of branching m, as shown in Equation (2):

$$k^*_{FBn} = k^*_{Pm} \tag{2}$$

where k^*_{Pm} is the output exergy flow exergy cost of branching m, kW/kW.

2.2.2. Exergy Cost Modeling Criteria of Output Exergy Flow

(1) For the energy conversion component, the output exergy flow exergy cost is equal to the input exergy flow exergy cost of components multiplied by unit exergy consumption, as shown in Equation (3):

$$k^*_{pn} = kB_n k^*_{FBn} \tag{3}$$

where k^*_{pn} is the output exergy flow exergy cost of the energy conversion component, kW/kW; kB_n is the unit exergy consumption, kW/kW.

(2) For the collection component, the output exergy flow exergy cost is equal to the sum of the input exergy flow exergy cost of each collection tributary multiplied by its exergy flow rate, as shown in Equation (4):

$$k^*_{pn} = \sum r_i k^*_{FBn} \tag{4}$$

where r_i is the exergy flow rate.

(3) For the branch component, the output exergy flow exergy cost is equal to the input exergy flow exergy cost k^*_{FB}, as shown in Equation (5):

$$k^*_{pn} = k^*_{FBn} \tag{5}$$

3. Calculation of Irreversible Cost Based on Exergy Flow Attribute Classification

3.1. Existing Irreversible Loss Index Based on Exergy Analysis

Irreversibility exists in all thermal processes and is an important factor for the decrease in system performance. In previous literature, the exergy loss coefficient has been commonly used as the evaluation index to complete the quantitative analysis of the process irreversible loss.

The exergy loss coefficient of the system is

$$\Omega_i = I_i / F \tag{6}$$

where Ω_i is the exergy loss coefficient, kW/kW; I_i is the loss of a certain process, kW; F is the total fuel input into the system, kW.

3.2. Calculation of Irreversible Loss Cost Based on Exergy Attribute Classification

In thermoeconomics, energy and funds are collectively referred to as resources. The generation of energy flow in an IES requires the consumption of external resources. The exergy cost modeling method based on exergy attribute classification has some new extensions to the exergy cost theory. The focus of this research is to analyze the influence of irreversible loss accumulation and transfer across the thermal cycle process on the formation of exergy cost.

3.2.1. Analysis of the Relationship between Exergy Cost and Irreversible Cost Based on Exergy Attribute Classification

Based on disturbance theory, the impact of irreversible increases in external resource consumption by each component needs to be evaluated [16] (Valero et al. (1993)), and the fuel effect formula was proposed for the first time in [27–35] (Valero and Torres et al.). Further research has been performed on this basis. The basic idea is to obtain the incremental form of external resource consumption from the mathematical form of overall resource consumption. The study focuses on the analysis of irreversible loss increments and further obtains the expression of irreversible cost by linear transformation <KP> matrix.

When the exergy attribute classification method is adopted, the coefficient of technical products is equal to the unit exergy consumption of components, after the exergy of the same attribute is collected. At this time, the diagonal matrix $(n \times n)$ $\langle \mathrm{KP} \rangle$ containing the coefficient of technical product k_{ij} in the system becomes an n-dimensional vector containing the unit exergy consumption of the exergy conversion component.

According to the exergy balance principle of the system, the increased fuel consumption is converted into an irreversible increase inside the system and an increase in the final product of the system. When the exergy attribute classification method is adopted, after the exergy of the same attribute is collected, the $(\mathrm{n} \times 1)$ vector $\mathrm{P_s}$ containing the final product of the system in the system becomes a scalar ω. From this, an expression of irreversible cost can be obtained, as shown in Equations (7) and (10):

$$I^* = |\mathrm{I}\rangle = \mathrm{I}/\mathrm{P_s} \tag{7}$$

where $\mathrm{P_s}$ is the $(\mathrm{n} \times 1)$ vector containing the final product B_{i0} of the system. The irreversible cost and exergy cost of the classified product have the following relationship:

$$k^*_{p,i} = 1 + \sum_{i=1}^{n} I^*_i \tag{8}$$

where $k^*_{p,i}$ is the exergy cost of the ith exergy flow, kW/kW; I^*_i is the irreversible cost of obtaining the ith exergy flow, kW/kW.

A detailed demonstration process is shown in Appendix A.

This equation characterizes the relationship between the exergy cost and the irreversible process cost necessary to obtain the exergy. The formation of the exergy cost of the entire energy conversion process reflects the accumulation of the irreversible cost of the process. With the change in the progressive exergy properties of the system energy conversion process, the exergy cost presents a gradually increasing trend.

3.2.2. Calculation of Irreversible Cost Based on Exergy Flow Attribute Classification

From the analysis of the relationship between exergy cost and irreversible cost based on exergy attribute classification, it can be seen that the exergy cost based on exergy attribute classification can be used to quantify the process impact of process irreversibility on the formation of the final cost of the system. In the same way, through equation deformation, the exergy cost difference can be calculated to obtain the process's irreversible loss, which converted to external fuel conversion cost (hereafter, this term will be abbreviated as "irreversible cost"). From Equation (8), we can see that

$$k_{p,i+1}^{*} = 1 + \sum_{=1}^{n} I_{i+1}^{*} = 1 + \sum_{i=1}^{n} I_{i}^{*} + I_{i+1}^{*} = k_{p,i}^{*} + I_{i+1}^{*} \quad (9)$$

and we can then obtain the following:

$$I_{i+1}^{*} = k_{p,i+1}^{*} - k_{p,i}^{*} \quad (10)$$

Equation (10) shows that the irreversible loss cost is equal to the cost difference between adjacent exergy flows after classification, and the irreversible loss cost can be obtained through the calculation of exergy cost.

4. Analysis of Examples

4.1. Typical Integrated Energy System Parameters under Rated Conditions

Take the typical integrated energy system as an example, as shown in Figure 3. It is mainly composed of a gas turbine unit cycle system (topping cycle), Rankine cycle system (bottom cycle), flue gas absorption chiller and hot water heat exchanger, with rated power generation of 389 MW, rated cooling capacity of 11,000 kW and rated heat capacity of 800 kW. After the air is compressed by the air compressor (AC), the air is burned in the combustion chamber (CC) with natural gas. The fuel releases chemical energy to form high-temperature flue gas. The flue gas drives the gas turbine (GT) to perform work, convert heat energy into mechanical energy, and drive the generator (GEN) to generate power. The temperature of the gas discharged from the gas turbine is as high as 583 °C and still has high energy. A portion of these high-temperature gases is sent to the waste heat boiler (HRSG) to heat water into steam to drive the steam turbine (ST) to do work, then drive the generator to generate electricity. The other part enters the absorption refrigerator (REF) for refrigeration, then enters the hot water heat exchanger (HWHE) to recover the waste heat and produce hot water, and is finally discharged into the atmosphere.

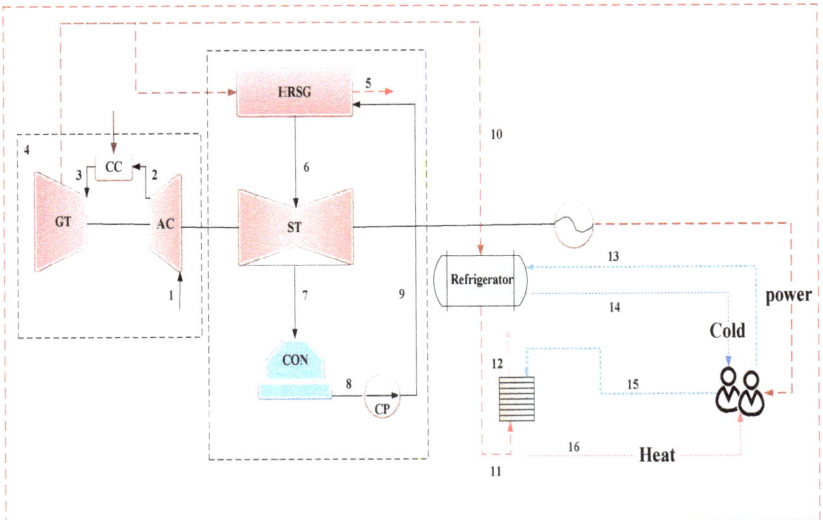

Figure 3. Thermal system diagram of the integrated energy system.

The design working condition parameters and calculation conditions are shown in Table 1. The main parameters of each stream under rated conditions are shown in Table 2. The composition of natural gas is shown in Table 3.

Table 1. Design working condition parameters and calculation conditions.

Item	Parameters	Values
Apical circulation (gas turbine device cycle)	Atmospheric pressure (kPa)	101.1
	Ambient temperature (°C)	17.4
	Compressor pressure ratio	15.4
	Low calorific value of natural gas (kJ/kg)	48,686.3
	Isentropic efficiency of compressor (%)	85
	Isentropic efficiency of gas turbine (%)	80.7
Bottom circulation (Rankine cycle)	Steam turbine inlet temperature (°C)	565.5
	Steam turbine inlet pressure (kPa)	9563
	Steam turbine exhaust pressure (kPa)	5.96
	Generator efficiency (%)	98
REF	Inlet flue gas temperature (°C)	583.6
	Outlet flue gas temperature (°C)	175
	Inlet temperature of refrigerant water (°C)	12
	Outlet temperature of refrigerant water (°C)	7
	Cooling water inlet temperature (°C)	32
	Cooling water outlet temperature (°C)	36
	Condensation temperature (°C)	39
	Energy efficiency of heat exchanger (%)	85
HWHE	Inlet flue gas temperature (°C)	175
	Outlet flue gas temperature (°C)	90.6
	Inlet water temperature (°C)	50
	Outlet water temperature (°C)	65
System	Rated power generation (MW)	389
	Rated cooling capacity (kW)	11,000
	Rated heat capacity (kW)	800

Table 2. Main parameters of each stream under rated conditions.

Number	m/ (t/h)	P/ (kPa)	T/ (°C)	h/ (kJ/kg)	e_x/ (kJ/kg)
1	2270.2	101.1	17.4	42.3	0
2	2270.2	1556.9	422.2	456.9	398.716
3	2329.9	1533.6	1273.2	1494.3	1237.621
4	2329.9	104.4	607.1	693.8	368.278
5	2307.4	101.1	83.8	64.8	26.998
6	280.9	9563	565.5	3542.2	1557.956
7	361	5.857	35.7	2418.6	134.616
8	395	5.856	35.7	149.7	0.168
9	395	2460	36.1	153.2	2.351
10	22.49	101.1	583.6	631.8	257.32
11	22.49	101.1	175	94.8	59.314

Table 2. *Cont.*

Number	m/ (t/h)	P/ (kPa)	T/ (°C)	h/ (kJ/kg)	e_x/ (kJ/kg)
12	22.49	101.1	90.6	66.12	29.32
13	1939.32	101.1	12	50.506	0.067
14	1939.32	101.1	7	29.525	0.997
15	43.54	101.1	50	209.42	7.117
16	43.54	101.1	65	272.18	14.719

Table 3. Composition of natural gas.

Name	Mole Percentage (%)
Methane	97.6%
Ethane	0.62%
Propane	0.41%
Butane	0.21%
Pentane	0.01%
Hexane	0.05%
Carbon Dioxide	0.65%
Nitrogen	0.45%

4.2. Establishment of the Integrated Calculation Model of Exergy Cost and Irreversible Loss Cost Based on Exergy Flow Attribute Classification

According to Section 2, the integrated calculation model of exergy cost and irreversible loss cost based on exergy flow attribute classification was established, and the cost equations are as shown in Table 4.

Table 4. Exergy cost equations.

	Exergy Cost Modeling of Exergy Flow	Exergy Flow Attribute Classification	Exergy Cost Equations
Topping Cycle (Gas Turbine Device Cycle)	Input exergy flow exergy cost modeling	Input fuel exergy components CC	$k^*_{FB} = 1$
		Input heat exergy components GT	$k^*_{FBn} = k^*_{PbTC1}$
		Input mechanics exergy components AC	$k^*_{FBn} = k^*_{PbTC2}$
	Output exergy flow exergy cost modeling	Components AC, CC and GT	$k^*_{pn} = kB_n k^*_{FBn}$
		Join components J(TC)1 and J(TC)2	$k^*_{pn} = \sum r_i k^*_{FBn}$
		Branch components B(TC)1 and B(TC)2	$k^*_{pn} = k^*_{FBn}$
Bottoming Cycle (Rankine Cycle)	Input exergy flow exergy cost modeling	Input fuel exergy components HRSG	$k^*_{FBn} = k^*_{Pbs1}$
		Input heat exergy components ST	$k^*_{FBn} = k^*_{PbBC1}$
	Output exergy flow exergy cost modeling	Components HRSG, CND, ST and GEN	$k^*_{pn} = kB_n k^*_{FBn}$
		Join components J(BC)1 and J(BC)2	$k^*_{pn} = \sum r_i k^*_{FBn}$
		Branch components B(BC)1 and B(BC)2	$k^*_{pn} = k^*_{FBn}$

Table 4. *Cont.*

	Exergy Cost Modeling of Exergy Flow	Exergy Flow Attribute Classification	Exergy Cost Equations
System	Input exergy flow exergy cost modeling	Input mechanics exergy components GEN	$k^*_{FBn} = k^*_{Pbs2}$
		Input heat exergy components REF	$k^*_{FBn} = k^*_{Pbs1}$
		Input heat exergy components HWHE	$k^*_{FBn} = k^*_{Pbs4}$
	Output exergy flow exergy cost modeling	Join components J(S)1, J(S)2, J(S)3 and J(S)4	$k^*_{pn} = \sum r_i k^*_{FBn}$
		Branch components B(S)1, B(S)2, B(S)3 and B(S)4	$k^*_{pn} = k^*_{FBn}$
		Output electric exergy components GEN	$k^*_{pn} = kB_n k^*_{FBn}$
		Output cooling exergy components REF	$k^*_{pn} = kB_n k^*_{FBn}$
		Output heat exergy components HWHE	$k^*_{pn} = kB_n k^*_{FBn}$

4.3. Analysis of Exergy Cost and Irreversible Loss Cost Distribution

Using the above modeling method, the cost of the integrated energy system is calculated. The fuel-product calculation results for the integrated energy system under rated conditions are shown in Table 5.

Table 5. Fuel-product calculation for the integrated energy system under rated conditions.

Component *		Fuel Exergy Consumption *FB* (kW)	Product *P* (kW)	Unit Exergy Consumption kB_n (kW/kW)	Exergy Flow Rate r
1	AC	261,432.5	251,434.8	1.04	0.060
2	CC	718,204.1	529,022.4	1.36	0.940
3	GT	562,633.6	518,127.6	1.09	0.654
4	HRSG	219,357	168,359.1	1.30	0.999
5	ST	151,210.12	141,167.47	1.07	0.346
6	CND	6280.34	214,884.5	0.03	-
7	CP	384.03	239.52	1.60	0.001
8	GEN	395,678.8	387,765.3	1.02	-
9	REF	1237	500.991	2.47	-
10	HWHE	187.41	91.95	2.03	-

Note *: Common component fuel-product definitions are given in references [36].

The exergy cost equations given in Table 4 are solved by Seidel iterative method; the calculation results are shown in Figures 4 and 5 and Table 6.

As shown in Figure 4, Figure 5 and Table 6, in the process of conversion from fuel exergy to electric exergy, heat exergy and cold exergy, the cost of exergy flow with different attributes increases with the increase in irreversible links. Fuel exergy cost 1 kW/kW increases to terminal mechanical exergy cost 1.513 kW/kW in the topping cycle. Heat exergy cost 1.392 kW/kW increases to terminal mechanical exergy cost 1.955 kW/kW in the bottoming cycle.

In the system, after the mechanical exergy of the topping cycle and the bottoming cycle is collected, the mechanical exergy cost of the system is 1.666 kW/kW, increasing to terminal electric exergy cost 1.699 kW/kW. Heat exergy (REF) cost 1.392 kW/kW increases to terminal cold exergy cost 3.438 kW/kW. Heat exergy (HWHE) cost 1.392 kW/kW increases to terminal heat exergy cost 2.825 kW/kW.

As shown in Figure 6 and Table 7, the exergy cost difference of adjacent different attribute exergy flows in the thermal cycle is the irreversible loss of the production process itself. The exergy cost is composed of the accumulated irreversible costs of each link.

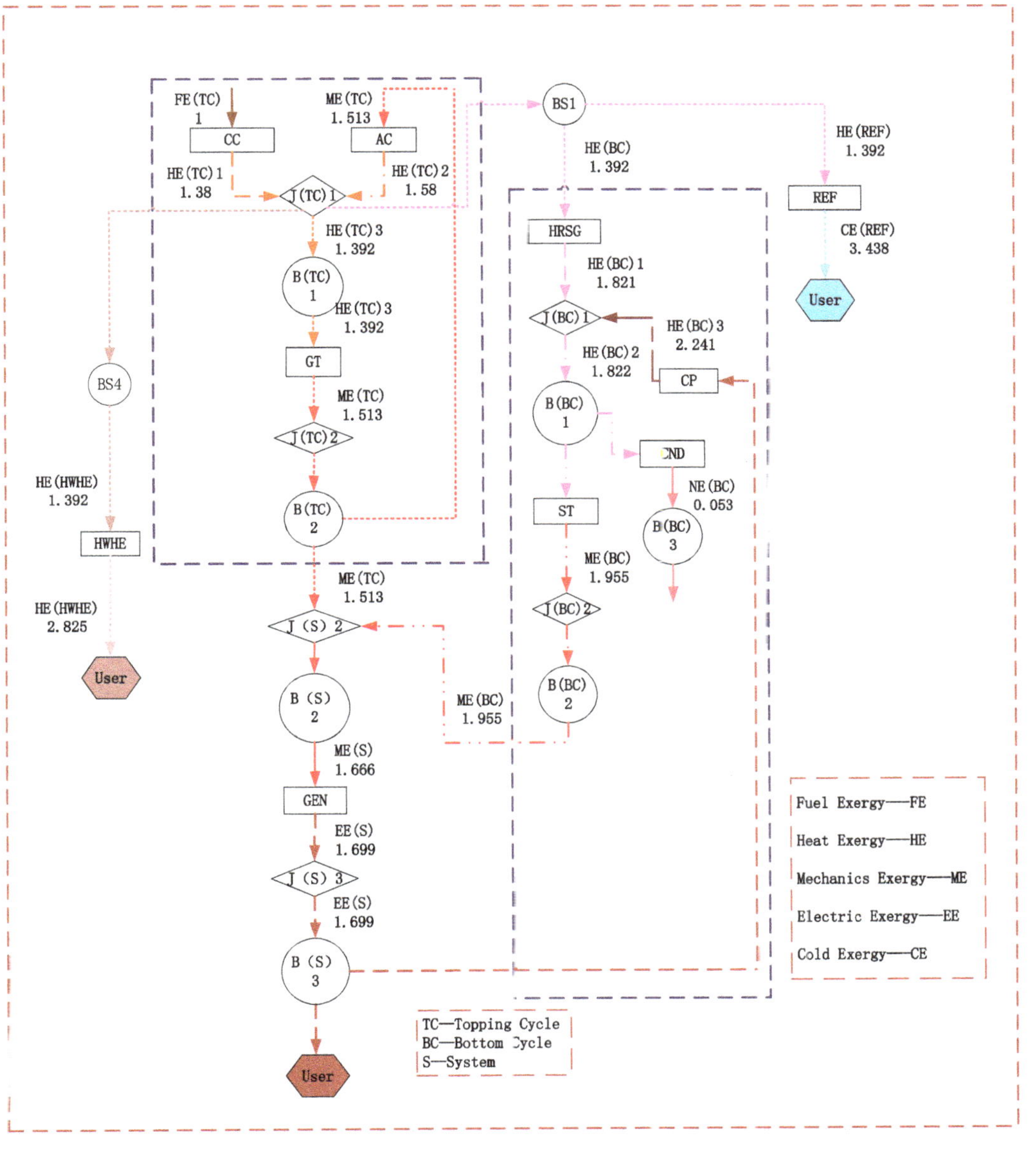

Figure 4. Exergy cost distribution of the IES.

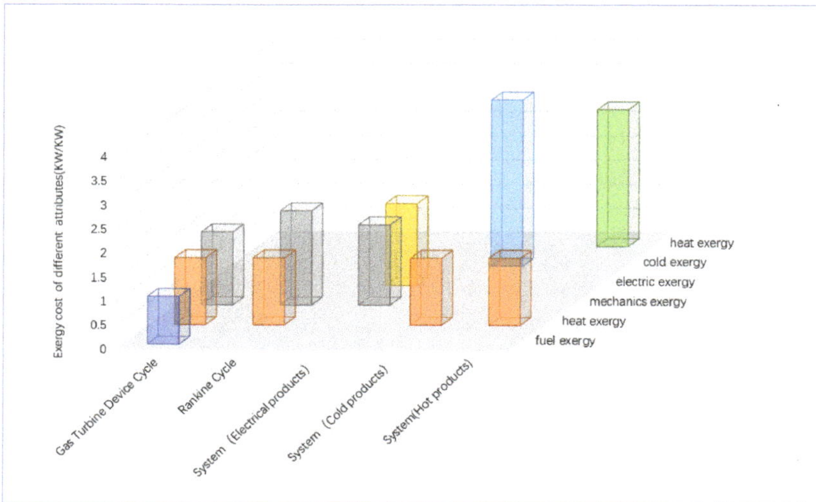

Figure 5. Exergy cost of exergy flow with different attributes.

Table 6. Exergy cost of exergy flow with different attributes.

Exergy Cost k_p^* (KW/KW)		**FE**	**HE**	**ME**	**EE**	**CE**	**HE**
Topping Cycle		1	1.392	1.513			
Bottoming Cycle			1.392	1.955			
System	GEN			1.666	1.699		
	REF		1.392			3.438	
	HWHE		1.392				2.825

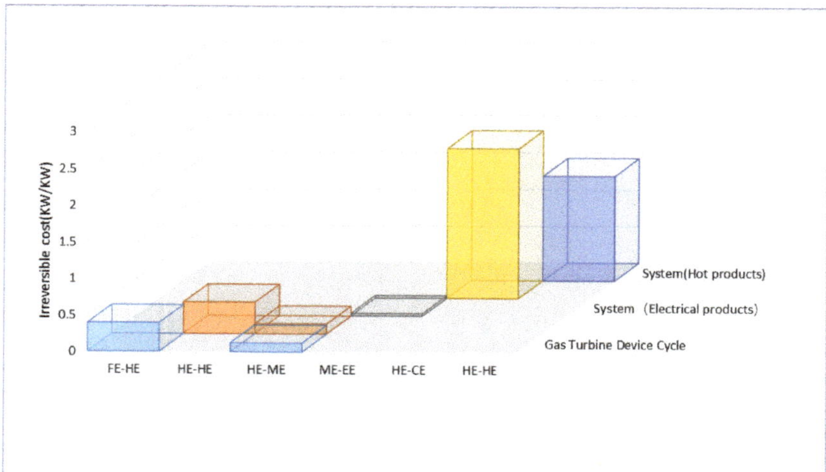

Figure 6. Irreversible loss cost of exergy flow with different attributes.

Table 7. Irreversible loss cost of exergy flow with different attributes.

Irreversible Loss Cost I^* (KW/KW)		FE-HE	HE-HE	HE-ME	ME-EE	HE-CE	HE-HE
Topping Cycle		0.392		0.121			
Bottoming Cycle			0.429	0.134			
System	GEN				0.033		
	REF					2.046	
	HWHE						1.433

The energy conversion process from fuel to heat FE(TC)—HE(TC), HE(BC)—HE(BC)1, HE(REF)—CE(REF) and HE(HWHE)—HE(HWHE) produces a higher irreversible cost. Due to the inequivalence of process irreversibility loss, we should pay attention to the energy conversion processes with a high irreversible cost

In the process of energy conversion, with the accumulation of irreversible loss, the exergy cost of exergy flow with different attributes increases. An example is given to analyze the relationship between the increase in irreversible loss and increase in cost, and the transfer process of irreversible loss across thermal cycles is given. Due to the transmission and coupling of the irreversible cost in each energy conversion link, the mechanism of exergy cost formation can be obtained by the distribution diagram of the exergy flow cost and the irreversible cost.

5. Conclusions

An IES provides an important pathway to realize energy cascade utilization. Developing suitable thermodynamic analysis methodologies is important for IES assessment and optimization. This paper proposes an irreversible cost and exergy cost integrated analysis method for IESs based on improved thermoeconomics, in which linear transformation of the <KP> matrix is applied to unify the process irreversible loss and the exergy cost of the system together. A case study on a 389MW combined cooling, heating and power IES shows that: (1) IES exergy costs increase with the accumulation of irreversible losses. The combustion process and the conversion process from high-quality energy into low-quality energy have the highest irreversible losses, which lead to a substantial increase in exergy costs. (2) In the process of energy cascade utilization, the accumulation of irreversible losses in the upstream cycle will be transferred to the downstream through the interaction, which will influence the exergy costs of downstream products. The proposed approach can provide deep insight into the formation of exergy costs within a polygeneration system and characterize processes of irreversible loss across thermal cycles and energy cascade utilization, thus revealing the weak links in an IES and offering directions for system performance improvements. This study focused on energy loss and energy grade decline, thus analyzing the position of the maximum irreversible loss of the system. If more advanced equipment is used, the equipment costs should not be ignored. Considering the cost of the equipment, thermoeconomic costs in price terms will be our future research interest.

Author Contributions: S.L.: conceptualization, formal analysis, methodology, writing—original draft, investigation, data curation, software, and validation. J.S.: conceptualization, supervision, and methodology. All authors have read and agreed to the published version of the manuscript.

Funding: This research was funded by the Natural Science Foundation of Institutions of Higher Education of Jiangsu Province, China under Grant 19KJD470002 and the National Natural Science Foundation of China (NSFC) under Grant 51936003.

Institutional Review Board Statement: Not applicable.

Informed Consent Statement: Not applicable.

Conflicts of Interest: The authors declare no conflict of interest.

Nomenclature

k^*_F	The exergy cost of input exergy flow (kW/kW)
k^*_P	The exergy cost of output exergy flow (kW/kW)
r_i	Exergy flow rate (kW/kW)
F	Exergy of a fuel stream (kW)
P	Exergy of a product stream (kW)
I	Exergy loss (kW)
kB	Unit exergy consumption (kW/kW)
k_{ij}	Technical product coefficient (kW/kW)
F_T	Total external resource consumption of the system, kW
Δk_i	Exergy consumption change of the *i*th unit
$k^*_{p,i}$	Exergy cost of the *i*th exergy flow after classification (kW/kW)
I^*_i	Irreversible cost of the *i*th exergy flow after classification (kW/kW)
Vectors and Matrices	
P_s	Vector containing the final product B_{i0} of the system (n × 1)
$<KP>$	Diagonal matrix containing technical product coefficients k_{ij} (n × n)
U_D	Unit diagonal matrix (n × n)
K_D	Diagonal matrix containing unit exergy consumption k
^tk_e	N-dimensional vector containing the unit consumption of system input resources (n × 1)
$\Delta\omega$	The change in the system's external output products (n × 1)
I^*	N-dimensional vector containing the process irreversible cost (n × 1)
Abbreviations	
J	Join
B	Branch
TC	Topping Cycle (Gas Turbine Device Cycle)
BC	Bottoming Cycle (Rankine Cycle)
S	System
J(TC)	Join components in the topping cycle
B(TC)	Branch components in the topping cycle
J(BC)	Join components in the bottoming cycle
B(BC)	Branch components in the bottoming cycle
J(S)	Join components in the system
B(S)	Branch components in the system
FE	Fuel Exergy
HE	Heat Exergy
ME	Mechanics Exergy
EE	Electric Exergy
CE	Cold Exergy
CC	Combustion Chamber
AC	Air Compressor
HRSG	Waste Heat Boiler
ST	Steam Turbine
GT	Gas Turbine
CND	Condenser
GEN	Generator
REF	Refrigerator
HWHE	Hot Water Heat Exchanger
Subscripts, superscripts and accents	
n	The component n
m	The branching m
i	The ith exergy flow
*	Related to exergy costs

Appendix A

In exergy cost theory [36] (Valero and Torres et al.), the concept of marginal cost in economics is introduced into thermoeconomics. The unit marginal cost is a derivative that characterizes the additional resources required to produce one more units of product under certain conditions, as shown in Equation (A1):

$$k_i^* = \left[\frac{\partial B_0}{\partial B_i}\right]_{conditions} \tag{A1}$$

where k_i^* is the unit marginal cost of the ith exergy; conditions denote the partial derivative under a certain condition; B_0 reflects the exergy value of the resources needed to obtain the ith exergy, kW; B_i is the exergy value of the ith product, kW.

Exergy loss (irreversible loss) occurs in all processes, and part of the fuel exergy is lost during the production process. Considering the definitions of fuel and product, the process component exergy balance equation is shown in Equation (A2):

$$F = P + I \tag{A2}$$

where F is the fuel exergy, kW; P represents the product exergy, kW; I refers to irreversible loss (exergy loss), kW.

Therefore, fuel exergy consumption is also related to the irreversible loss of the process. It can be characterized by unit exergy consumption k, as shown in Equation (A3):

$$kB = \frac{F}{P} = 1 + \frac{I}{P} \tag{A3}$$

where kB is the unit exergy consumption, kW/kW.

On the basis of the unit exergy consumption, the concept of the technical product coefficient is extended. The technical product coefficient measures the unit exergy consumption of a single fuel flow in the process component, as shown in Equation (A4):

$$k_{ij} = \frac{F_{ij}}{P_j} \tag{A4}$$

where k_{ij} is the technical product coefficient of the ith fuel flow of the jth process component, kW/kW; F_{ij} is the ith fuel of the jth process component, P_j is the jth process component product.

The sum of the technical product coefficients of each fuel flow of the process component is the unit exergy consumption of the component, as shown in Equation (A5):

$$kB_j = \sum k_{ij} = \sum \frac{F_{ij}}{P_j} = \frac{F_j}{P_j} \tag{A5}$$

In thermoeconomics, a logical topological structure diagram enables the resource allocation of the entire system to be represented graphically. This diagram determines how energy is distributed and transformed among process components. The exergy cost model (the mathematical expression of logical topology) will use a series of characteristic equations to describe the allocation of resources among process components.

Fuel and product operators are used to characterize the characteristic equation of the system, as shown in Equation (A6):

$$P_i = B_{i0} + \sum_{j=1}^{n} B_{ij} \; i = 0, 1, 2, \ldots, n \tag{A6}$$

where P_i is the product of the ith process component; B_{i0} indicates the product of the ith component that is output to the environment as the total product of the system; B_{ij} represents the product of the jth component used as fuel to provide the exergy of the jth component.

This equation reveals how a certain component product is distributed within the system as part of the system or as fuel for other components. The equation is expressed in unit exergy consumption, as shown in Equation (A7):

$$P_i = B_{i0} + \sum_{j=1}^{n} k_{ij} P_j \quad i = 0, 1, 2, \ldots, n \tag{A7}$$

This is expressed in matrix form in Equation (A8):

$$\mathrm{P} = \mathrm{P_s} + \langle \mathrm{KP} \rangle \mathrm{P} \tag{A8}$$

where $\mathrm{P_s}$ is the (n × 1) vector containing the final product B_{i0} of the system. $\langle \mathrm{KP} \rangle$ is the $n \times n$ diagonal matrix containing the technical product coefficient.

The above equation can be rewritten as follows:

$$\mathrm{P} = |\mathrm{P}\rangle \mathrm{P_s} \tag{A9}$$

where $|\mathrm{P}\rangle \equiv (\mathrm{U_D} - \langle \mathrm{KP} \rangle)^{-1}$, and $\mathrm{U_D}$ denotes the $(n \times n)$ unit diagonal matrix.

The matrix expression form of the irreversible loss can be obtained by analysis using the exergy balance equation:

$$\mathrm{I} = (\mathrm{K_D} - \mathrm{U_D})\mathrm{P} = |\mathrm{I}\rangle \mathrm{P_S} \tag{A10}$$

where $|I\rangle \equiv (K_D - U_D)|P\rangle$, and $\mathrm{K_D}$ is the $(n \times n)$ diagonal matrix containing unit exergy consumption k.

From the definition of unit exergy consumption in Equation (A3), the expression form of the overall system resource consumption can be analyzed, as shown in Equation (A11):

$$F_T = {}^t k_e |\mathrm{P}\rangle \mathrm{P_s} = {}^t k_e \mathrm{P} \tag{A11}$$

where F_T is the total external resource consumption of the system, kW; ${}^t k_e$ is an n-dimensional vector containing the unit consumption of system input resources.

As mentioned above, since marginal cost expresses the increase in external resources required for each additional unit of product, we studied the relationship between the superposition of irreversible losses and the cost formation by calculating the increase in external resources and the increase in irreversible losses.

Based on Equation (A8), the unit product increment was analyzed using Equation (A12):

$$\Delta \mathrm{P} = \Delta \mathrm{P_s} + \Delta \langle \mathrm{KP} \rangle \mathrm{P}^0 + \langle \mathrm{KP} \rangle \mathrm{P} \tag{A12}$$

Equation (A13) can be obtained through mathematical derivation:

$$\Delta \mathrm{P} = |\mathrm{P}\rangle (\Delta \mathrm{P_s} + \Delta \langle \mathrm{KP} \rangle \mathrm{P}^0) \tag{A13}$$

Based on Equation (A11), the external resource increment was analyzed using Equation (A14):

$$\Delta F_T = \Delta {}^t k_e \mathrm{P}^0 + {}^t k_e \Delta \mathrm{P} \tag{A14}$$

Substituting Equation (A13) into Equation (A14), the matrix form relationship between the product unit exergy cost vector and the external resource increment can be obtained, as shown in Equation (A15):

$$\begin{aligned} \Delta F_T &= \Delta {}^t k_e \mathrm{P}^0 + {}^t k_e |\mathrm{P}\rangle \Delta \langle \mathrm{KP} \rangle \mathrm{P}^0 + {}^t k_e |\mathrm{P}\rangle \Delta \mathrm{P_s} \\ &= \Delta {}^t k_e \mathrm{P}^0 + {}^t k_p^* \Delta \langle \mathrm{KP} \rangle \mathrm{P}^0 + {}^t k_p^* \Delta \mathrm{P_s} \end{aligned} \tag{A15}$$

where ${}^t k_p^* = {}^t k_e |\mathrm{P}\rangle = F_T / \mathrm{P_s}$, which is an n-dimensional vector containing the unit exergy cost of the product.

When the exergy attribute classification method is adopted, after the exergy of the same attribute is collected, the diagonal matrix ($n \times n$) $\langle KP \rangle$ containing the coefficient k_{ij} of the technical product in the system becomes an n-dimensional vector containing the unit exergy consumption of the exergy conversion component. If Equation (A15) is decomposed into scalar form, Equation (A16) is obtained:

$$\Delta F_T = \sum_{i=1}^{n} k_{pi}^{*} \Delta k_i P^0 + k_{pi}^{*} \Delta \omega \tag{A16}$$

where $\Delta\omega$ is the change in the system's external output products; Δk_i indicates the change in exergy consumption of the ith component unit.

The increase of irreversible loss based on Equation (A10) can be analyzed to obtain Equation (A17):

$$\mathrm{I} = \Delta \mathrm{K_D P^0} + (\mathrm{K_D} - \mathrm{U_D})\Delta \mathrm{P} \tag{A17}$$

Substituting Equation (A13) into Equation (A17), the matrix form relationship between unit exergy consumption and irreversible loss increment can be obtained, as shown in Equation (A18):

$$\begin{aligned} \Delta I &= (\Delta K_D + |I\rangle \Delta \langle KP \rangle) P^0 + |I\rangle \Delta P_s \\ &= \Delta K_D \mathrm{P}^0 + I^* \Delta \langle \mathrm{KP} \rangle \mathrm{P}^0 + I^* \Delta \mathrm{P_s} \end{aligned} \tag{A18}$$

where $I^* = |\mathrm{I}\rangle = \mathrm{I}/\mathrm{P_s}$, which is an n-dimensional vector containing the irreversible cost of the process.

When the exergy attribute classification method is adopted, after the exergy of the same attribute is collected, the diagonal matrix ($n \times n$) $\langle \mathrm{KP} \rangle$ containing the technical product coefficient k_{ij} in the system becomes an n-dimensional vector containing the unit exergy consumption of the exergy conversion component. Equation (A18) can be decomposed into scalar form, as shown in Equation (A19):

$$\Delta I_i = \Delta k_i P^0 + \sum_{i=1}^{n} I_i^* \Delta k_i P^0 + I_i^* \Delta \omega \tag{A19}$$

At constant production, the increased fuel consumption is converted into an inside and outside (waste) irreversible increase in the system. When the exergy attribute classification method is adopted, after the exergy of the same attribute is collected, the ($\mathrm{n} \times 1$) vector $\mathrm{P_s}$ containing the final product of the system in the system becomes a scalar ω, so Equation (A20) can be established:

$$\Delta F_T = \Delta P_S + \Delta I = \Delta \omega + \sum_{i=1}^{n} \Delta I_i \tag{A20}$$

Substituting Equation (A19) into Equation (A20) and rearranging the above equation can obtain

$$\Delta F_T = (1 + \sum_{i=1}^{n} I_i^*) \Delta \omega + \sum_{i=1}^{n} (1 + \sum_{i=1}^{n} I_i^*) k_i P^0 \tag{A21}$$

Comparing Equation (A21) with Equation (A16), the irreversible cost and exergy cost have the following relationship:

$$k_{p,i}^{*} = 1 + \sum_{i=1}^{n} I_i^* \tag{A22}$$

where $k_{p,i}^{*}$ is the exergy cost of the ith exergy flow after classification, kW/kW; I_i^* is the irreversible cost of the ith exergy flow, kW/kW.

References

1. Köse, Ö.; Koç, Y.; Yağlı, H. Performance improvement of the bottoming steam Rankine cycle (SRC) and organic Rankine cycle (ORC) systems for a triple combined system using gas turbine (GT) as topping cycle. *Energy Convers. Manag.* **2020**, *211*, 112745. [CrossRef]
2. Xi, H.; Wu, X.; Chen, X.; Sha, P. Artificial intelligent based energy scheduling of steel mill gas utilization system towards carbon neutrality. *Appl. Energy* **2021**, *295*, 117069. [CrossRef]
3. Memon, A.G.; Memon, R.A.; Qureshi, S. Thermo-environmental and economic analyses of combined cycle power plants with regression modelling and optimization. *Appl. Therm. Eng.* **2017**, *125*, 489–512. [CrossRef]
4. Wu, X.; Xi, H.; Ren, Y.; Lee, K.Y. Power-Carbon Coordinated Control of BFG-Fired CCGT Power Plant Integrated with Solvent-based Post-Combustion CO_2 Capture. *Energy* **2021**, *226*, 120435. [CrossRef]
5. Postnikov, I.; Stennikov, V.; Penkovskii, A. Integrated Energy Supply Schemes on Basis of Cogeneration Plants and Wind Power Plants. *Energy Procedia* **2019**, *158*, 154–159. [CrossRef]
6. Chen, X.; Wu, X.; Lee, K.Y. The mutual benefits of renewables and carbon capture: Achieved by an artificial intelligent scheduling strategy. *Energy Convers. Manag.* **2021**, *233*, 113856. [CrossRef]
7. Kler, A.M.; Potanina, Y.M.; Marinchenko, A.Y. Co-optimization of thermal power plant flowchart, thermodynamic cycle parameters, and design parameters of components. *Energy* **2020**, *193*, 116679. [CrossRef]
8. Zhu, M.; Wu, X.; Shen, J.; Lee, K. Dynamic modeling, validation and analysis of direct air-cooling condenser with integration to the coal-fired power plant for flexible operation. *Energy Convers. Manag.* **2021**, *245*, 114601. [CrossRef]
9. Chen, Q.; Han, W.; Zheng, J.-J.; Sui, J.; Jin, H.-G. The exergy and energy level analysis of a combined cooling, heating and power system driven by a small scale gas turbine at off design condition. *Appl. Therm. Eng.* **2014**, *66*, 590–602. [CrossRef]
10. Jie, P.; Yan, F.; Wen, Z.; Li, J. Evaluation of the biomass gasification-based combined cooling, heating and power system using the maximum generalized entropy principle. *Energy Convers. Manag.* **2019**, *192*, 150–160. [CrossRef]
11. Wang, J.; Chen, Y.; Dou, C.; Gao, Y.; Zhao, Z. Thermodynamic performance analysis and comparison of a combined cooling heating and power system integrated with two types of thermal energy storage. *Energy* **2018**, *163*, 475–489. [CrossRef]
12. Khan, M.S.; Abid, M.; Ratlamwala, T.A.H. Energy, Exergy and Economic Feasibility Analyses of a 60 MW Conventional Steam Power Plant Integrated with Parabolic Trough Solar Collectors Using Nanofluids. *Iran. J. Sci. Technol. Trans. Mech. Eng.* **2019**, *43*, 193–209. [CrossRef]
13. Mohammadi, A.; Ahmadi, M.H.; Bidi, M.; Joda, F.; Valero, A.; Uson, S. Exergy analysis of a Combined Cooling, Heating and Power system integrated with wind turbine and compressed air energy storage system. *Energy Convers. Manag.* **2017**, *131*, 69–78. [CrossRef]
14. Adibhatla, S.; Kaushik, S.C. Energy, exergy and economic (3E) analysis of integrated solar direct steam generation combined cycle power plant. *Sustain. Energy Technol. Assess.* **2017**, *20*, 88–97. [CrossRef]
15. Cai, Y.; Li, J.; Liu, H.; He, Q. Exergy Analysis of Compressed Air Energy Storage System Combined with Absorption Chiller. *Zhongguo Dianji Gongcheng Xuebao/Proc. Chin. Soc. Electr. Eng.* **2018**, *38*, 186–194.
16. Lozano, M.A.; Valero, A. Theory of the exergetic cost. *Energy* **1993**, *18*, 939–960. [CrossRef]
17. Erlach, B.; Serra, L.; Valero, A. Structural theory as standard for thermoeconomics. *Energy Convers. Manag.* **1999**, *40*, 1627–1649. [CrossRef]
18. Haydargil, D.; Abuşoğlu, A. A comparative thermoeconomic cost accounting analysis and evaluation of biogas engine-powered cogeneration. *Energy* **2018**, *159*, 97–114. [CrossRef]
19. Marques, A.S.; Carvalho, M.; Ochoa, Á.A.V.; Souza, R.J.; dos Santos, C.A.C. Exergoeconomic assessment of a compact electricity-cooling cogeneration unit. *Energies* **2020**, *13*, 5417. [CrossRef]
20. Boyaghchi, F.A.; Chavoshi, M. Multi-criteria optimization of a micro solar-geothermal CCHP system applying water/CuO nanofluid based on exergy, exergoeconomic and exergoenvironmental concepts. *Appl. Therm. Eng.* **2017**, *112*, 660–675. [CrossRef]
21. Ghaffarpour, Z.; Mahmoudi, M.; Mosaffa, A.; Farshi, L.G. Thermoeconomic assessment of a novel integrated biomass based power generation system including gas turbine cycle, solid oxide fuel cell and Rankine cycle. *Energy Convers. Manag.* **2018**, *161*, 1–12. [CrossRef]
22. Villafana, E.D.S.; Bueno, J.P.V.M. Thermoeconomic and environmental analysis and optimization of the supercritical CO_2 cycle integration in a simple cycle power plant. *Appl. Therm. Eng.* **2019**, *152*, 1–12. [CrossRef]
23. Ameri, M.; Mohammadzadeh, M. Thermodynamic, thermoeconomic and life cycle assessment of a novel integrated solar combined cycle (ISCC) power plant. *Sustain. Energy Technol. Assess.* **2018**, *27*, 192–205. [CrossRef]
24. Wang, J.; Liu, W.; Meng, X.; Liu, X.; Gao, Y.; Yu, Z.; Bai, Y.; Yang, X. Study on the Coupling Effect of a Solar-Coal Unit Thermodynamic System with Carbon Capture. *Energies* **2020**, *13*, 4779. [CrossRef]
25. Zhai, R.; Liu, H.; Wu, H.; Yu, H.; Yang, Y. Analysis of Integration of MEA-Based CO_2 Capture and Solar Energy System for Coal-Based Power Plants Based on Thermo-Economic Structural Theory. *Energies* **2018**, *11*, 1284. [CrossRef]
26. Arena, A.P.; Borchiellini, R. Application of different productive structures for thermoeconomic diagnosis of a combined cycle power plant. *Int. J. Therm. Sci.* **1999**, *38*, 601–612. [CrossRef]
27. Tsatsaronis, G.; Park, M.-H. On avoidable and unavoidable exergy destructions and investment costs in thermal systems. *Energy Convers. Manag.* **2002**, *43*, 1259–1270. [CrossRef]

28. Valero, A.; Serra, L.; Uche, J. Fundamentals of exergy cost accounting and thermoeconomics. Part I: Theory. *J. Energy Resour. Technol. Trans. ASME* **2006**, *128*, 1–8. [CrossRef]
29. Torres, C.; Valero, A.; Rangel, V.; Zaleta, A. On the cost formation process of the residues. *Energy* **2008**, *33*, 144–152. [CrossRef]
30. Valero, A.; Usón, S.; Torres, C.; Stanek, W. Theory of exergy cost and thermo-ecological cost. In *Thermodynamics for Sustainable Management of Natural Resources*; Stanek, W., Ed.; Springer: Berlin/Heidelberg, Germany, 2017; ISBN 978-3-319-48648-2.
31. Torres, C.; Valero, A.; Serra, L.; Royo, J. Structural theory and thermoeconomic diagnosis: Part I. On malfunction and dysfunction analysis. *Energy Convers. Manag.* **2002**, *43*, 1503–1518. [CrossRef]
32. Valero, A.; Correas, L.; Zaleta, A.; Lazzaretto, A.; Verda, V.; Reini, M.; Rangel, V. On the thermoeconomic approach to the diagnosis of energy system malfunctions. Part 1: The TADEUS problem and Part 2: Malfunction definitions and assessment. *Energy* **2004**, *29*, 1875–1907. [CrossRef]
33. Usón, S.; Uche, J.; Martínez, A.; del Amo, A.; Acevedo, L.; Bayod, A. Exergy assessment and exergy cost analysis of a renewable-based and hybrid trigeneration scheme for domestic water and energy supply. *Energy* **2019**, *168*, 662–683. [CrossRef]
34. Valero, A.; Lerch, F.; Serra, L.; Royo, J. Structural theory and thermoeconomic diagnosis: Part II: Application to an actual power plant. *Energy Convers. Manag.* **2002**, *43*, 1519–1535. [CrossRef]
35. Cuadra, C.T. Symbolic thermoeconomic analysis of energy systems. In *Exergy, Energy System Analysis and Optimization: Thermoeconomic Analysis Modeling, Simulation and Optimization in Energy Systems*; EOLSS Publishers: Oxford, UK, 2006.
36. Torres, C.; Valero, A. The Exergy Cost Theory Revisited. *Energies* **2021**, *14*, 1594. [CrossRef]

processes MDPI

Article

Optimization of Pipeline Network Layout for Multiple Heat Sources Distributed Energy Systems Considering Reliability Evaluation †

Ziyuan Cui [1], Hai Lin [1], Yan Wu [2], Yufei Wang [1,*] and Xiao Feng [2]

1 School of Chemical Engineering and Technology, China University of Petroleum, Beijing 102249, China; 2018010581@student.cup.edu.cn (Z.C.); 2018010588@student.cup.edu.cn (H.L.)

2 School of Chemical Engineering & Technology, Xi'an Jiaotong University, Xi'an 710049, China; wuyan91@stu.xjtu.edu.cn (Y.W.); xfeng@mail.xjtu.edu.cn (X.F.)

* Correspondence: wangyufei@cup.edu.cn

† This paper is an extended version of our paper published in Chemical Engineering Transactions for the 23rd Conference on Process Integration, Modelling and Optimisation for Energy Saving and Pollution Reduction (PRES'20), Xi'an, Shaanxi, China, 17–21 August 2020.

Abstract: Due to the target of carbon neutrality, energy saving has become more important than ever. At the same time, the widespread use of distributed energy systems and the regional utilization of industrial waste heat leads to the existence of multiple heat sources in an area. Therefore, how to design an economical and reliable pipeline network to meet energy-saving demand under multiple heat source conditions becomes a problem. In this work, an optimization method is established to determine the optimal pipeline network topology with minimum total annual cost. In this optimization method, Star tree algorithm, Kruskal algorithm and GeoSteiner algorithm are combined with a linear programming model to establish a distributed energy pipeline network for multiple heat sources. The model incorporates Euclidean Steiner Minimum Tree and Rectilinear Steiner Minimum Tree in the consideration of the topology optimization of Distributed Energy System pipeline networks. Four pipeline network topologies, STAR, Minimum Spanning Tree, Euclidean Steiner Minimum Tree and Rectilinear Steiner Minimum Tree, are evaluated in this paper from economic and reliability perspectives. A case extracted from a real industrial park where steam is the medium is used to prove the validity of the model. The optimization results show that a Euclidean Steiner Minimum Tree pipeline network has a lower total annual cost than three other types of pipeline network and ranks second in reliability. Considering the comprehensive economy and reliability, ESMT is the optimal pipeline network type of distributed energy system with steam as the medium.

Keywords: distributed energy system; pipeline network layout; reliability; GeoSteiner algorithm; Kruskal algorithm; Star tree algorithm

Citation: Cui, Z.; Lin, H.; Wu, Y.; Wang, Y.: Feng, X. Optimization of Pipeline Network Layout for Multiple Heat Sources Distributed Energy Systems Considering Reliability Evaluation. *Processes* **2021**, *9*, 1308. https://doi.org/10.3390/pr9081308

Academic Editors: Pei Liu, Ming Liu and Xiao Wu

Received: 9 July 2021
Accepted: 26 July 2021
Published: 28 July 2021

Publisher's Note: MDPI stays neutral with regard to jurisdictional claims in published maps and institutional affiliations.

Copyright: © 2021 by the authors. Licensee MDPI, Basel, Switzerland. This article is an open access article distributed under the terms and conditions of the Creative Commons Attribution (CC BY) license (https://creativecommons.org/licenses/by/4.0/).

1. Introduction

Based on the proposal of carbon neutrality target, optimization for energy-saving and emission-reduction in energy system field attracts various academic research. As a high efficiency and promising technology [1] for energy conservation, a distributed energy system (DES) can use natural gas and renewable clean energy as primary energy source, including biomass, solar energy and hydrogen, and achieve the step utilization of the energy. Therefore, it has become very popular recently compared with traditional centralized energy systems.

DES can be improved from several aspects and a number of progress have been made. Firstly, DES can be improved by optimizing the process within the station. A number of works focus on applying renewable energy in DES. For instance, Ren et al. [2]

combined solar thermochemical fuel production technology with cogeneration of cold, heat and power to improve energy utilization efficiency. Multi-objective linear programming was used to optimize the operation of DES. Wu et al. [3] established a multi-objective optimization model considering cogeneration and renewable energy. The best combination of relative technologies is determined to reduce total annual cost. Jing et al. [4] proposed a comprehensive framework of multi-objective optimization and multi-criteria evaluation for the optimization of DES. In addition, solar-assisted solid oxide fuel cell technology was introduced into the system. Besides considering renewable energy, some other elements are also considered to improve the performance of DES. For example, Yuan et al. [5] considered thermal energy storage system and proposed a control strategy to improve the performance of the DES. Wang et al. [6] used different waste heat recovery technologies to recover heat with different temperatures for further improving the system efficiency. A MILP model was used to optimize the system. In some other studies, more evaluation criterions are used rather than single economic performance. For example, Buoro et al. [7] developed a multi-objective MILP model to determine the optimal operation mode of DES with the goal of minimizing the emissions and total cost of the system. Wang et al. [8] compared different DES modeling methods under various uncertainties, and optimized the operation of the system by combining Monte Carlo simulation and multi-objective optimization (economic and environmental objectives).

Beyond the work of improving energy generation process, how to effectively transport energy to consumers is also a hot spot of research interests. During transportation, it is desirable to reduce heat loss with an acceptable investment, so Keçebaş et al. [9] optimized the thickness of insulation layer of pipeline with the objective of optimal economy. Salem et al. [10] employed heating degree-days method to optimize the insulation thickness of pipelines. The diameter of pipeline is also a key element that has a significant impact on pipeline investment and power consumption. Li et al. [11] determined the pipeline diameter by considering the variation of cold and thermal loads based on fixed road topology constraint using GA algorithm. Wang et al. [12] considered the siting and sizing of the station considering pipeline network design using a cost-based function model. The topology of pipeline network will also affect the performance of energy transportation process. Currently, tree pipeline network topology is frequently studied, for instance, Zeng et al. [13] established an optimization method to determine the most economic pipeline network based on substation loads. Improved GA is used to find the optimal pipeline diameter for the tree pipeline network topology. Mehleri et al. [14] developed a MILP model to optimize the heating pipeline network considering the storage tank capacity and the size of the tree pipeline network topology. Khir et al. [15] proposed a MIP model to optimize the tree pipeline network size and layout based on the investment and operating costs in a district cooling system. Chan et al. [16] combined GA with local search techniques to optimize tree pipeline network layout of district cooling systems. Carl et al. [17] established a linear model to optimize the pipeline network layout considering the variation in the needs of consumers in different periods. Studies on pipeline network topology can be found in some other area except DES, for example, Sanaye et al. [18] adopted GA to optimize the gas pipeline network layout with MST topology structure. Su et al. [19] considered the structure of tree pipeline network and adopted NSGA-II algorithm to optimize the gas pipeline network. Sokolov et al. [20] proposed a new dynamic programming methodology to optimize the tree-shaped water network. Liu [21] considered the problem of pipeline network layout on the road and proposed a pipeline network optimization method based on Kruskal algorithm to optimize the direction of load access to the pipeline network. However, most of the above-mentioned studies are only based on tree network topology for optimization, with little consideration given to other topologies of the pipeline network.

Reliability is an important aspect in pipeline network design. Some studies have been done in this subject, for instance, Rimkevicius et al. [22] proposed probabilistic mathematical analysis, deterministic hydrothermal force analysis and deterministic probabilistic structural integrity analysis to evaluate the reliability of pipeline network energy system. Shan et al. [23] evaluated the dynamic reliability of heating pipeline network. Alsharqawi et al. [24] used the network fault tree method to determine the work and fault state of pipeline networks. Chen et al. [25] considered the probability distribution of gas shortages under the normal scenario and proposed an integrated methodology to assess the gas supply reliability of a gas pipeline network. Yu et al. [26] proposed an integrated method considering hydraulic and pressure constrains to provide a new way to assess the reliability of natural gas pipeline network. Zhu et al. [27] proposed an operation state determination model to assess an actual pipeline network based on failure and repair processes.

At present, large amount of low degree waste heat is discharged in industries, and such heat can be used in the region to save energy for lowering carbon emission. When industry and DES are combined to provide energy, several heat sources are existed in the region and such situation requires more consideration on how to effectively establish the pipeline network topology, but this point has not been well studied. Moreover, it is important to find the best pipe connection between heat source and heat users to cut down the investment and energy consumption. Aim to solve such problem, the new method is proposed in this work to optimize pipeline network under multiple heat sources situation. The aim of the methodology is to find the optimal pipeline network considering both economic and reliable factors. Moreover, different pipeline network topologies have different performance in economic and reliability, but the research considered reliability in multiple heat sources condition were rare. To study the above-mentioned problems, this work firstly uses clustering algorithm to divide consumers into different heat sources. Then, through coupling clustering algorithm combining Star tree algorithm, Kruskal algorithm, GeoSteiner algorithm and Linear Programming (LP) model, different pipeline network topology and flow rate of each branch pipeline are solved. This work takes the minimum value of total annual cost (TAC) as the objective function, including construction cost, pressure loss cost and heat loss cost of pipeline network. Additionally, then, the reliability of pipeline network is assessed. Finally, both the economy and reliability aspects of different types of pipeline network are evaluated to provide a reference for the optimization of pipeline network layout of DES.

2. Methodology

2.1. Problem Statement

For the energy supply problem in a region with industries, it is necessary to connect multiple heat sources with a number of consumers scattered in different parts of the region to form an energy transportation pipeline network. How to construct a pipeline network topology with the low cost and the high reliability should be considered.

2.1.1. Assumption

(1) The pipeline length is the distance between the two vertices that are connected.
(2) The steam in the pipeline network system is an incompressible fluid.
(3) The temperature of the fluid in each pipeline is constant.
(4) The flow rate of the fluid in each pipeline is constant.

2.1.2. Given

(1) The number of consumers and heat sources.
(2) The coordinates and heat demand of consumers.
(3) The coordinates and the heat supply of heat source.

2.1.3. Determine

(1) The pipeline network topology of multiple heat sources DES.
(2) Total annual cost and reliability of multiple heat sources DES.

As shown in Figure 1, this work first classifies the consumers into clusters and applies the Star tree algorithm, Kruskal algorithm or GeoSteiner algorithm to obtain different pipeline network topologies. LP model is then applied to find the optimal flow rate of each branch pipeline. After that, parameters such as the length and flow rate of branch pipelines are embedded in the pipeline network system, and the economic model with minimum TAC as the objective function is solved. Finally, reliability assessment model is solved.

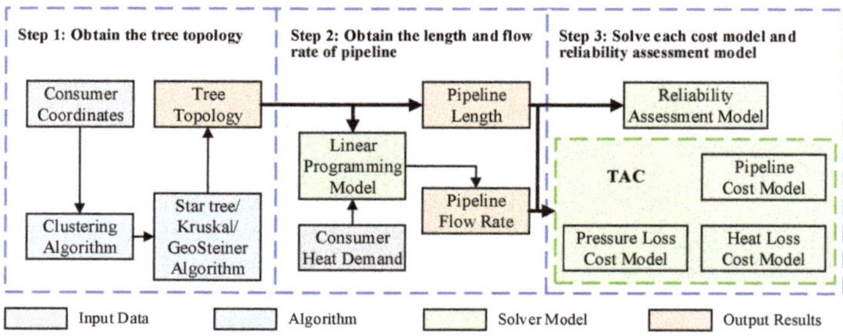

Figure 1. Schematic diagram of proposed methodology.

Four different types of pipeline network topologies are STAR, Minimum Spanning Tree-MST, Rectilinear Steiner Minimum Tree-RSMT and Euclidean Steiner Minimum Tree-ESMT. Among them, STAR is solved by Star tree algorithm, MST is solved by Kruskal algorithm, and RSMT and ESMT are solved by GeoSteiner algorithm.

Figure 2 presents the optimization schematic diagram of a DES pipeline network layout consisting of 3 heat sources and 14 consumers. The optimal design of the pipeline network layout consists of four parts. Firstly, clustering algorithm (see Section 3.1 for details) is applied to cluster all consumers, which is able to avoid too many pipelines with high flow rates to reduce investment costs, as shown in Figure 2a. Secondly, each heat source is connected to obtain a tree (see Sections 3.2–3.4 for details) to achieve the energy supply–demand balance between heat source, as shown in Figure 2b. Thirdly, the consumers within each cluster are connected to obtain a tree (see Sections 3.2–3.4 for details). Figure 2c takes the ESMT pipeline network as an example, and Steiner Point can be obtained by GeoSteiner algorithm in Section 3.4. Finally, the complete pipeline network layout of the multiple heat sources DES is obtained, as shown in Figure 2d.

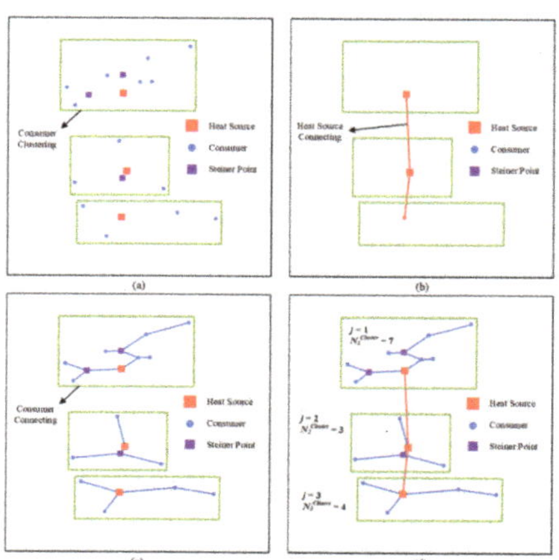

Figure 2. Schematic of optimization of multiple heat sources DES network system: (**a**) cluster consumers; (**b**) generate heat source–heat source tree; (**c**) generate heat source–consumers tree; (**d**) obtain the layout of multiple heat sources DES pipeline network.

2.2. Mathematical Model

2.2.1. Objective Function

The objective function in this paper is to minimize the total annual cost $TAC(¥ \cdot a^{-1})$, including construction cost $C^{Pipeline}(¥ \cdot a^{-1})$, pressure loss cost $C^{Pressure\ Loss}(¥ \cdot a^{-1})$ and heat loss cost $C^{Heat\ Loss}(¥ \cdot a^{-1})$, of the pipeline network, as shown Equation (1). It is noted that steam is used as heat transfer medium in this work, as the studied DES providing steam to industry consumer. For the DES providing hot water, the method can be easily adopted by changing properties from steam to hot water.

$$\min TAC = C^{Pipeline} + C^{Pressure\ Loss} + C^{Heat\ Loss} \tag{1}$$

$$C^{Pipeline} = \sum_{j=1}^{N^{Cluster}} \sum_{i=1}^{N_j^{Cluster}} C_{i,j}^{Pipeline} \tag{2}$$

$$C^{Pressure\ Loss} = \sum_{j=1}^{N^{Cluster}} \sum_{i=1}^{N_j^{Cluster}} C_{i,j}^{Pressure\ Loss} \tag{3}$$

$$C^{Heat\ Loss} = \sum_{j=1}^{N^{Cluster}} \sum_{i=1}^{N_j^{Cluster}} C_{i,j}^{Heat\ Loss} \tag{4}$$

where $N^{Cluster}$ is the number of clusters of pipeline network system and $N_j^{Cluster}$ is the number of branch pipelines in cluster j. For instance, in Figure 2, the number of clusters of pipeline network system $N^{Cluster} = 3$, and the number of branch pipelines within each cluster is $N_1^{Cluster} = 7$, $N_2^{Cluster} = 3$, $N_3^{Cluster} = 4$.

2.2.2. Linear Programming Model

Besides the length of pipeline, the diameter of pipeline also has a great impact on the total cost of pipeline network. To consider diameter of pipeline, flowrate in each branch of pipeline network should be optimized.

After determining the connection of the pipeline network, the sum of the mass flow of steam in all branch pipelines of the steam pipeline network W^S can be obtained by solving the LP model shown in Equation (5).

$$min\ W^S = \sum_{z}^{N^{V^T}} \sum_{x}^{N^{V^T}} \sum_{y}^{N^{V^T}} b_{x,y,z}^{Binary} \times w_{x,y,z} \tag{5}$$

The constraints of the LP model are shown in Equations (6)–(10).

$$\sum_{x}^{N^{V^T}} \sum_{y}^{N^{V^T}} \left((-1)^D b_{x,y,z}^{Binary} \times w_{x,y,z} + (-1)^D b_{x,y,z}^{Binary} \times w_{y,x,z} \right) - W_z^S = 0 \tag{6}$$

$$b_{x,y,z}^{Binary} = \begin{cases} 1, (x,y) \in \mathbf{E_z^T} \\ 0, \text{otherwise} \end{cases}, \forall x,y,z \in \mathbf{V^T} \tag{7}$$

$$D = \begin{cases} 1, x < y \\ 2, x \geq y \end{cases} \tag{8}$$

$$w_{x,y,z} \geq 0 \tag{9}$$

$$w_{y,x,z} \geq 0 \tag{10}$$

where N^{V^T} is the number of vertices in the pipeline network topology. The binary variables $b_{x,y,z}^{Binary}$ are obtained from Equation (7) based on the result of the pipeline network topology.

$b_{x,y,z}^{Binary}$ indicates whether there is an edge (x, y), which belongs to the connected tree, is directly connected to vertex z. Equation (6) accounts for the material balance for each vertex in the pipeline network topology. $w_{x,y,z}(\text{kg} \cdot \text{s}^{-1})$ is the mass flow rate of steam in the edge (x, y) connected to vertex z when the material balance is performed on the vertex z. D is an exponent, which is determined by Equation (8). $W_z^S(\text{kg} \cdot \text{s}^{-1})$ is the mass flow rate at vertex z. If vertex z produces steam, W_z^S is negative, otherwise, W_z^S is positive. The flow in each branch line has two directions from x to y and from y to x. In Equation (7), for each vertex z, there is a set of edges $\mathbf{E}_\mathbf{z}^\mathbf{T}$ connected to it, each set $\mathbf{E}_\mathbf{z}^\mathbf{T}$ is a subset of the connected tree, and the number of edge sets is N^{V^T}, so that one of the two vertices corresponding to each edge in $\mathbf{E}_\mathbf{z}^\mathbf{T}$ must be z. If the edge (x, y) belongs to $\mathbf{E}_\mathbf{z}^\mathbf{T}$, $b_{x,y,z}^{Binary}$ is equal to 1, otherwise, it is equal to 0.

In Equation (6), edge (x, y) calculates the mass flow rate twice from x to y $(w_{x,y,z})$ and from y to x $(w_{y,x,z})$. The two mass flow rates are independent with each other due to the constraints (Equations (9) and (10)). After solving the optimization, one of $w_{x,y,z}$ and $w_{y,x,z}$ must be 0, and the other is a positive number. The aim of solving Equation (5) is to find the value of $w_{x,y,z}$ and $w_{y,x,z}$ and let one of them to be 0.

2.2.3. Pipeline Cost Model

The construction cost [28] of the pipeline $C_{i,j}^{Pipeline}(\text{¥} \cdot \text{a}^{-1})$ and the unit price of the pipeline $a_{i,j}(\text{¥} \cdot \text{m}^{-1})$ are obtained by Equations (11) and (12).

$$C_{i,j}^{Pipeline} = a_{i,j} L_{i,j} \frac{I(1+I)^{N^{year}}}{(1+I)^{N^{year}} - 1} \tag{11}$$

$$a_{i,j} = 5.74 Wt_{i,j} + 1295\left(D_{i,j}^{outer,pipeline}\right)^{0.48} + 47.6 + 2065 D_{i,j}^{outer,pipeline} \tag{12}$$

$$Wt_{i,j} = 644.3\left(D_{i,j}^{inner,pipeline}\right)^2 + 72.5 D_{i,j}^{inner,pipeline} + 0.4611 \tag{13}$$

$$D_{i,j}^{outer,pipeline} = 1.052 D_{i,j}^{inner,pipeline} + 0.005251 \tag{14}$$

$$D_{i,j}^{inner,pipeline} = \sqrt{\frac{4W_{i,j}}{\pi \rho u}} \tag{15}$$

where, $N^{year}(\text{a})$ is the life cycle of the pipeline network system, I is the annual interest rate, and $L_{i,j}(\text{m})$ is the length of the branch pipeline.

The weight per unit length of the pipeline $Wt_{i,j}(\text{kg} \cdot \text{m}^{-1})$, the outer diameter $D_{i,j}^{outer,pipeline}(\text{m})$ of the pipeline and the inner diameter $D_{i,j}^{inner,pipeline}(\text{m})$ of the pipeline are obtained from Equations (13)–(15), where $W_{i,j}(\text{kg} \cdot \text{s}^{-1})$ is the mass flow rate of steam in the pipeline, $\rho(\text{kg} \cdot \text{m}^{-3})$ is the density of the fluid, and $u(\text{m} \cdot \text{s}^{-1})$ is the velocity of the fluid in the pipeline.

2.2.4. Pressure Loss Cost Model

The pressure loss cost $C_{i,j}^{Pressure\ Loss}(\text{¥} \cdot \text{a}^{-1})$ of the pipeline network, the shaft power $N_{i,j}(\text{W})$ to overcome the resistance during steam transportation and the effective power $Ne_{i,j}(\text{W})$ of the transportation equipment are calculated by Equations (16)–(19). To consider the local resistance, it is assumed an average of four standard elbows (90°) per 100 m, then $\zeta_{i,j}^{Elbow}$ can be calculated by Equation (20).

$$C_{i,j}^{Pressure\ Loss} = \frac{a^E T^{Operating} N_{i,j}}{1000} \tag{16}$$

$$N_{i,j} = \frac{Ne_{i,j}}{\eta} \tag{17}$$

$$Ne_{i,j} = gH_{i,j}^{f}W_{i,j} \tag{18}$$

$$H_{i,j}^{f} = \left(\sigma\frac{L_{i,j}}{D_{i,j}^{inner,pipeline}} + \zeta_{i,j}^{Elbow} + \zeta_{i,j}\right)\frac{u^2}{2g} \tag{19}$$

$$\zeta_{i,j}^{Elbow} = \frac{L_{i,j}}{25}\zeta_{i,j}^{E} \tag{20}$$

$$\zeta_{i,j} = \begin{cases} 0, & [i=0] \vee \left[S_{i-1,j}^{inner,pipeline} = S_{i,j}^{inner,pipeline}\right] \\ \left(1 - \frac{S_{i-1,j}^{inner,pipeline}}{S_{i,j}^{inner,pipeline}}\right)^2, & S_{i-1,j}^{inner,pipeline} < S_{i,j}^{inner,pipeline} \\ 0.5\left(1 - \frac{S_{i,j}^{inner,pipeline}}{S_{i-1,j}^{inner,pipeline}}\right)^2, & S_{i-1,j}^{inner,pipeline} > S_{i,j}^{inner,pipeline} \end{cases} \tag{21}$$

$$S_{i,j}^{inner,pipeline} = \frac{\pi}{4}\left(D_{i,j}^{inner,pipeline}\right)^2 \tag{22}$$

where $a^E\left(¥ \cdot kW^{-1} \cdot h^{-1}\right)$ is the power cost, $T^{Operating}(h)$ is the number of annual operating hours of the device, and η is the efficiency of the conveying equipment. The head loss $H_{i,j}^{f}(m)$ caused by the resistance along the pipeline is calculated by Equation (19), where $\zeta_{i,j}$ is the local resistance coefficient when the cross section suddenly increases or decreases, $\zeta_{i,j}^{E}$ is the local resistance coefficient at the elbow, $S_{i,j}^{inner,pipeline}\left(m^2\right)$ is the cross-sectional area of the branch pipeline, and σ is the pipeline friction coefficient.

2.2.5. Heat Loss Cost Model

The heat loss cost $C_{i,j}^{Heat\ Loss}\left(¥ \cdot a^{-1}\right)$ can be calculated by Equation (23). Considering the insulation, the heat loss $Q_{i,j}\left(kJ \cdot m^{-1} \cdot s^{-1}\right)$ of the steam pipeline network is calculated by Equation (24).

$$C_{i,j}^{Heat\ Loss} = \frac{a^S Q_{i,j} L_{i,j} T^{Operating}}{q} \tag{23}$$

$$Q_{i,j} = \frac{\varepsilon\pi(T - T^a)}{1000\left(\frac{1}{\lambda}ln\frac{D_{i,j}^{outer,insulation}}{D_{i,j}^{inner,insulation}} + \frac{2}{\varepsilon D_{i,j}^{outer,insulation}}\right)} \tag{24}$$

where $a^S\left(¥ \cdot kg^{-1}\right)$ is the unit price of steam, $q\left(kJ \cdot kg^{-1}\right)$ is the latent heat of steam, $T(°C)$ is the temperature of the fluid medium in the pipeline, $T^a(°C)$ is the ambient temperature, $D_{i,j}^{inner,insulation}(m)$ and $D_{i,j}^{outer,insulation}(m)$ are the inner diameter and outer diameter of the insulation layer, $\lambda\left(W \cdot m^{-1} \cdot K^{-1}\right)$ is the thermal conductivity of the insulation material at the average temperature, $\varepsilon\left(W \cdot m^{-2} \cdot K^{-1}\right)$ is the heat transfer coefficient between the outer surface of the insulation layer and the surroundings. The inner diameter of the insulation layer $D_{i,j}^{inner,insulation}$ is taken as the outer diameter of the steam pipeline $D_{i,j}^{outer,pipeline}$. Due to the difference in fluid temperatures and pipeline diameters, the thickness of the insulation layer is different. In this work, the thickness of insulation layer is selected according to Table 1.

Table 1. Table for selecting glass wool insulation layer thickness.

DN (mm)	Pipeline Surface Temperature (°C)				
	≤60	≤150	≤250	≤300	≤350
	Insulation Layer Thickness (mm)				
15	30	30	40	50	50
20	30	30	40	50	50
25	30	30	50	50	60
40	30	50	50	60	60
50	30	50	50	60	70
80	30	50	60	70	70
100	30	50	60	70	80
150	30	60	70	70	80
200	30	60	70	80	90
250	30	60	70	80	90
300	30	60	70	80	90
350	30	50	70	80	90
400	30	50	70	80	90
450	30	50	70	80	90
500	30	50	70	80	90
600	30	50	70	80	90
700	30	50	70	80	90
800	30	50	70	80	90
900	30	50	70	80	100
1000	30	50	70	80	100
1100	30	50	70	80	100
1200	30	50	70	80	100

2.2.6. Reliability Assessment Model

To calculate the reliability of the pipeline network system $R^{Reliability}$, the connected probability between the consumer α and the heat source $P_\alpha^{Probability}$ should be known first. Connected probability is used to describe when some part of the pipeline does not work, the probability that the consumer can still connect with the heat source. Reliability $R^{Reliability}$ and connected probability $P_\alpha^{Probability}$ can be calculated by the Equations (25) and (26). $P_\alpha^{Probability}$ is the product value of the probability of connectivity of each pipeline between consumer α to the heat source.

$$R^{Reliability} = \frac{\sum_{\alpha=1}^{N^{VT}} \left(P_\alpha^{Probability} \times Q_\alpha^{Heat\ Demand}\right)}{\sum_{\alpha=1}^{N^{VT}} Q_\alpha^{Heat\ Demand}} \tag{25}$$

$$P_\alpha^{Probability} = \prod_\beta \left(p_{\alpha,\beta}^{Path}\right)^{\frac{L_{\alpha,\beta}}{1000}} \tag{26}$$

where $Q_\alpha^{Heat\ Demand}$ is the heat demand of consumer α, $p_{\alpha,\beta}^{Path}\left(\mathrm{km}^{-1}\right)$ is the probability of connecting the βth pipeline in the path between consumer α and the heat source, and the size of the array $p_{\alpha,\beta}^{Path}$ is not exactly the same for different consumers, and $L_{\alpha,\beta}(\mathrm{m})$ is the length of the βth pipeline between consumer α and the heat source.

3. Algorithm

This method uses the concept of graph theory. In graph theory, each graph consists of some preset points and the lines connecting them. Such a graph is used to describe some certain relation between some things. The concept of graph theory is very suitable to solve the problem raised in this work. The algorithms chose in this work are all well accepted algorithms to solve such problems. This work aims to find the best connection between heat sources and heat users. The computational complexity is the combination of clustering, LP, and different graph theory algorithms.

3.1. Clustering Algorithm

The clustering algorithm is used in this paper to obtain a set of clusters $\mathbf{C} = (C_1, C_2, \cdots, C_{N^{Cluster}})$ by assigning different consumers to the heat source with the shortest Euclidean distance. The specific process of the clustering algorithm is shown in Algorithm 1.

Algorithm 1: Clustering Algorithm

Input: The coordinates of consumers $[x, y] = \{(x_1, y_1), (x_2, y_2), \ldots, (x_{N^{Consumer}}, y_{N^{Consumer}})\}$, the number of clusters $N^{Cluster}$, and the coordinates of heat sources $[\mu, \lambda] = \{(\mu_1, \lambda_1), (\mu_2, \lambda_2), \ldots, (\mu_{N^{Consumer}}, \lambda_{N^{Consumer}})\}$.
Output: A set of clusters $\mathbf{C} = (C_1, C_2, \cdots, C_{N^{Cluster}})$.

Let $C_j = \{\text{Heat source}_j\}, j = 1, 2, \cdots, N^{Cluster}$
for all $i = 1, 2, \cdots, N^{Consumer}$ **do**
$d_{ij} = \sqrt{(x_i - \mu_j)^2 + (y_i - \lambda_j)^2}, j = 1, 2, \cdots, N^{Cluster}$
$\xi_i = argmin d_{ij}, j = 1, 2, \cdots, N^{Cluster}$
$C_{\xi_i} = C_{\xi_i} \cup \{(x_i, y_i)\}$
end for
return $\mathbf{C} = (C_1, C_2, \cdots, C_{N^{Cluster}})$

3.2. Star Tree Algorithm

Star tree algorithm [29] is a graph theory algorithm that directly connects the consumer to a selected center. The specific flow of the algorithm is shown in Algorithm 2.

Algorithm 2: Star tree algorithm

Input: A graph $\mathbf{G} = (\mathbf{V}, \mathbf{E})$.
Output: A Star tree $\mathbf{T} = (\mathbf{V^T}, \mathbf{E^T})$.

Let $\mathbf{E^T} = \Phi$
choose the heat source as the center of all consumers
for all $i = 1, 2, \cdots, N^V$ **do**
generate e_i^T by connecting consumer i and the center straightly
$\mathbf{E^T} = \mathbf{E^T} \cup \{e_i^T\}$
end for
return $\mathbf{T} = (\mathbf{V^T}, \mathbf{E^T})$

3.3. Kruskal Algorithm

Kruskal algorithm [30] is a classical algorithm for solving MST of weighted connected graphs, which gradually transforms the connected components into an MST connecting all vertices by finding the edges with minimum weights. The specific process of Kruskal algorithm is shown in Algorithm 3.

Algorithm 3: Kruskal algorithm

Input: A connected graph $\mathbf{G} = (\mathbf{V}, \mathbf{E}, \mathbf{W})$.
Output: A minimum spanning tree $\mathbf{T} = \left(\mathbf{V^T}, \mathbf{E^T}, \mathbf{W^T}\right)$.

Let $\mathbf{E^T} = \Phi$
Sort the edges such that $\mathbf{W}(e_1) \leq \mathbf{W}(e_2) \leq \cdots \leq \mathbf{W}(e_m)$
for all $i = 1, 2, \cdots, N^E$ **do**
$\mathbf{E^T} = \mathbf{E^T} \cup \{e_i^T\}$
if cycle is generated in $\mathbf{T}$ **then**
delete e_i^T from the $\mathbf{E^T}$
else
maintain the $\mathbf{E^T}$ unchanged
end if
if $N^{E^T} = N^V - 1$ **then**
break
else
continue
end if
end for
return $\mathbf{T} = \left(\mathbf{V^T}, \mathbf{E^T}, \mathbf{W^T}\right)$

3.4. GeoSteiner Algorithm

GeoSteiner algorithm is a fast and accurate algorithm for solving ESMT and RSMT. It has been improved through the work of Zachariasen [31] and Warme [32]. The geometric Steiner tree problem is known to be NP-hard for the Euclidean metric, and NP-complete for the rectilinear metric.

For the problem of generating Steiner trees, GeoSteiner algorithm includes two phases: First, a small but sufficient set of full Steiner trees (FSTs) is generated and then a Steiner minimum tree (SMT) is constructed from this set. These phases are called FSTs generation and FSTs concatenation. Before generating FSTs, a preprocessing phase is required to reduce the complexity of the calculation process.

The input to the FSTs generation algorithm is the set of terminal points, and the output is an embedded hypergraph. The embedding of each hyperedge (or FSTs) is the geometric tree structure of the FSTs. FSTs generation in GeoSteiner is performed by enumerating branch trees, and only the results after passing distance tests, long-leg segment tests, branch tree tests and FSTs tests will be retained to the final concatenation phase.

The FSTs concatenation problem can be modelled as an instance of the MST in hypergraph (MSTHG) problem [33]: given a hypergraph $\mathbf{H} = \left(\mathbf{V^H}, \mathbf{E^H}\right)$, and a weight function $w^{\mathbf{H}} : \mathbf{E^H} \mapsto \mathbb{R}$ find a subset $\mathbf{T^H} \subseteq \mathbf{E^H}$ such that $\mathbf{T^H}$ is a spanning tree of $\mathbf{H}$ that minimizes $w^{\mathbf{H}}\left(\mathbf{T^H}\right) = \sum_{e \in \mathbf{E^H}} w^{\mathbf{H}}(e)$. The MSTHG problem can be solved by integer programming, as shown in Equations (27)–(31):

$$\text{Minimize } \sum\nolimits_{e \in E^H} w^{\mathbf{H}}(e) x_e \tag{27}$$

Subject to

$$\sum_{e \in \mathbf{E^H}} (|e| - 1) x_e = \left|\mathbf{V^H}\right| - 1 \tag{28}$$

$$\sum_{e \in \mathbf{E^H}} max(|e \cap S| - 1, 0)x_e \leq |S| - 1, \forall S \subset \mathbf{V^H}, |S| \geq 2 \quad (29)$$

$$0 \leq x_e \leq 1, \forall e \in \mathbf{E^H} \quad (30)$$

$$x_e \in \mathbb{Z}, \forall e \in \mathbf{E^H} \quad (31)$$

where $w^{\mathbf{H}} \in \mathbb{R}^{|\mathbf{E^H}|}$ is the cost vector for all $e \in \mathbf{E^H}$, x_e is the solution vector, and $x \in \mathbb{R}^{|\mathbf{E^H}|}$. The constraint demonstrated in Equation (28) ensures that the spanning tree has the right number and sizes of hyperedges, while the subtour constraint demonstrated in Equation (29) ensures that the tree has no cycles.

For the Euclidean metric, the edges at a Steiner point in ESMT must meet exactly at an angle of 120°, and all other edges must meet at an angle of 120° or more, otherwise the tree can be easily shortened. Additionally, all edges are straight line segments directly connected by terminals and/or Steiner points. Figure 3 shows a schematic topology consisting of 3 terminals, where Terminal-1, Terminal-2, and Terminal-3 are the three vertices of the unit square triangle. From Figure 3, it can be calculated that the total length of MST is 2 and the total length of ESMT is $\sqrt{3}$. In terms of total length, ESMT is shorter than MST.

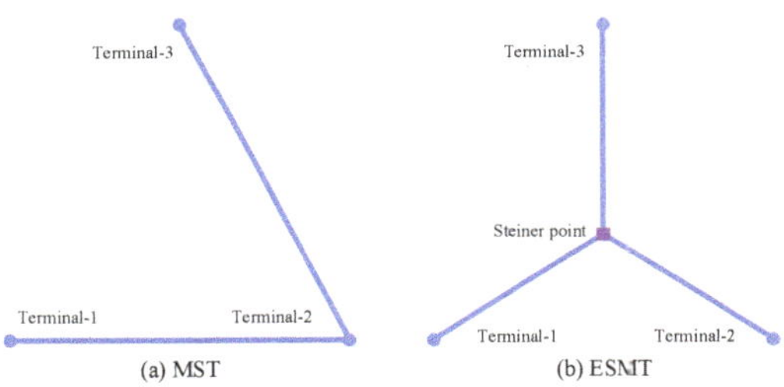

Figure 3. A schematic diagram of the topology consisted of 3 terminals.

For the rectilinear metric [33], all line segments meet at a common corner point $w = (u_x, v_y)$, which is obtained by intersecting the vertical line from point u and the horizontal line from point v. Additionally, the backbone, for Hwang topologies [34], (u, v) is consisted of line segments uw and vw. The points u, v, and w may be any combination of terminals, Steiner points, and corner points.

4. Case Study

4.1. Data Acquisition

In this case, the number of heat sources is 3. By clustering, consumers are assigned to these heat sources. The whole optimization process is realized by C++ programming. This case is simplified through a real industrial park, all the energy consumers require steam rather than hot water. The input data of consumers and heat sources is shown in Table 2, and if the heat demand is positive, then it is a consumer, on the contrary it is a heat source. The required parameters in the mathematical model are shown in Table 3.

4.2. Optimal Results and Analysis

The results are summarized in Table 4. Figure 4 shows the topological schemes of four types of pipeline network. The red line segment in Figure 4 represents the pipeline between the heat sources, and the other line segments represent the pipeline between the consumers and the heat sources. The thickness of the line segment in Figure 4 indicates the diameter of the branch pipeline.

From the perspective of economy, the TAC of STAR is the highest while that of ESMT pipeline network is the lowest, as shown in Table 4. The following analysis is carried out according to the pipeline construction cost, pressure loss cost and heat loss cost in the objective function.

In terms of pipeline construction cost, as a result of the existence of multiple heat sources, as shown in Figure 4, a single heat source no longer covers all the consumers, so the pipeline length is reduced, especially for those long, thick pipelines. This is beneficial for reducing pipeline costs.

Table 2. The table of input data of consumers and heat sources.

Name	Coordinate X (m)	Coordinate Y (m)	Heat Demand ($t{\cdot}h^{-1}$ Steam)
Consumer-01	7097	9542	15.0
Consumer-02	8800	4024	6.0
Consumer-03	9602	5124	0.5
Consumer-04	12,013	7072	4.0
Consumer-05	13,392	11,430	3.0
Consumer-06	13,949	14,857	5.0
Consumer-07	3384	24,093	10.0
Consumer-08	25,483	0	25.0
Consumer-09	11,914	2235	3.0
Consumer-10	11,369	3893	4.0
Consumer-11	6561	6546	5.0
Consumer-12	12,437	10,805	12.0
Consumer-13	12,454	11,538	0.3
Consumer-14	16,227	13,461	10.0
Consumer-15	18,914	14,341	1.2
Consumer-16	7171	12,650	0.3
Consumer-17	13,255	20,390	4.0
Consumer-18	15,489	16,736	7.0
Consumer-19	17,153	17,937	1.0
Consumer-20	22,789	24,165	2.0
Consumer-21	24,416	24,203	1.2
Consumer-22	6206	20,012	10.0
Consumer-23	5818	10,120	4.0
Consumer-24	3465	24,915	10.0
Heat source-01	0	3144	−45.8
Heat source-02	12,200	21,944	−45.2
Heat source-03	19,200	7544	−52.5

Table 3. The table of mathematical model parameters.

Parameter	Number	Unit	Parameter	Number	Unit
N^{year}	10	a	a^S	0.1945	$\cdot\mathrm{kg}^{-1}$
I	0.02		q	1999.9	$\mathrm{kJ}\cdot\mathrm{kg}^{-1}$
ρ	0.60	$\mathrm{kg}\cdot\mathrm{m}^{-3}$	T^a	3.5	°C
u	30.00	$\mathrm{m}\cdot\mathrm{s}^{-1}$	λ	0.06	$\mathrm{W}\cdot\mathrm{m}^{-1}\cdot\mathrm{K}^{-1}$
a^E	0.21	$\cdot\mathrm{kW}\cdot\mathrm{h}^{-1}$	v	3.50	$\mathrm{m}\cdot\mathrm{s}^{-1}$
$T^{Operating}$	8760	h	ε	11.63	$\mathrm{W}\cdot\mathrm{m}^{-2}\cdot\mathrm{K}^{-1}$
η	0.8		$p^{Path}_{\alpha,\beta}$	0.98	km^{-1}
σ	0.015				

Table 4. The table of results of optimization cases.

Topology Type of Pipeline Networks	STAR	MST	RSMT	ESMT
Total pipeline cost ($\times 10^7 \cdot \mathrm{a}^{-1}$)	2.706 (100%)	2.035 (75.2%)	2.152 (79.5%)	1.962 (72.5%)
Pressure loss cost ($\times 10^7 \cdot \mathrm{a}^{-1}$)	0.668 (100%)	0.504 (75.4%)	0.533 (79.8%)	0.486 (72.8%)
Heat loss cost ($\times 10^7 \cdot \mathrm{a}^{-1}$)	0.990 (100%)	0.623 (62.9%)	0.654 (66.0%)	0.605 (61.1%)
Total annual cost ($\times 10^7 \cdot \mathrm{a}^{-1}$)	4.364 (100%)	3.162 (72.5%)	3.339 (76.5%)	3.053 (70.0%)
$R^{Reliability}$	0.848 (100%)	0.815 (96.1%)	0.802 (94.6%)	0.822 (96.9%)

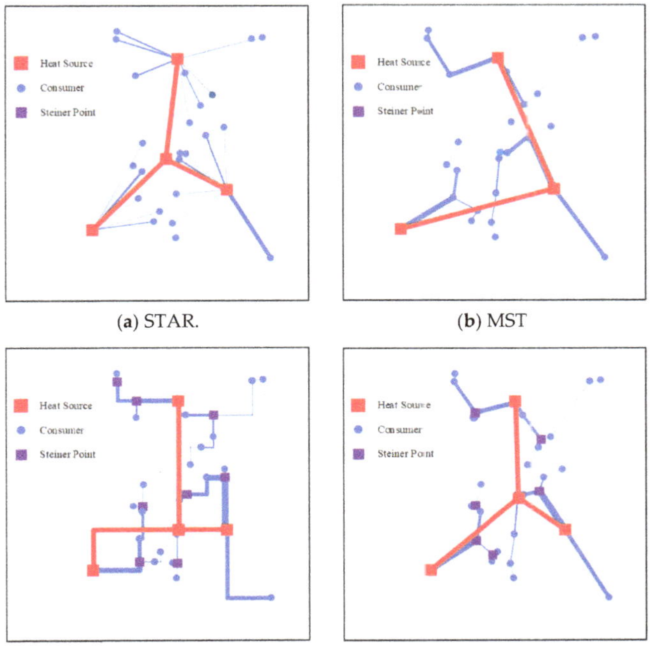

(**a**) STAR. (**b**) MST

(**c**) RSMT. (**d**) ESMT.

Figure 4. Pipeline network layouts.

On the basis of multiple heat sources, different pipeline network topologies affect pipeline costs through pipeline length, as shown in Figure 5. For the STAR pipeline network, since each consumer needs to be connected to the heat source in the region, there is no shared pipeline, leading to a significant increase in the pipeline length. As a result, this type is not applicable in large-scale projects. The MST pipeline network takes into account the situation of shared pipelines to reduce the pipeline length. Compared with STAR and MST, RSMT and ESMT pipeline networks introduce Steiner points to optimize the pipeline network connection mode. When the limitation of the actual path is considered, the RSMT pipeline network can better adapt to the actual road layout through rectangular connection, as shown in Figure 4c. However, this rectangular connection increases the total length of the network to a certain extent. Compared with RSMT pipeline network, ESMT pipeline network is not limited by the actual path, and the linear connection between two points makes its pipeline length the shortest.

In addition, in this model, pipeline construction cost is not only related to pipeline length, but also related to pipeline diameter. Since the diameter of pipeline is directly related to flow rate, the total flow rate of the pipeline network is defined as the sum of the flow rate of each piecewise pipeline. The total flow rate can reflect the overall diameter of the pipeline network and to further reflect the economy of the pipeline network. Figure 6 also demonstrates the relationship between TAC and the total flow rate.

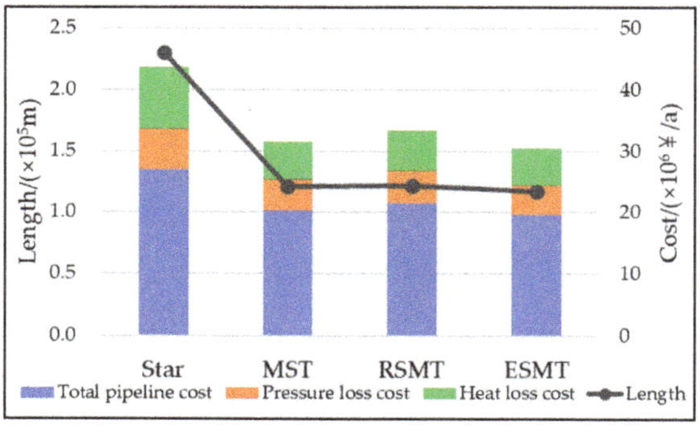

Figure 5. The relationship between the pipeline length and each cost of the four topologies.

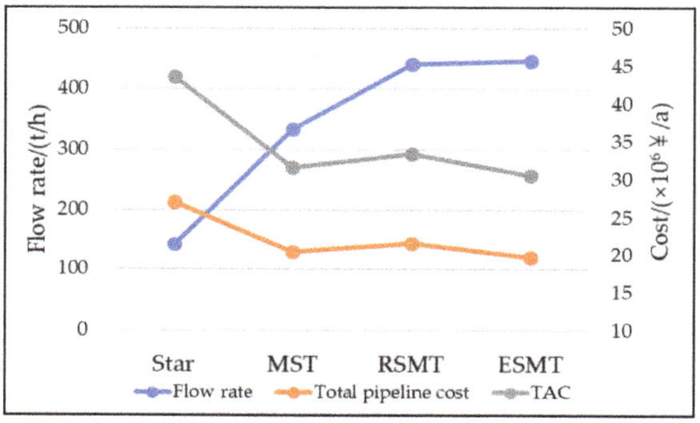

Figure 6. The relationship between the flow rate, the total pipeline cost and TAC of the four topologies.

On the basis of multiple heat sources, a comprehensive analysis of the influence of pipeline length and pipeline diameter on pipeline cost shows that, when the main pipeline length is almost constant, the increase of total flow in pipeline network will lead to the increase of total cost, as shown in Figures 5 and 6 (MST and RSMT). This indicates that increasing the flow of the pipeline and making the pipeline diameter larger will result in increased costs. For the STAR and MST networks, the overall cost of the MST is significantly reduced, indicating that pipeline length is a more sensitive factor compared with flow rate For ESMT network and RSMT network, where the total flow is similar, the economy of ESMT is better because the former has shorter pipeline length.

The STAR pipeline network also has the highest cost in terms of pressure loss and heat loss. This is because the cost of pressure loss and heat loss are related to the pipeline length. As shown in Figure 5, the longer pipeline length of STAR pipeline network causes more pressure loss and heat loss. On the other hand, for ESMT pipelines, the pressure loss and heat loss are significantly reduced due to shorter pipeline.

From the perspective of reliability, the reliability of Star is the highest, while that of RSMT is the lowest, as shown in Table 4. In this model, the reliability of the pipeline network is related to the connection mode between the consumers and the heat sources as well as the length of the main pipeline. The existence of multiple heat sources can shorten the long-distance transportation and reduce the number of shared pipelines, so that the system reliability can be improved.

In this case, the STAR pipeline network is more reliable than the other three structures because it does not contain shared pipelines. This means that the consumers are independent from each other. In the process of energy transportation, the damage of a single pipeline will not cause a large-scale failure of the pipeline network system. In order to meet the actual road layout, some consumers are connected by vertical/horizontal pipelines in RSMT structure, leading to a longer pipeline compared to MST, so that the reliability of RSMT is lower than that of MST. Compared to RSMT and MST, ESMT is the shortest, so that the reliability is higher, making ESMT superior to the other two in terms of reliability.

It can be found from the results that the STAR pipeline network has the best reliability, but at the same time it is the most expensive one. MST pipeline network, as a common pipeline network topology, has better economy compared with the STAR structure. Compared with MST, the economy and reliability of RSMT becomes worse, but it may fit the actual road layout better. The performance of ESMT pipeline network is the best one in economic aspect and the second best one in reliability.

4.3. Analysis of Small-Scale DES

From the analysis above it can be found that the distance is a key factor in the problem. The case used in last section is a large-scale problem. To identify the performance of the four types in small scale problem, this section will discuss DES that are scaled down in equal proportions, where the heat sources are more geographically concentrated with the consumers.

By comparing the data of Tables 4 and 5, it can be seen that in terms of the economics of DES, the TAC of STAR compared with that of ESMT changed from 142.9% in the large-scale case to 137.5% in the small-scale case, and the gap narrowed by 5.4%; when the pipeline network topologies are MST and RSMT, none of the gap changes by more than 1% in this ratio, and the TAC of ESMT is still the smallest. In terms of the reliability of DES, STAR has the highest reliability and ESMT has the second highest reliability, they are also in accordance with large-scale case.

Table 5. The table of results of small-scale DES case.

Topology Type of Pipeline Networks	STAR	MST	RSMT	ESMT
Total pipeline length ($\times 10^4$ m)	4.603	2.421	2.433	2.345
Total pipeline cost ($\times 10^6 \cdot a^{-1}$)	5.412	4.069	4.304	3.925
Pressure loss cost ($\times 10^6 \cdot a^{-1}$)	1.313	1.002	1.062	0.967
Heat loss cost ($\times 10^6 \cdot a^{-1}$)	1.313	1.002	1.062	0.967
Total annual cost ($\times 10^6 \cdot a^{-1}$)	6.725	5.072	5.366	4.892
$R^{Reliability}$	0.967	0.959	0.956	0.961

From the analysis, it can be concluded that the scale of the region does not have a great impact on the relative performance of the four structures.

5. Conclusions

In this paper, the layout of the DES pipeline network is optimized for a multiple heat sources scenario to save energy during the energy transportation process. In this work, a topology optimization model for the pipeline network of DES containing multiple heat sources is developed by coupling clustering algorithm, Star tree algorithm, Kruskal algorithm, GeoSteiner algorithm and LP model. In the context of multiple heat sources, this paper investigates the effects of four different topologies (STAR, MST, ESMT, RSMT) on the pipeline network system in terms of both economy and reliability. The following conclusions are obtained based on the case studies.

1. Compared with the single heat source scenario, the multiple heat sources system will reduce the long-distance and high-flow pipelines in the system, so that both economy and reliability of the pipeline network system is improved.

2. Compared with the traditional pipeline network obtained using MST, an ESMT pipeline network can reduce the total annual cost by 3% and increase reliability by 1%.
3. When considering the actual path constraints, the RSMT pipeline network can be better adapted to the road layout.
4. The geographically scale of the problem does not have a great impact on the relative performance of the four structures.
5. By using the proposed method, both economic and reliability can be improved for the DES system.

In the work, the model developed can optimize the pipeline network layout under multiple heat sources conditions. However, the study in this paper has not taken into account the uncertainty of customer demand changes, which limits the optimization results to some extent. In the future, this work will be extended to consider reliability and economic under uncertainty factors. In addition, to make the optimization results more practical, in future, more practical factors should be concerned, e.g., the layout of existing road.

Author Contributions: Conceptualization, Y.W. (Yufei Wang) and Y.W. (Yan Wu); methodology, Y.W. (Yan Wu); software, Z.C. and H.L.; validation, Z.C. and H.L.; formal analysis, Z.C.; investigation, Z.C. and H.L.; resources, X.F.; writing—original draft preparation, Z.C. and H.L.; writing—review and editing, Y.W. (Yufei Wang); supervision, Y.W. (Yufei Wang) and X.F.; funding acquisition, Y.W. (Yufei Wang). All authors have read and agreed to the published version of the manuscript.

Funding: Financial support from the National Natural Science Foundation of China under Grant (No. 22022816 and 22078358) are gratefully acknowledged.

Institutional Review Board Statement: Not applicable.

Informed Consent Statement: Not applicable.

Conflicts of Interest: The authors declare no conflict of interest.

Nomenclature

Abbreviations	
DES	Distributed Energy System
ESMT	Euclidean Steiner Minimum Tree
FSTs	Full Steiner trees
GA	Genetic algorithm
LP	Linear programming
MILP	Mixed Integer Linear Programming
MIP	Mixed Integer Programming
MST	Minimum Spanning Tree
MSTHG	The MST in hypergraph
NSGA-II	Non-dominated Sorting Genetic Algorithm II
RSMT	Rectilinear Steiner Minimum Tree
SMT	Steiner Minimum Tree
STAR	Star tree
TAC	Total annual cost, $\cdot a^{-1}$
Indices and Sets	
$\mathbf{C}$	The set of clusters, denoted by index j
D	An index, which is used to control the positive and negative of $w_{x,y,z}$ by $(-1)^D$
e	An index, referring to the edge in $\mathbf{E^H}$
$\mathbf{E}$	The set of edges in a graph
$\mathbf{E^H}$	The set of edges in the hypergraph
$\mathbf{E^T}$	The set of edges in the connected tree, denoted by indices i,β
$\mathbf{E_z^T}$	The set of edges in the connected tree which are connected directly to the vertex z.

$\mathbf{G}$	The graph of empty set in Star tree algorithm, or the weighted connected graph without direction in Kruskal algorithm
$\mathbf{H}$	A hypergraph
$\mathbf{V}$	The set of all vertices in a graph, denoted by indices x, y, z
$\mathbf{V^H}$	The set of vertices in the hypergraph, denoted by indices x, y, z $\mathbf{V^H} \subseteq \mathbf{V}$
$\mathbf{V^T}$	The set of vertices in MST, denoted by indices x, y, z, α, $\mathbf{V^T} \subseteq \mathbf{V}$
$\mathbf{W}$	The set of weights of edges in a graph, denoted by indices x, y, z
$\mathbf{W^T}$	The set of weights of edges in the connected tree, denoted by indices x, y, z, $\mathbf{W^T} \subseteq \mathbf{W}$
Variables	
a^E	The electricity cost, $\cdot\text{kW}\cdot\text{h}^{-1}$
$a_{i,j}$	The unit price of the ith pipeline of the cluster j, $\cdot\text{m}^{-1}$
a^S	The unit price of steam, $\cdot\text{kg}^{-1}$
$b_{x,y,z}^{Binary}$	Binary variables indicating whether the edge (x, y) in the connected tree is connected directly to the vertex z.
$C^{Heat\ Loss}$	The heat loss cost of pipeline, $\cdot\text{a}^{-1}$
$C_{i,j}^{Heat\ Loss}$	The heat loss cost of the ith pipeline of the cluster j, $\cdot\text{a}^{-1}$
$C^{Pipeline}$	The construction cost of pipeline, $\cdot\text{a}^{-1}$
$C_{i,j}^{Pipeline}$	The construction cost of the ith pipeline of the cluster j, $\cdot\text{a}^{-1}$
$C^{Pressure\ Loss}$	The pressure loss cost of pipeline, $\cdot\text{a}^{-1}$
$C_{i,j}^{Pressure\ Loss}$	The pressure loss cost of the ith pipeline of the cluster j, $\cdot\text{a}^{-1}$
$d_{i,j}$	The Euclidean distance between vertex i and vertex j
$D_{i,j}^{inner,insulation}$	The inner diameter of the insulation layer of the ith pipeline of the cluster j, m
$D_{i,j}^{inner,pipline}$	The inner diameter of the ith pipeline of the cluster j, m
$D_{i,j}^{outer,insulation}$	The outer diameter of the insulation layer of the ith pipeline of the cluster j, m
$D_{i,j}^{outer,pipline}$	The outer diameter of the ith pipeline of the cluster j, m
$e_i^{\mathbf{H}}$	The ith edge in the hypergraph
$e_i^{\mathbf{T}}$	The ith edge in the connected tree
g	Gravitational acceleration
$H_{i,j}^f$	The head loss of the ith pipeline of the cluster j, m
I	The annual interest rate
$L_{i,j}$	The length of the ith pipeline of the cluster j, m
$L_{\alpha,\beta}$	The length of the βth pipeline between consumer α and the heat source in the area where this consumer is located, m
$N^{Cluster}$	The number of clusters of the pipeline network system
$N_j^{Cluster}$	The number of branch pipelines in cluster j
$N^{Consumer}$	The number of consumers of the pipeline network system
$N^{\mathbf{V^T}}$	The number of vertices in the connected tree
$N^{\mathbf{E^T}}$	The number of vertices in the connected tree
N^{year}	The life cycle of the pipeline network system, a
$N_{i,j}$	The shaft power of the ith pipeline of the cluster j, W
$Ne_{i,j}$	The effective power of the ith pipeline of the cluster j, W
$p_{\alpha,\beta}^{Path}$	The probability of connecting the βth pipeline in the path connected between the consumer α and the heat source in its area
$P_\alpha^{Probability}$	The connected probability between the consumer α and the heat source of the area where this consumer is located
q	The latent heat of steam, $\text{kJ}\cdot\text{kg}^{-1}$
$Q_\alpha^{Heat\ Demand}$	The heat demand of consumer α
$Q_{i,j}$	The heat loss of the ith pipeline of the cluster j, $\text{kJ}\cdot\text{m}^{-1}\cdot\text{s}^{-1}$
$R^{Reliability}$	The reliability of the pipeline network system
$S_{i,j}^{inner,pipeline}$	The head loss of the ith pipeline of the cluster j, m^2
T	The outer surface temperature of the pipeline, °C
T^a	The ambient temperature, °C
$T^{Operating}$	The number of annual operating hours of the device, h

u	The flow rate of the steam, $m \cdot s^{-1}$
W^S	The sum of the steam mass flow rate in all branch pipelines in the entire pipeline network
$W_{i,j}$	The mass flow rate of the ith pipeline of the cluster j, $kg \cdot s^{-1}$
W_z^S	The mass flow rate of steam at vertex z, $kg \cdot s^{-1}$
$w^{\mathbf{H}}$	A weight function of the edges in a hypergraph
$w^{\mathbf{H}}(e)$	The weight of edge e in a spanning tree in the hypergraph
$w^{\mathbf{H}}(\mathbf{T}^{\mathbf{H}})$	The sum of weights of the edges in a spanning tree in the hypergraph
$w_{x,y,z}$ and $w_{y,x,z}$	The mass flow rate within the branch pipeline connecting vertices x and y when using vertex z as a reference for material accountancy, $kg \cdot s^{-1}$
$Wt_{i,j}$	The weight per unit length of the ith pipeline of the cluster j, $kg \cdot m^{-1}$
x_e	The solution vector in the hypergraph
Greek letters	
ε	Heat transfer coefficient between the outer surface of the insulation and the atmosphere, $W \cdot m^{-2} \cdot K^{-1}$
$\zeta_{i,j}^{E}$	The local resistance coefficient at the standard elbows (90°)
$\zeta_{i,j}^{Elbow}$	The local resistance coefficient of the ith pipeline of the cluster j
$\zeta_{i,j}$	The resistance coefficient of the ith pipeline of the cluster j
η	The efficiency of the conveying equipment
λ	The thermal conductivity of insulating material products at average temperature, $W \cdot m^{-1} \cdot K^{-1}$
ξ_i	The index used to determine which heat source consumer i is assigned to
ρ	The density of the steam, $kg \cdot m^{-3}$
σ	The friction coefficient of the pipeline
Φ	The empty set

References

1. Fichera, A.; Marrasso, E.; Sasso, M.; Volpe, R. Environmental and Economic Performance of an Urban Community Hybrid Distributed Energy System. *Energy* **2020**, *13*, 2545.
2. Ren, H.; Zhou, W.; Nakagami, K.I.; Gao, W.; Wu, Q. Multi-objective optimization for the operation of distributed energy systems considering economic and environmental aspects. *Appl. Energy* **2010**, *87*, 3642–3651. [CrossRef]
3. Wu, Q.; Ren, H.; Gao, W.; Ren, J. Multi-objective optimization of a distributed energy network integrated with heating interchange. *Energy* **2016**, *109*, 353–364. [CrossRef]
4. Jing, R.; Zhu, X.; Zhu, Z.; Wang, W.; Meng, C.; Shah, N.; Li, N.; Zhao, Y. A multi-objective optimization and multi-criteria evaluation integrated framework for distributed energy system optimal planning. *Energy Convers. Manag.* **2018**, *166*, 445–462. [CrossRef]
5. Yuan, J.; Xiao, Z.; Zhang, C.; Gang, W. A control strategy for distributed energy system considering the state of thermal energy storage. *Sustain. Cities Soc.* **2020**, *63*, 102492. [CrossRef]
6. Wang, X.; Tian, H.; Yan, F.; Feng, W.; Wang, R.; Pan, J. Optimization of a distributed energy system with multiple waste heat sources and heat storage of different temperatures based on the energy quality. *Appl. Therm. Eng.* **2020**, *181*, 115975. [CrossRef]
7. Buoro, D.; Casisi, M.; De Nardi, A.; Pinamonti, P.; Reini, M. Multicriteria optimization of a distributed energy supply system for an industrial area. *Energy* **2013**, *58*, 128–137. [CrossRef]
8. Wang, M.; Yu, H.; Lin, X.; Jing, R.; He, F.; Li, C. Comparing stochastic programming with posteriori approach for multi-objective optimization of distributed energy systems under uncertainty. *Energy* **2020**, *210*, 118571. [CrossRef]
9. Keçebaş, A.; Ali Alkan, M.; Bayhan, M. Thermo-economic analysis of pipe insulation for district heating piping systems. *Appl. Therm. Eng.* **2011**, *31*, 3929–3937. [CrossRef]
10. Salem, E.A.; Farid Khalil, M.; Sanhoury, A.S. Optimization of insulation thickness and emissions rate reduction during pipeline carrying hot oil. *Alex. Eng. J.* **2021**, *60*, 3429–3443. [CrossRef]
11. Li, X.-l.; Duanmu, L.; Shu, H.-w. Optimal design of district heating and cooling pipe network of seawater-source heat pump. *Energy Build.* **2010**, *42*, 100–104. [CrossRef]
12. Wang, Y.; Wang, J.; Gao, M.; Zhang, D.; Liu, Y.; Tan, Z.; Zhu, J. Cost-based siting and sizing of energy stations and pipeline networks in integrated energy system. *Energy Convers. Manag.* **2021**, *235*, 113958. [CrossRef]
13. Zeng, J.; Han, J.; Zhang, G. Diameter optimization of district heating and cooling piping network based on hourly load. *Appl. Therm. Eng.* **2016**, *107*, 750–757. [CrossRef]
14. Mehleri, E.D.; Sarimveis, H.; Markatos, N.C.; Papageorgiou, L.G. A mathematical programming approach for optimal design of distributed energy systems at the neighbourhood level. *Energy* **2012**, *44*, 96–104. [CrossRef]
15. Khir, R.; Haouari, M. Optimization models for a single-plant District Cooling System. *Eur. J. Oper. Res.* **2015**, *247*, 648–658. [CrossRef]

16. Chan, A.L.S.; Hanby, V.I.; Chow, T.T. Optimization of distribution piping network in district cooling system using genetic algorithm with local search. *Energy Convers. Manag.* **2007**, *48*, 2622–2629. [CrossRef]
17. Haikarainen, C.; Pettersson, F.; Saxén, H. A model for structural and operational optimization of distributed energy systems. *Appl. Therm. Eng.* **2014**, *70*, 211–218. [CrossRef]
18. Sanaye, S.; Mahmoudimehr, J. Optimal design of a natural gas transmission network layout. *Chem. Eng. Res. Des.* **2013**, *91*, 2465–2476. [CrossRef]
19. Su, H.; Zio, E.; Zhang, J.; Li, X.; Chi, L.; Fan, L.; Zhang, Z. A method for the multi-objective optimization of the operation of natural gas pipeline networks considering supply reliability and operation efficiency. *Comput. Chem. Eng.* **2019**, *131*, 106584. [CrossRef]
20. Sokolov, D.V.; Barakhtenko, E.A. Optimization of transmission capacity of energy water pipeline networks with a tree-shaped configuration and multiple sources. *Energy* **2020**, *210*, 118469. [CrossRef]
21. Liu, X. Energy stations and pipe network collaborative planning of integrated energy system based on load complementary characteristics. *Sustain. Energy Grids Netw.* **2020**, *23*, 100374. [CrossRef]
22. Rimkevicius, S.; Kaliatka, A.; Valincius, M.; Dundulis, G.; Janulionis, R.; Grybenas, A.; Zutautaite, I. Development of approach for reliability assessment of pipeline network systems. *Appl. Energy* **2012**, *94*, 22–33. [CrossRef]
23. Shan, X.; Wang, P.; Lu, W. The reliability and availability evaluation of repairable district heating networks under changeable external conditions. *Appl. Energy* **2017**, *203*, 686–695. [CrossRef]
24. Alsharqawi, M.; Zayed, T.; Parvizsedghy, L.; Senouci, A.; Al-Derham, H. Reliability Assessment Model for Water Distribution Networks. *J. Pipeline Syst. Eng. Pract.* **2020**, *11*, 04019059. [CrossRef]
25. Chen, Q.; Zuo, L.; Wu, C.; Cao, Y.; Bu, Y.; Chen, F.; Sadiq, R. Supply reliability assessment of a gas pipeline network under stochastic demands. *Reliab. Eng. Syst. Saf.* **2021**, *209*, 107482. [CrossRef]
26. Yu, W.; Huang, W.; Wen, Y.; Li, Y.; Liu, H.; Wen, K.; Gong, J.; Lu, Y. An integrated gas supply reliability evaluation method of the large-scale and complex natural gas pipeline network based on demand-side analysis. *Reliab. Eng. Syst. Saf.* **2021**, *212*, 107651. [CrossRef]
27. Zhu, Y.; Wang, P.; Wang, Y.; Tong, R.; Yu, B.; Qu, Z. Assessment method for gas supply reliability of natural gas pipeline networks considering failure and repair. *J. Nat. Gas Sci. Eng.* **2021**, *88*, 103817. [CrossRef]
28. Chang, C.; Chen, X.; Wang, Y.; Feng, X. An efficient optimization algorithm for waste Heat Integration using a heat recovery loop between two plants. *Appl. Therm. Eng.* **2016**, *105*, 799–806. [CrossRef]
29. Zhou, J.; Li, Z.; Liang, G.; Zhou, L.; Zhou, X. General Models for Optimal Design of Star–Star Gathering Pipeline Network. *J. Pipeline Syst. Eng. Pract.* **2021**, *12*, 04021024. [CrossRef]
30. Weiss, M.A. *Data Structures and Algorithm Analysis in C++*, 4th ed.; Pearson Education India: London, UK, 2014.
31. Zachariasen, M. Rectilinear full Steiner tree generation. *Networks* **1999**, *33*, 125–143. [CrossRef]
32. Warme, D.M. Spanning Trees in Hypergraphs with Applications to Steiner Trees. Ph.D. Thesis, University of Virginia, Charlottesville, VA, USA, 1998.
33. Juhl, D.; Warme, D.M.; Winter, P.; Zachariasen, M. The GeoSteiner software package for computing Steiner trees in the plane: An updated computational study. *Math. Prog. Comp.* **2018**, *10*, 487–532. [CrossRef]
34. Salowe, J.S.; Warme, D.M. Thirty-five-point rectilinear steiner minimal trees in a day. *Networks* **1995**, *25*, 69–87. [CrossRef]

processes

MDPI

Article

Research on Dynamic Modeling of the Supercritical Carbon Dioxide Power Cycle

Ping Song [1,2], Zhenxing Zhao [1,2], Lie Chen [1,2], Chunhui Dai [1,2,*], Chonghai Huang [1,2], Mengran Liao [1,2], Xingsheng Lao [1,2], Yuansheng Lin [1,2] and Wei Wang [1,2]

1 Key Laboratory of Thermal Power Technology, Wuhan 430064, China; pingsong@stu.xjtu.edu.cn (P.S.); zzxxjtu@163.com (Z.Z.); hellowcrldum@163.com (L.C.); bingzhihai@126.com (C.H.); lmr@mail.ustc.edu.cn (M.L.); lacxingsheng@tsinghua.org.cn (X.L.); keylab_rndl@163.com (Y.L.); jchdlzhjs@126.com (W.W.)

2 Wuhan Second Ship Design and Research Institute, Wuhan 430064, China

* Correspondence: ch.dai@163.com

Abstract: The supercritical carbon dioxide (SCO_2) Brayton cycle, as a substitute for the steam cycle, can be widely used in a variety of power generation scenarios. However, most of the existing SCO_2 cycle studies are restricted to basic thermodynamics research, parameter optimizations, system design in different application fields, and even economic analysis. Considering the load variability and control flexibility of the power generation system, the dynamic performance research of the SCO_2 cycle is also crucial, but the work done is still limited. Based on the previous studies, Simulink software is used in this paper to develop a dynamic model of the 20 MW-SCO_2 recompression cycle, which specifically includes component models that can independently realize physical functions and an overall closed-loop cycle model. A series of comparative calculation are carried out to verify the models and the results are very positive. The SCO_2 recompression power system is built with the developed models and the dynamic model runs stably with a maximum error of 0.56%. Finally, the simulation of the dynamic switching conditions of the 20 MW-SCO_2 recompression cycle are performed and the analysis results supply instructive suggestions for the system operation and control.

Keywords: supercritical carbon dioxide brayton cycle; dynamic model; simulink

Citation: Song, P.; Zhao, Z.; Chen, L.; Dai, C.; Huang, C.; Liao, M.; Lao, X.; Lin, Y.; Wang, W. Research on Dynamic Modeling of the Supercritical Carbon Dioxide Power Cycle. *Processes* **2021**, *9*, 1946. https://doi.org/10.3390/pr9111946

Academic Editors: Pei Liu and Xiao Wu

Received: 2 September 2021
Accepted: 27 October 2021
Published: 29 October 2021

Publisher's Note: MDPI stays neutral with regard to jurisdictional claims in published maps and institutional affiliations.

Copyright: © 2021 by the authors. Licensee MDPI, Basel, Switzerland. This article is an open access article distributed under the terms and conditions of the Creative Commons Attribution (CC BY) license (https://creativecommons.org/licenses/by/4.0/).

1. Introduction

The supercritical carbon dioxide (SCO_2) cycle can be traced back to the 1960s, after which research on the cycle was temporarily shelved due to technical limitations [1]. As the fourth-generation nuclear reactor developed rapidly in the 21st century, the operating temperature now reaches as high as 500–900 °C. Considering that the material reliability of the ultra-supercritical steam cycle cannot be guaranteed [2], the SCO_2 Brayton cycle as an alternative cycle has attracted widespread attention from scholars. Additionally, comparative studies on the SCO_2 cycle, steam cycle, and helium cycle have also been carried out [3,4]. A large number of results show that the SCO_2 cycle is more suitable for higher temperature nuclear reactors due to the simple structure and high efficiency, and has the potential to reduce costs.

The SCO_2 cycle also has the advantages of a compact structure and good heat source matching performance, and can be widely used in nuclear energy [5–7], waste heat recovery [8–10], solar energy [11–13], and thermal power [14–16]. A lot of work has been done on the basic thermodynamic research of the SCO_2 cycle, configuration and parameter optimizations, system design in different application fields, and even economic analysis and optimization.

Sarkar [17] performed second-law analysis of the SCO_2 cycle, and the results showed that the minimum operating temperature has a greater impact on the optimal pressure ratio and cycle efficiency than the maximum operating temperature, and the irreversibility of heat exchangers is significantly higher than that of turbomachinery. AHN et al. [18] found that the recompression cycle showed the highest efficiency by comparing different cycle layouts. Moreover, the recompression cycle can effectively avoid the recuperator pinch point, which is more likely to occur in the simple recuperation cycle [19]. Ma et al. [20] analyzed the SCO_2 cycle with an intercooling system of the main compressor, and studied the influence of the pressure ratio and the pressure distribution of the two compression stages on the performance of the system. The results show that the coupled main compressor intercooling system can improve the cycle efficiency, save system cost, and improve cycle stability. Liu et al. [14] studied the effects of the reheat stages on the cycle efficiency. Li et al. [21] conceptually designed the power system of the SCO_2 cycle coupled with a small lead-cooled fast reactor, and analyzed the thermodynamic and economic performance. In addition, some scholars have conceptually designed the configurations of SCO_2 coal-fired power plants with different loads, and proposed a series of SCO_2 boiler and cycle-side temperature coupling methods to rationally utilize the flue gas heat [15,22,23]. Zhu et al. [24] used the weighted quality method to evaluate the cost of an SCO_2 boiler, and then evaluated the cost of SCO_2 coal-fired power generation, which shows the economic advantages of the SCO_2 cycle in the thermal power field. At the same time, a large quantity of related research has also been carried out in the fields of solar energy and waste heat recovery [25,26].

The above research has mainly focused on the steady state calculation. Since the SCO_2 cycle is used in the power generation system, attention needs to be focused on the performance of a variable load and flexibility. Thus, research on the performance and simulation of the SCO_2 cycle has also attracted attention from all over the world [27]. Dyreby [28] studied the method of SCO_2 Brayton cycles numerical modelling. Moisseytsev and Sienicki [29] simulated the performance characteristics of the 100 kW simple recuperated Brayton cycle configuration by developing The Plant Dynamics Code (PDC). Wu P. et al. [30] and Deng T. et al. [31] also developed the transient analysis code for the SCO_2 Brayton cycle.

As a supplement to the previous research, this paper studies the dynamic performance of the SCO_2 recompression cycle of a 20 MW sodium-cooled reactor with Simulink software.

2. System Configuration and Modeling Descriptions

The power generation system described in this paper adopts the SCO_2 recompression cycle, and the configuration is shown in Figure 1. The overall system includes a turbine (T), main compressor (MC), re-compressor (RC), high-temperature and low-temperature recuperators (HTRs and LTRs), intermediate heat exchanger (IHE), pre-cooler (PC), surge tank (ST), and heater.

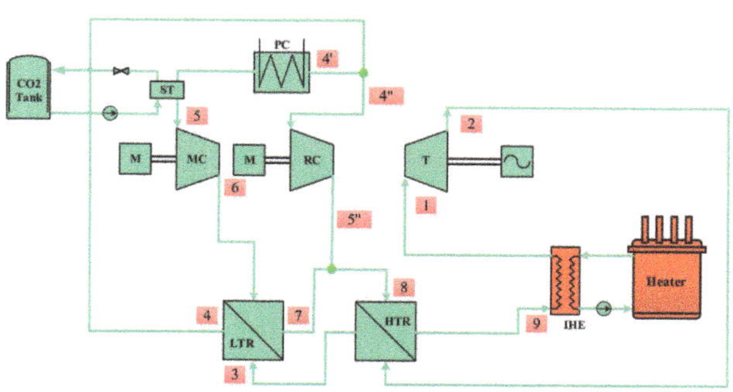

Figure 1. Supercritical carbon dioxide recompression cycle power generation system.

As shown in Figure 1, the high-temperature and high-pressure CO_2 from the IHE enters the turbine to do work and generates electricity. The exhausted CO_2 from turbine flows into the HTR and LTR and releases heat. The low-pressure fluid at the LTR outlet is then split into two streams: one stream is cooled in the PC and then compressed in the MC, and the other stream is directly compressed in the RC. Two streams converge at the high-pressure side outlet of LTR and are sent to the HTR and IHE for heating.

In order to meet the flexible control requirements, the MC, RC, and turbine adopt a split-axis layout. At the same time, the ST is arranged between the PC and the MC to facilitate the control of the MC inlet pressure and buffer the pressure change. The ST is connected with a CO_2 storage tank, and the MC inlet pressure is maintained above the critical point by controlling the amount of CO_2 in the ST.

This paper focuses on exploring the dynamic characteristics of the SCO_2 recompression cycle, and the mathematical model of each component is developed to reflect the dominant characteristics of the cycle. The model of each component could be able to realize independent physical functions and be mathematically independent. At the same time, there are clear boundaries and data interfaces between the modules. The internal parameters of the models can be changed according to actual needs, and the impact of each parameter on the overall circulatory system can be analyzed. Based on the above requirements, this paper develops the dynamic model of the SCO_2 system based on the Simulink platform, which is also convenient for system control strategy research.

3. Dynamic Model Development

3.1. Heat Exchanger Model

The heat exchangers involved in the SCO_2 power cycle, including HTR, LTR, PC, and IHE, adopt printed circuit heat exchangers (PCHEs), which have great advantages in dynamic modeling due to the regular layout structure. The flow and heat transfer model of PCHEs are based on the three conservation equations of mass, momentum, and energy, which can be simplified to one-dimensional models. The structure of PCHE is shown in Figure 2a.

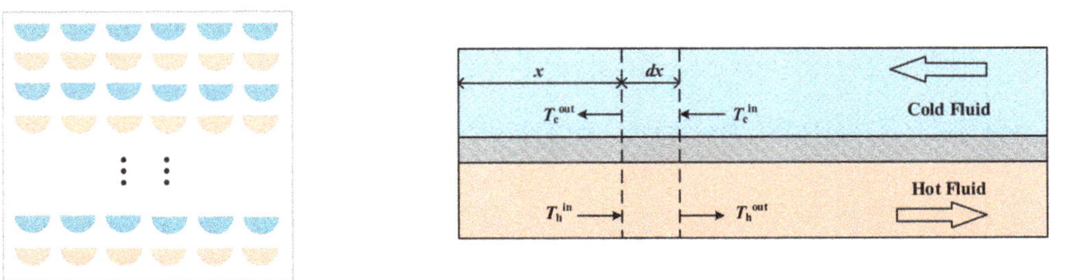

(a) (b)

Figure 2. The model of PCHE. (**a**) the structure of PCHE. (**b**) schematic diagram of PCHE partition calculation.

In this section, PCHE is divided into several parts, and the flow and heat transfer models are established for each part, as shown in Figure 2b. Assuming that the fluid in each part is incompressible, the steady-state model is adopted to describe the flow process. In each part, the lumped parameter method is applied to the calculation and the fluids and wall are treated as zero-dimensional nodes. The heat transfer processes between the cold fluid and the wall, the wall, and the hot fluid are considered, while the heat transfer in the direction of the fluid flow is ignored to ensure the accuracy and simplification of the calculation process. The specific calculation process is as follows [32,33].

Hot fluid:

$$V_h \rho_h c_{p,h} \frac{dT_h}{dt} + m_h (h_h^{out} - h_h^{in}) = -(kA_s)_h (T_h - T_w) \quad (1)$$

Cold fluid:

$$V_c\rho_c c_{p,c}\frac{dT_c}{dt} + m_c(h_c^{out} - h_c^{in}) = (kA_s)_c(T_w - T_c) \tag{2}$$

Wall:

$$m_w c_{p,w}\frac{dT_w}{dt} = (hA_s)_h(T_h - T_w) - (hA_s)_c(T_w - T_c) \tag{3}$$

The heat transfer coefficient calculation methods of different heat exchangers are selected based on the fluid type and Reynolds number (*Re*). For example, the Gnielinski formula is applied to calculate the Nusselt number (*Nu*) of the cold and hot side fluids in HTR and LTR, as shown in Equations (4) and (5). The cold and hot side of HTR and LTR are both CO_2, so the Nusselt number (*Nu*) can be calculated by the Gnielinski formula. Thus, the heat transfer coefficient can be obtained. The specific heat capacity of the tube wall is determined by the material type:

$$f = (1.82\lg Re - 1.64)^{-2} \tag{4}$$

$$Nu = \frac{(f/8)(Re - 1000)Pr}{1 + 12.7(Pr^{2/3} - 1)\sqrt{f/8}} \tag{5}$$

where Pr is the Prandtl number of the fluid.

The pressure drop of each part can be determined by Equation (6). The pressure drop is proportional to the square of the flow rate and inversely proportional to the fluid density:

$$\Delta P = \gamma \cdot \frac{m^2}{\rho} \tag{6}$$

where γ is the effective loss coefficient, which is determined by the pipeline characteristics and can be calculated with the parameters of the design condition.

3.2. Compressor Model

The dynamic response time of different processes and equipment in the SCO_2 power cycle is significantly different. The devices with a long response time adopt dynamic-state models, such as recuperators, coolers, etc. On the contrary, the devices and thermal processes with a short response time adopt steady-state models, such as turbine, compressors, working fluid separation, and mixing processes.

The compressor outlet parameters are determined by the inlet parameters and isentropic efficiency. Taking the main compressor (MC), for example, the specific calculation is shown as follows [34]:

$$h_{MC,in} = h(T_{MC,in}, P_{MC,in}) \tag{7}$$

$$s_{MC,in} = s(T_{MC,in}, P_{MC,in}) \tag{8}$$

$$h_{MC,out,is} = h(s_{MC,in}, P_{MC,out}) \tag{9}$$

$$h_{MC,out} = h_{MC,in} + (h_{MC,out,is} - h_{MC,in})/\eta_{MC,is} \tag{10}$$

where $h_{MC,in}$ is the MC inlet-specific enthalpy, kJ/kg; $h_{MC,out}$ is the MC outlet-specific enthalpy, kJ/kg; $h_{MC,out,is}$ is the isentropic-specific enthalpy of the MC outlet, kJ/kg; and $\eta_{MC,is}$ is the isentropic efficiency of the MC.

The outlet pressure and isentropic efficiency of the compressor are provided by the performance prediction curves. The curves describe the relationship between the outlet pressure and efficiency with the mass flow rate and rotating speed. The main compressor performance prediction curves are shown in Figure 3. The re-compressor performance prediction curves are shown in Figure 4, which are obtained through the simulation results based on the compressor design value [35].

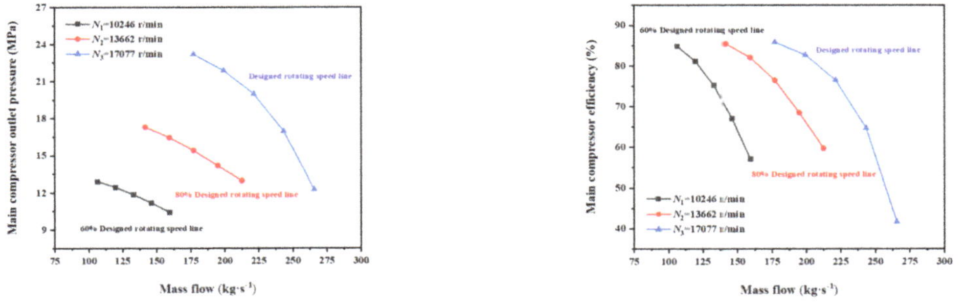

(a) (b)

Figure 3. The performance prediction curves of the main compressor. (**a**) Outlet pressure prediction of MC. (**b**) Efficiency prediction of MC.

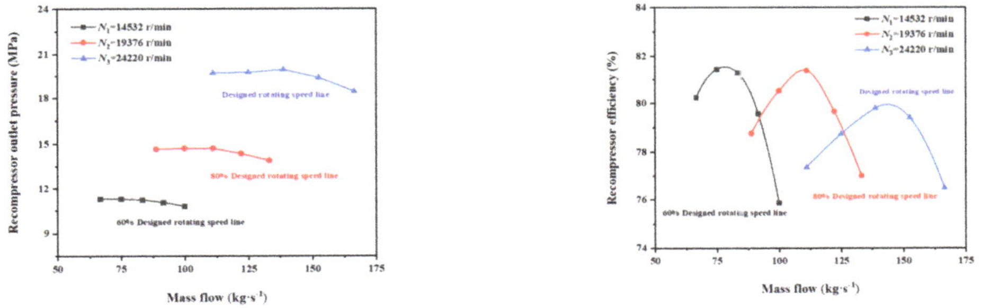

(a) (b)

Figure 4. The performance prediction curves of the recompressor. (**a**) Outlet pressure prediction of RC. (**b**) Efficiency prediction of RC.

It can be seen from the recompression configuration in Figure 1 that the low-pressure side outlet working fluid of the LTR is divided into two streams: one stream is cooled in the PC and then compressed in the MC, while the other stream is directly compressed in the RC. The split ratio (SR) is used to express the share of flow entering the recompressor, as shown in Equation (11):

$$SR = m_{RC}/m \tag{11}$$

Therefore, the power consumption of the main compressor is:

$$W_{MC} = m\cdot(1 - SR)\cdot(h_{MC,out} - h_{MC,in}) \tag{12}$$

The power consumption of the recompressor is:

$$W_{RC} = m\cdot SR\cdot(h_{RC,out} - h_{RC,in}) \tag{13}$$

where m is the mass flow rate of the circulation system, kg/s; W_{MC} is the MC power consumption, kW; $h_{RC,in}$ is the RC inlet-specific enthalpy, kJ/kg; $h_{RC,out}$ is the RC outlet-specific enthalpy, kJ/kg; and W_{RC} is the RC power consumption, kW.

3.3. Turbine Model

Similar to the compressor model, the turbine outlet parameters are also determined by the inlet parameters and isentropic efficiency. The specific calculation is as follows [29]:

$$h_{\text{tur,in}} = h(t_{\text{tur,in}}, p_{\text{tur,in}}) \tag{14}$$

$$s_{\text{tur,in}} = s(t_{\text{tur,in}}, p_{\text{tur,in}}) \tag{15}$$

$$h_{\text{tur,out,is}} = h(s_{\text{tur,in}}, p_{\text{tur,out}}) \tag{16}$$

$$h_{\text{tur,out}} = h_{\text{tur,in}} - (h_{\text{tur,in}} - h_{\text{tur,out,is}}) \cdot \eta_{\text{tur,is}} \tag{17}$$

where $h_{\text{tur,in}}$ is the turbine inlet-specific enthalpy, kJ/kg; $h_{\text{tur,out}}$ is the turbine outlet-specific enthalpy, kJ/kg; $h_{\text{tur,out}}$ is the turbine outlet isentropic-specific enthalpy, kJ/kg; and η_{tur} is the isentropic efficiency of the turbine.

The turbine output power is calculated as follows:

$$W_{\text{tur}} = m_{\text{tur}} \cdot (h_{\text{tur,in}} - h_{\text{tur,out}}) \tag{18}$$

where m_{tur} is the turbine flow rate, kg/s; and W_{tur} is turbine output power, kW.

The turbine performance prediction curves are shown in Figure 5 [35].

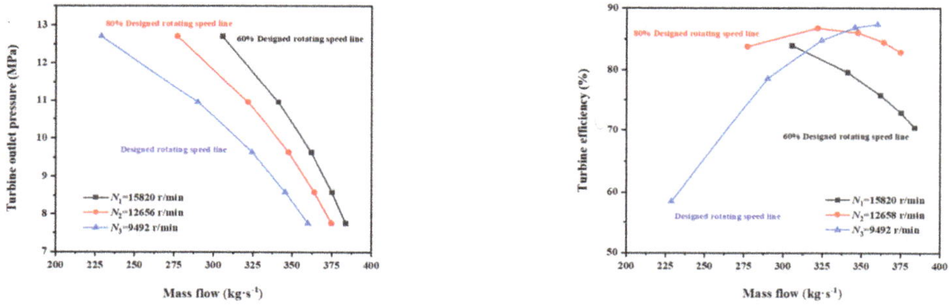

(a) (b)

Figure 5. The performance prediction curves of the turbine. (**a**) Outlet pressure prediction of turbine. (**b**) Efficiency prediction of turbine.

3.4. Other Models

3.4.1. Surge Tank (ST) Model

In this section, considering the rapid response of pressure, the volume in the entire loop composed of HTR hot end-LTR hot end-PC hot end-ST is lumped into the surge tank to calculate the pressure change. Additionally, it is considered that the working fluid flow rate before and after the HTR hot end, LTR hot end, and PC hot end remains unchanged.

According to the mass conservation equation, the density change of the working fluid in the surge tank is as follows:

$$\frac{d\rho_{\text{ST}}}{dt} = \frac{m_{\text{HTR,h}} - m_{\text{RC}} + m_{\text{CFP}} - m_{\text{PRV}} - m_{\text{MC}}}{V_{\text{HTR}} + V_{\text{LTR}} + V_{\text{PC}} + V_{\text{ST}}} \tag{19}$$

where ρ_{ST} is the working fluid density in the surge tank, kg/m^3; $m_{\text{HTR,h}}$ is the working fluid flow rate at the HTR hot end, kg/s; m_{CFP} is the flow rate from the CO_2 filling pump (CFP) into the ST, kg/s; m_{PRV} is the flow rate of the pressure-reducing valve (PRV), kg/s; V_{HTR} is the working fluid volume at the HTR hot end, m^3; V_{LTR} is the working fluid volume at the LTR hot end, m^3; V_{PC} is the working fluid volume at the PC hot end, m^3; and V_{ST} is the ST volume, m^3.

According to the energy conservation equation, the specific enthalpy change of the working fluid in the surge tank is shown as below [36]:

$$\frac{dh_{ST}}{dt} = \frac{m_{LTR,h}h_{PC,h} + m_{CFP}h_{CFP} - m_{PRV}h_{ST} - m_{MC}h_{ST}}{\rho_{ST}V_{ST}} \tag{20}$$

where h_{ST} is the specific enthalpy of the CO_2 in ST, kJ/kg; h_{PC} is the specific enthalpy of the working fluid at the PC hot end, kJ/kg; and h_{CFP} is the specific enthalpy of the CO_2 at the CFP outlet, kJ/kg. According to the density ρ_{ST} and specific enthalpy h_{ST}, the temperature T_{ST} and pressure P_{ST} of the working fluid in surge tank can be obtained by consulting the REFPROP software.

3.4.2. Working Fluid Mixing Process

In the recompression cycle system, there is a node where the working fluid at the LTR cold end is mixed with that at the outlet of RC. According to the equations of mass conservation and energy conservation, the parameters of the mixed working fluid can be calculated as follows [37]:

$$m_{HTR,c} = m_{LTR,c} + m_{RC} \tag{21}$$

$$h_{HTR,cin} = \frac{m_{LTR,c}h_{LTR,cout} + m_{RC}h_{RC,out}}{m_{HTR,c}} \tag{22}$$

where $h_{HTR,cin}$ is the specific enthalpy of the mixed working fluid, kJ/kg; $h_{LTR,cout}$ is the specific enthalpy of the working fluid at the outlet of the LTR cold end, kJ/kg; and $h_{RC,out}$ is the specific enthalpy of the working fluid at the RC outlet, kJ/kg. According to the pressure and specific enthalpy, the temperature of the mixed working fluid can be obtained by consulting the REFPROP software.

4. Model Verification

A series of comparative calculations were carried out to verify the models. Firstly, the main component models were verified with the design values to make sure that the models are built correctly. All component models were connected based on the CO_2 flow sequence and system structure, and constitute the overall dynamic model of the power generation system. Then, based on the system design value, we carried out the system steady-state simulation.

4.1. Component Verification

4.1.1. PCHE Model Verification

The structural parameters of the PCHE are shown in Table 1. In the Simulink model, the PCHE was divided into five nodes, and the model was built with the Simulink graphical modeling tool based on the above-mentioned unsteady heat transfer relationship. We obtained the calculation results with the structural parameters and the thermal parameters based on the Simulink model. According to the comparison with the design value, the model calculation accuracy is reliable. As shown in Table 2, the outlet temperature of the hot side error is as small as 0.065%, while that of the cold side is larger, with an error of 0.25%. The results prove that the model has high accuracy.

Table 1. Structural parameters of PCHE.

Parameters	Hot Side	Cold Side
Fluid	CO_2	CO_2
Diameter	2 mm	1.5 mm
Hole pitch	3 mm	2.5 mm
Plate thickness	2 mm	1.8 mm
Sheet number	667	667
volume	1.41 m^3	0.95 m^3
PCHE size	2.5 m × 2.5 m × 1.61 m	
PCHE mass	60,442 kg	

Table 2. The outlet temperature of PCHE.

	Design Value	Calculation Value	Error
Hot side outlet temperature	429.12 K	429.4 K	0.065%
Hot side outlet pressure	7740 kPa	7740 kPa	-
Cold side outlet temperature	627.15 K	628.7 K	0.25%
Cold side outlet pressure	19,950 kPa	19,950 kPa	-

We also performed the disturbance analysis, and the results were compared with other model data to verify the dynamic simulation capacity of the model. At 300 s, the inlet temperature of the hot fluid increased and decreased by 30 °C, respectively. The characteristics of the outlet temperature of the cold and hot fluids were analyzed and compared with the literature [38]. The comparison results are shown in Figure 6:

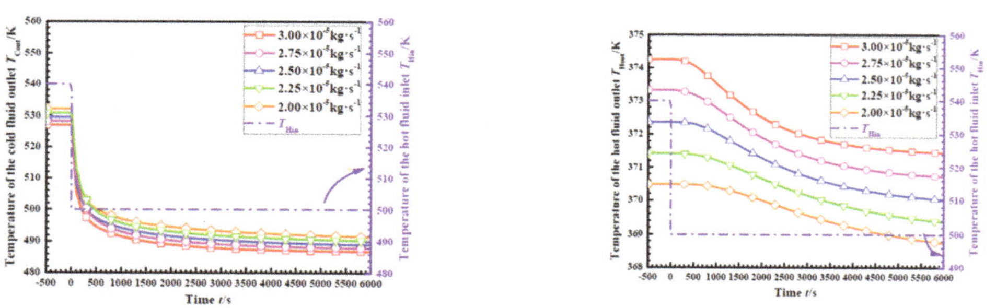

(**a**) (**b**)

Figure 6. The comparison of the temperature disturbance. (**a**) Temperature of the cold fluid outlet. (**b**) Temperature of the hot fluid outlet.

The authors of [38] carried out the PCHE thermodynamic analysis and performance prediction of a 1000 MW SCO_2 coal-fired power plant with CFD mesh simulation. Although the background and method were different from this paper, the model establishment was similar, being based on the basic conservation equations. As can be seen from the comparison, when affected by the disturbance of the step change in the hot fluid inlet temperature, the changes in the outlet temperature of the cold and hot fluids show similar dynamic characteristics:

(1) When the inlet temperature of the hot fluid is subject to a step change, the outlet temperature of the cold fluid responds faster. At the same time, the outlet temperature of the cold fluid varies widely, which is equivalent to the change of the disturbance temperature;
(2) The outlet temperature of the hot fluid is less affected by the inlet temperature and changes slowly after being disturbed. The temperature change is smaller as well.

4.1.2. Compressor Model Verification

According to the mathematical calculation formula and performance curve of the compressor in the steady state, the main compressor and recompressor models were built in Simulink. The calculation of the performance curve in the model adopted the interpolation calculation of the two-dimensional Look up table. The steady state calculation results were compared with the design value to verify the model accuracy. As shown in Table 3, the main compressor outlet temperature error is as large as 1.1%, which is considered acceptable.

Table 3. Comparison of the compressor.

	Main Compressor			Recompressor		
	Design Value	**Model Value**	**Error**	**Design Value**	**Model Value**	**Error**
Outlet temperature\°C	58.99	58.35	1.1%	153.99	154.35	0.23%
Outlet pressure\MPa	20	20	-	19.97	19.97	-
Power consumption\MW	5.07	5.05	0.4%	8.48	8.47	0.11%

4.1.3. Turbine Model Verification

For the turbine, the model build process is similar to the compressor model. The design value was also applied to the verification of the model and the comparison results are shown in Table 4. As can be seen, the largest error is the turbine power with a value of 0.74%, which is totally acceptable. Thus, it can be considered that the model is reliable.

Table 4. Comparison of the turbine.

	Design Value	Calculation Value	Error
Outlet temperature\°C	394.73	394.35	0.1%
Outlet pressure\MPa	7760	7733	0.35%
Power\MW	40.68	40.38	0.74%

4.2. System Verification

The closed-loop system model was debugged, and the initial operating parameters were calculated. With the problems of the algebraic loop and initial values solved, the SCO_2 recompression power system dynamic model can be run stably. The Simulink configuration diagram is shown in Figure 7. Similar to the system configuration and modeling description in Section 2, the Simulink configuration also mainly includes the turbine (T), main compressor (MC), recompressor (RC), high-temperature and low-temperature recuperators (HTRs and LTRs), intermediate heat exchanger (IHE), pre-cooler (PC), surge tank (ST), and heater.

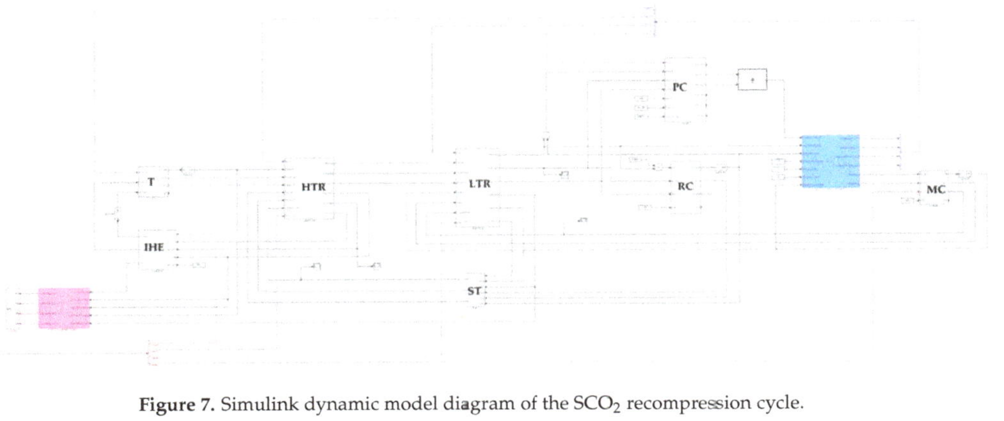

Figure 7. Simulink dynamic model diagram of the SCO_2 recompression cycle.

The heat duty was input to the system to simulate the heater in this paper at a value of 64.3 MW. The SCO_2 in the IHE absorbs the heat and flows into the turbine to do work. After the heat release in the HTR and LTR, with a given split ratio (SR) at 0.3852, about 38.52% of the SCO_2 flows into the RC to be compressed directly while the remaining is cooled in the PC and compressed in the MC. Then, the two streams converge at the outlet of the high-pressure side of LTR and continually flow into the HTR and IHE. Thus, an integral process is completed. The recompression model was verified under design condition. The input parameters of the design condition are shown in Table 5, and the model results and error analysis are shown in Table 6.

Table 5. Input parameters under the design condition of the recompression cycle model.

Input Parameters	Values
Rotating speed of MC\% Design rotating speed	100
Rotating speed of RC\% Design rotating speed	100
Spilt ratio	0.3853
Cooling water temperature\K	304.4
Cooling water mass flow rate\kg/s	1400
Cooling water pressure\MPa	0.101
Heater duty\MW	64.3

Table 6. Verification of the operation results of the SCO_2 recompression cycle design condition.

Node Number	Temperature\K			Pressure\MPa		
	Design Value	Model	Errors	Design Value	Model	Errors
1	773.90	774.80	0.12%	19.93	19.92	0.050%
2	667.88	669.25	0.21%	7.76	7.758	0.026%
3	429.12	426.95	0.51%	7.74	7.738	0.026%
4	337.14	335.45	0.50%	7.71	7.708	0.026%
4′	337.14	335.45	0.50%	7.71	7.708	0.026%
5	304.40	304.40	-	7.7	7.698	0.026%
6	332.14	331.50	0.19%	20.0	19.99	0.050%
7	421.85	419.70	0.51%	19.97	19.96	0.050%
4″	337.14	335.45	0.50%	7.71	7.708	0.026%
5″	427.14	424.98	0.51%	19.97	19.96	0.050%
8	424.12	421.75	0.56%	19.95	19.94	0.050%
9	627.15	628.50	0.22%	19.95	19.94	0.050%
Efficiency	41.7%	41.87%	0.41%	-	-	-

As shown in Table 6, the node number corresponds to the note number in Figure 1 in Section 2. Specifically, node 1 and node 2 are the turbine inlet and outlet, and node 3, node 4′, and node 5–6 refer to the flow process from the LTR inlet passing through PC to the MC outlet. The split SCO_2 to the RC outlet is noted by node 4″ and node 5″. The two streams concentrate in the HTR inlet noted as node 8 and then flow into IHE at node 9. It can be seen from the table that the maximum error appears on the HTR inlet temperature at a value of 0.56% with a temperature difference of 2.2 °C. It is easy to note that the MC inlet temperature is kept at 304.4 °C, which is a benefit of the control of the ST. The temperature of the IHE inlet and the turbine is lower by about 1 °C, while the SCO_2 temperature from the inlet of LTR to the inlet of the HTR high-pressure side is generally higher than the design value. In the preliminary analysis, the split ratio may contribute to the error and the strong coupling of the closed-loop system would also amplify the errors of the heat transfer coefficient and physical property. However, the maximum error is only 0.56%, showing that the calculation results of this model can be considered reliable.

4.3. Experiment Verification Plan

Considering the study of the CO_2 Brayton circle is not mature enough, the detailed experimental data of a real power plant are very hard to obtain to support the experiment verification. Actually, our laboratory is preparing to carry out an experiment on the supercritical carbon dioxide system. The experimental loop is based on a simple regenerative cycle, and the schematic diagram is shown in Figure 8. The experiment loop includes a turbine, compressor, printed circuit heat exchanger (PCHE), intermediate heat exchanger (IHE), cooler, surge tank (ST), and heater. The design output power scale of the cycle is 1 MW and the circle efficiency is as large as 20%. The designed maximum temperature is 500 °C at the outlet of IHE, and the maximum pressure is 20 MW at the compressor outlet. At present, the experiment has entered the equipment manufacture stage, and the experiment will be carried out within one year. At this time, we will validate the developed equipment models and circle system based on the experimental results. In fact, the system model we developed will also serve the calculation of the experiment. Then, we will first validate the steady state operation system. Additionally, a series of transient-condition verification will also be carried out, including the working load increase and decrease process. Especially, the working state at a very low power state will also be performed, which will be our main focus.

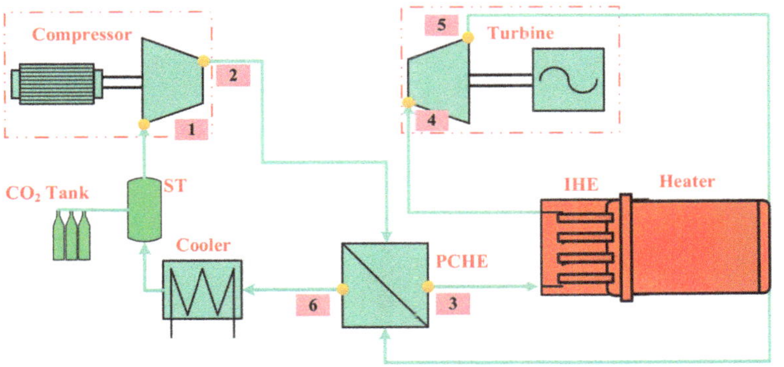

Figure 8. Simple regenerative cycle experiment system schematic diagram.

5. System Dynamic Simulation of the Switching Working Condition

This section mainly studies the 20 MW-SCO_2 system dynamic characteristics in the process of switching between various operating conditions of 100%, 80%, 60%, 40%, and 20% load. The thermal parameters of each key node of the system under non-design operating conditions were estimated according to the performance curve of the equipment. The system mass flow rate was adjusted by changing the opening of the control valve in front of the turbine, so as to achieve the purpose of a variable load. Each working load had the corresponding compressor speed, heat duty, cooling water mass flow rate, and opening degree of the control valve, as shown in Table 7.

Table 7. Corresponding parameters of each working load.

Working load	20%	40%	60%	80%	100%
Control valve opening degree	0.78	0.78	0.88	0.95	100
Rotating speed of MC	80	100	100	100	100
Rotating speed of RC	80	100	100	100	100
Heat duty (MW)	28.11	41.13	49.59	57.03	64.3
Cooling water mass flow rate (kg/s)	1070	1400	1400	1400	1400

In the simulation, the system initially operated stably under 20% operating conditions. When increasing the working load to the 40% load condition, the control valve opening degree was unchanged, and the compressor rotating speed, heat duty, and cooling water mass flow increased in a ramp manner. Then, the compressor and cooler maintained at the design condition and the control valve opening degree were increased continually to meet the load increase demand as shown in Figure 9.

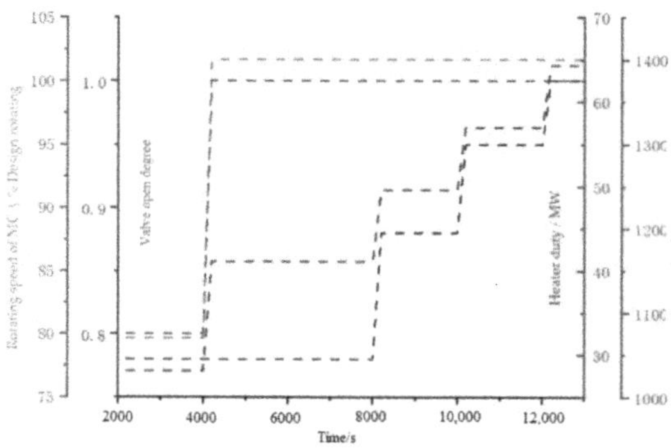

Figure 9. Changes of the parameters in the working load increase process.

The system dynamic characteristics of the load-increasing process are shown in Figure 10. As can be seen, this operating mode can switch operating conditions by adjusting the amount of turbine distribution, while the compressor operating environment remains unchanged. This method can avoid major changes in the compressor operating environment during variable operating conditions and is beneficial to the stable operation of the system. At the same time, this method has a fast rate of changing working conditions, which can meet the demand of a fast changing load. However, this method has certain limitations for a variable load: when the load demand is below 40%, the compressor speed and cooling water mass flow experience great changes. If switching working conditions at this time, the system parameters will fluctuate greatly, and the stability time will be long. Furthermore, the compressor power consumption has to be maintained at a relatively high level, so the cycle efficiency is very low when operating at a low load.

The load-decreasing process is the opposite of the load-increasing process as shown in Figure 11, and the system dynamic characteristics are shown in Figure 12. Assuming that the system initially operates stably under 100% operating conditions, the compressor and cooler operating conditions remained unchanged at first when reducing the load. The turbine power reduced with the decrease of the turbine gas flow rate and heat duty. When the load was below 40%, reducing the compressor rotating speed, and the cooling water mass flow rate at the same time caused fluctuation of the outlet temperature in the heat exchanger and displayed a longer stable time.

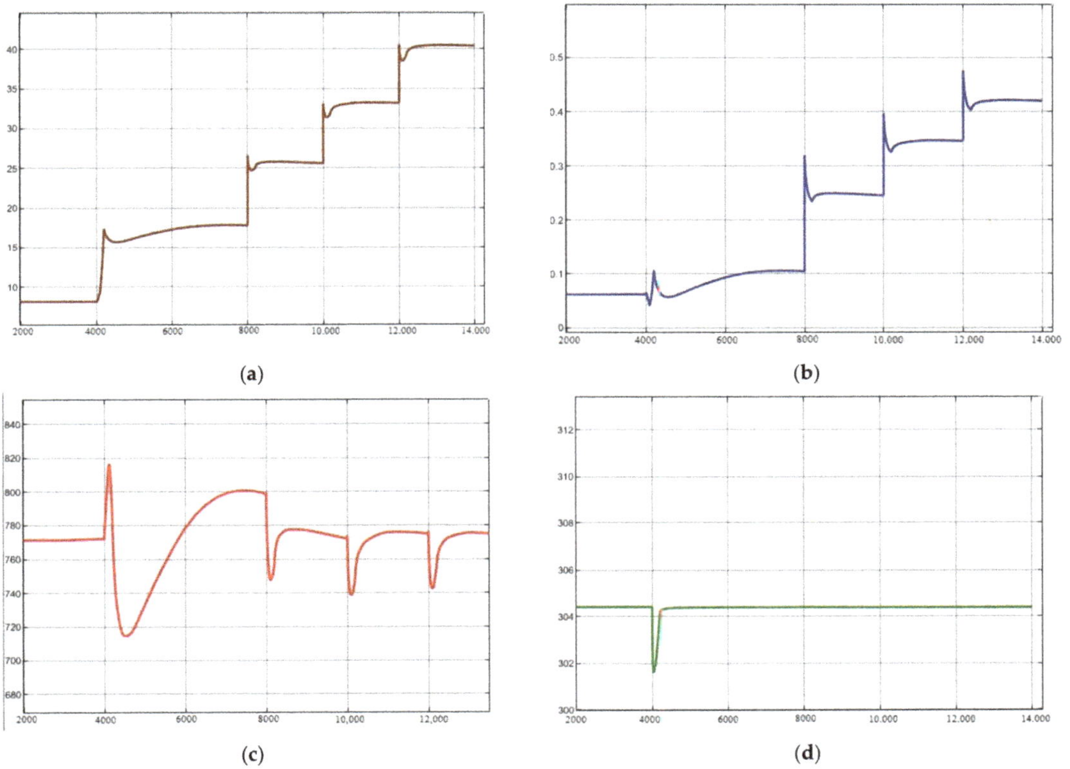

Figure 10. The system dynamic characteristics of the load-increasing process. (**a**) Turbine output power. (**b**) circle efficiency. (**c**) Turbine inlet temperature. (**d**) Compressor inlet temperature.

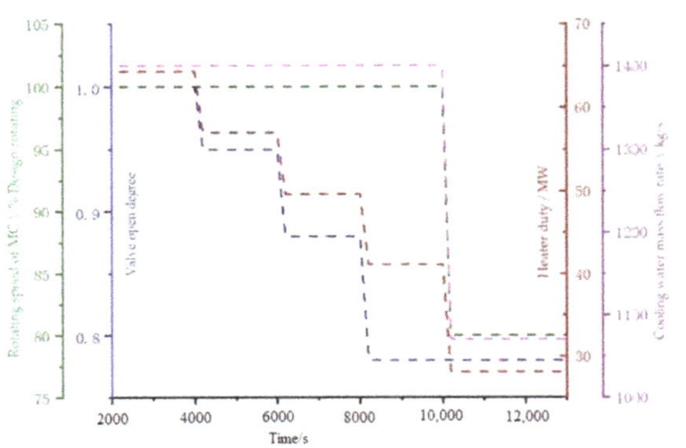

Figure 11. Changes of the parameters in the working load decrease process.

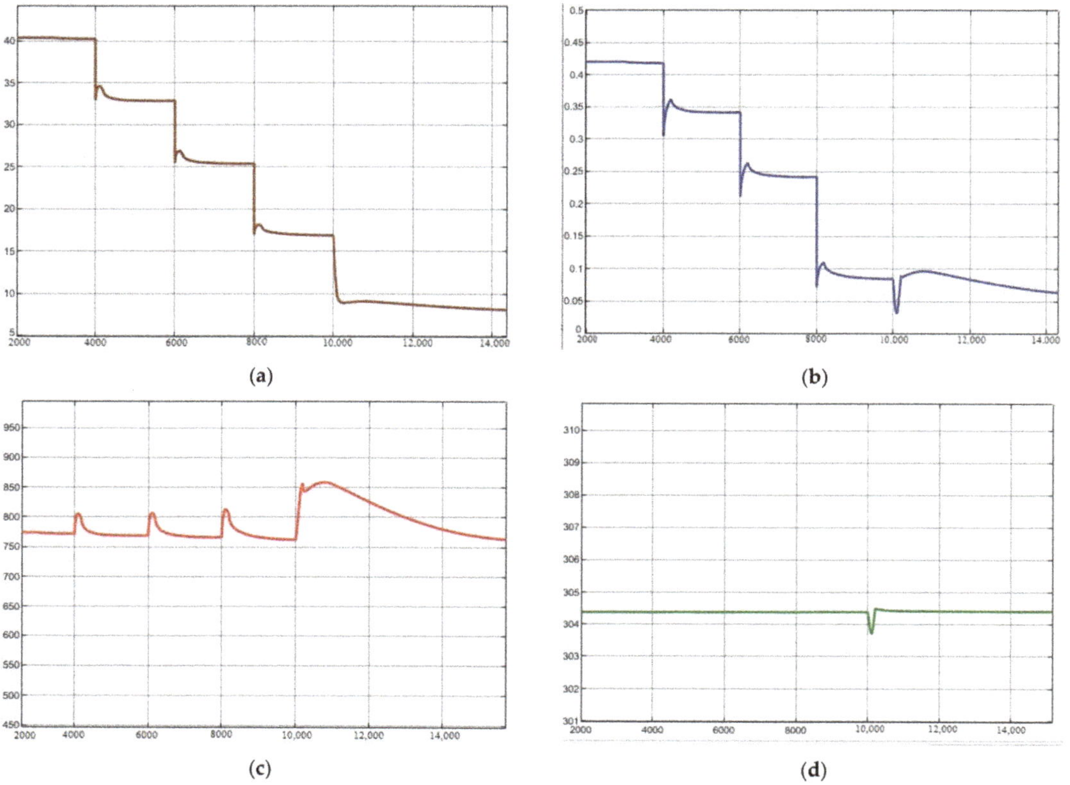

Figure 12. The system dynamic characteristics of the load-decreasing process. (**a**) Turbine output power. (**b**) circle efficiency. (**c**) Turbine inlet temperature. (**d**) Main Compressor inlet temperature.

6. Conclusions

In this study, Simulink software was applied to develop a dynamic model of the 20 MW-SCO_2 recompression cycle modularly. Corresponding component models were developed. A series of comparative calculations were carried out to verify the models. Additionally, the switching condition test of the 20 MW-SCO_2 recompression cycle was conducted. The main conclusions of the study are as follows:

- The models of a 20 MW-SCO_2 recompression cycle were developed based on Simulink software, including heat exchanger models, compressor models, turbine models, and surge tank models.
- The developed main component models were verified with the design values. The comparison results show that calculation errors are acceptable. The disturbance analysis results of the PCHE model were compared with PCHE in 1000 MW S-CO_2 coal-fired power plant data to verify the dynamic simulation capacity of the model. Additionally, the prediction trends were consistent.
- The SCO_2 recompression power system was built with the developed models and the dynamic model ran stably and the maximum error was 0.56% compared to the design value.
- The system dynamic characteristics were studied in the process of switching between various operating conditions of 100%, 80%, 60%, 40%, and 20% load. The analysis is instructive for the system operation and control.

Based on the analytical investigations, the results are very promising in that the developed model has the ability to evaluate the thermal-hydraulic characteristics of the SCO_2 Brayton circulation system, but further experimental data are still needed to validate the code.

Author Contributions: Conceptualization, Z.Z.; Formal analysis, X.L.; Methodology, C.H. and M.L. Project administration, Y.L.; Software, P.S. and L.C.; Supervision, W.W.; Writing—original draft, P.S.; Writing—review and editing, C.D. All authors have read and agreed to the published version of the manuscript.

Funding: This research was funded by National Key R&D Program of China, grant number 2019YFB1901201.

Conflicts of Interest: The authors declare no conflict of interest.

Nomenclature

Abbreviations	
CFP	CO_2 filling pump
HTR	high temperature recuperator
IHE	intermediate heat exchanger
LTR	low temperature recuperator
MC	main compressor
PC	pre-cooler
PCHE	printed circuit heat exchanger
PRV	pressure reducing valve
RC	re-compressor
SCO_2	supercritical carbon dioxide
ST	surge tank
T	Turbine
SR	split ratio
Symbols	
c_p	specific heat, kJ/(kg·K)
h	specific enthalpy, kJ/kg
m	mass flow, kg/s
P	pressure, kPa
V	volume, m^3
f	friction loss coefficient
T	temperature, K
γ	effective loss coefficient
ρ	density, kg/m^3
s	specific entropy, kJ/(kg·K)
η	efficiency
k	heat transfer coefficient, $W/(m^2 \cdot K)$
A_s	Cross-sectional area, m^2
Super- and Sub- scripts	
h	hot fluid
c	cold fluid
w	wall surface
in	inlet
out	outlet
is	isentropic

References

1. Liu, M.; Yang, K.; Zhang, X.; Yan, J. Design and optimization of waste heat recovery system for supercritical carbon dioxide coal-fired power plant to enhance the dust collection efficiency. *J. Clean. Prod.* **2020**, *275*, 122523. [CrossRef]
2. Ahn, Y.; Lee, J.; Kim, S.G.; Lee, J.I.; Cha, J.E.; Lee, S.W. Design consideration of supercritical CO_2 power cycle integral experiment loop. *Energy* **2015**, *86*, 115–127. [CrossRef]

3. Ishiyama, S.; Muto, Y.; Kato, Y.; Nishio, S.; Hayashi, T.; Nomoto, Y. Study of steam, helium and supercritical CO_2 turbine power generations in prototype fusion power reactor. *Prog. Nucl. Energy* **2008**, *50*, 325–332. [CrossRef]
4. Dostal, V.; Hejzlar, P.; Driscoll, M.J. The Supercritical Carbon Dioxide Power Cycle: Comparison to Other Advanced Power Cycles. *Nucl. Technol.* **2006**, *154*, 283–301. [CrossRef]
5. Li, H.; Fan, G.; Cao, L.; Yang, Y.; Yan, X.; Dai, Y.; Zhang, G.; Wang, J. A comprehensive investigation on the design and off-design performance of supercritical carbon dioxide power system based on the small-scale lead-cooled fast reactor. *J. Clean. Prod.* **2020**, *256*, 120720. [CrossRef]
6. Du, Y.; Yang, C.; Hu, C.; Zhang, C. Thermoeconomic analysis and inter-stage pressure ratio optimization of nuclear power supercritical CO_2 multi-stage recompression. *Int. J. Energy Res.* **2020**, *45*, 2367–2382. [CrossRef]
7. Wu, P.; Ma, Y.; Gao, C.; Liu, W.; Shan, J.; Huang, Y.; Wang, J.; Zhang, D.; Ran, X. A review of research and development of supercritical carbon dioxide Brayton cycle technology in nuclear engineering applications. *Nucl. Eng. Des.* **2020**, *368*, 110767. [CrossRef]
8. Pan, M.; Zhu, Y.; Bian, X.; Liang, Y.; Lu, F.; Ban, Z. Theoretical analysis and comparison on supercritical CO_2 based combined cycles for waste heat recovery of engine. *Energy Convers. Manag.* **2020**, *219*, 113049. [CrossRef]
9. Liu, L.; Yang, Q.; Cui, G. Supercritical Carbon Dioxide(s-CO_2) Power Cycle for Waste Heat Recovery: A Review from Thermodynamic Perspective. *Processes* **2020**, *8*, 1461. [CrossRef]
10. Sharma, O.P.; Kaushik, S.C.; Manjunath, K. Thermodynamic analysis and optimization of a supercritical CO_2 regenerative recompression Brayton cycle coupled with a marine gas turbine for shipboard waste heat recovery. *Therm. Sci. Eng. Prog.* **2017**, *3*, 62–74. [CrossRef]
11. Ma, Y.; Morosuk, T.; Liu, M.; Liu, J. Investigation of off-design characteristics of an improved recompression supercritical carbon dioxide cycle for concentrated solar power application. *Int. J. Energy Res.* **2021**, *45*, 1818–1835. [CrossRef]
12. Qiao, Z.; Cao, Y.; Li, P.; Wang, X.; Romero, C.E.; Pan, L. Thermoeconomic analysis of a CO_2 plume geothermal and supercritical CO_2 Brayton combined cycle using solar energy as auxiliary heat source. *J. Clean. Prod.* **2020**, *256*, 120374. [CrossRef]
13. Trevisan, S.; Guédez, R.; Laumert, B. Thermo-economic optimization of an air driven supercritical CO_2 Brayton power cycle for concentrating solar power plant with packed bed thermal energy storage. *Sol. Energy* **2020**, *211*, 1373–1391. [CrossRef]
14. Liu, M.; Zhang, X.; Yang, K.; Ma, Y.; Yan, J. Optimization and comparison on supercritical CO_2 power cycles integrated within coal-fired power plants considering the hot and cold end characteristics. *Energy Convers. Manag.* **2019**, *195*, 854–865. [CrossRef]
15. Guo, J.; Li, M.; Xu, J.; Yan, J.; Ma, T. Energy, exergy and economic (3E) evaluation and conceptual design of the 1000 MW coal-fired power plants integrated with S-CO_2 Brayton cycles. *Energy Convers. Manag.* **2020**, *211*, 112713. [CrossRef]
16. Park, S.; Kim, J.; Yoon, M.; Rhim, D.; Yeom, C. Thermodynamic and economic investigation of coal-fired power plant combined with various supercritical CO_2 Brayton power cycle. *Appl. Therm. Eng.* **2018**, *130*, 611–623. [CrossRef]
17. Sarkar, J. Second law analysis of supercritical CO_2 recompression Brayton cycle. *Energy* **2009**, *34*, 1172–1178. [CrossRef]
18. Ahn, Y.; Bae, S.J.; Kim, M.; Cho, S.K.; Baik, S.; Lee, J.I.; Cha, J.E. Review of supercritical CO_2 power cycle technology and current status of research and development. *Nucl. Eng. Technol.* **2015**, *47*, 647–661. [CrossRef]
19. Michael, P.; Alex, K.; Edward, Z.; Timothy, H. *Supercritical CO_2 Power Cycle Developments and Commercialization: Why sCO_2 Can Displace Steam Ste*; Power-Gen India & Central Asia: New Delhi, India, 2012.
20. Ma, Y.; Liu, M.; Yan, J.; Liu, J. Thermodynamic study of main compression intercooling effects on supercritical CO_2 recompression Brayton cycle. *Energy* **2017**, *140*, 746–756. [CrossRef]
21. Li, M.; Xu, J.; Cao, F.; Guo, J.; Tong, Z.; Zhu, H. The investigation of thermo-economic performance and conceptual design for the miniaturized lead-cooled fast reactor composing supercritical CO_2 power cycle. *Energy* **2019**, *173*, 174–195. [CrossRef]
22. Xu, J.; Sun, E.; Li, M.; Liu, H.; Zhu, B. Key issues and solution strategies for supercritical carbon dioxide coal fired power plant. *Energy* **2018**, *157*, 227–246. [CrossRef]
23. Sun, E.; Xu, J.; Li, M.; Liu, G.; Zhu, B. Connected-top-bottom-cycle to cascade utilize flue gas heat for supercritical carbon dioxide coal fired power plant. *Energy Convers. Manag.* **2018**, *172*, 138–154. [CrossRef]
24. Zhu, M.; Zhou, J.; Su, S.; Xu, J.; Li, A.; Chen, L.; Wang, Y.; Hu, S.; Jiang, L.; Xiang, J. Study on supercritical CO_2 coal-fired boiler based on improved genetic algorithm. *Energy Convers. Manag.* **2020**, *221*, 113163. [CrossRef]
25. Yin, J.M.; Zheng, Q.Y.; Peng, Z.R.; Zhang, X.R. Review of supercritical CO_2 power cycles integrated with CSP. *Int. J. Energy Res.* **2019**, *44*, 1337–1369. [CrossRef]
26. Liang, Y.; Chen, J.; Luo, X.; Chen, J.; Yang, Z.; Chen, Y. Simultaneous optimization of combined supercritical CO_2 Brayton cycle and organic Rankine cycle integrated with concentrated solar power system. *J. Clean. Prod.* **2020**, *266*, 121927. [CrossRef]
27. Liu, Y.; Wang, Y.; Huang, D. Supercritical CO_2 Brayton cycle: A state-of-the-art review. *Energy* **2019**, *189*, 115900. [CrossRef]
28. Dyreby, J.J. *Modeling the Supercritical Carbon Dioxide Brayton Cycle with Recompression*; The University of Wisconsin-Madison: Madison, WI, USA, 2014.
29. Moisseytsev, A.; Sienicki, J.J. Simulation of S-CO_2 integrated system test with anl plant dynamics code. In Proceedings of the 5th International Symposium–Supercritical CO_2 Power Cycles, San Antonio, TX, USA, 28–31 March 2016.
30. Wu, P.; Gao, C.; Shan, J. Development and verification of a transient analysis tool for reactor system using supercritical CO_2 brayton cycle as power conversion system. *Sci. Technol. Nucl. Install.* **2018**, *2018*. [CrossRef]
31. Deng, T.; Li, X.; Wang, Q.; Ma, T. Dynamic modelling and transient characteristics of supercritical CO_2 recompression Brayton cycle. *Energy* **2019**, *180*, 292–302. [CrossRef]

32. Chen, M.; Kim, I.H.; Sun, X.; Christensen, R.; Utgikar, V.; Sabharwall, P. Transient analysis of an FHR coupled to a helium Brayton power cycle. *Prog. Nucl. Energy* **2015**, *83*, 283–293. [CrossRef]
33. Nielsen, T.R.; Rose, J.; Kragh, J. Dynamic model of counter flow air to air heat exchanger for comfort ventilation with condensation and frost formation. *Appl. Therm. Eng.* **2009**, *29*, 462–468. [CrossRef]
34. Liu, M.; Zhang, X.; Yang, K.; Wang, B.; Yan, J. Comparison and sensitivity analysis of the efficiency enhancements of coal-fired power plants integrated with supercritical CO_2 Brayton cycle and steam Rankine cycle. *Energy Convers. Manag.* **2019**, *198*, 111918. [CrossRef]
35. Ma, C.; Wang, W.; Wu, J.; Dai, L. Analysis of unsteady flow in a supercritical carbon dioxide radial compressor stage. International Conference on Nuclear Engineering. *Am. Soc. Mech. Eng.* **2018**, *51494*, V06BT08A037.
36. Ying, Q.; Zhuge, W.; Zhang, Y.; Ma, C.; Gou, J.; Wang, W. Vortex Patterns Investigation and Enstrophy Analysis in a Small Scale S-CO_2 Axial Turbine. *Energies* **2021**, *14*, 6112. [CrossRef]
37. Sun, J. *Characteristic Simulation and Control of Supercritical Carbon Dioxide Cycle Power Generation System*; Harbin Institute of Technology: Harbin, China, 2018.
38. Ma, T.; Li, M.; Xu, J.; Cao, F. Thermodynamic analysis and performance prediction on dynamic response characteristic of PCHE in 1000 MW S-CO_2 coal fired power plant. *Energy* **2019**, *175*, 123–138. [CrossRef]

processes

MDPI

Article

Research on Response Characteristics and Control Strategy of the Supercritical Carbon Dioxide Power Cycle

Chunhui Dai [1,2], Ping Song [1,2,*], Can Ma [1,2], Kelong Zhang [1,2], Wei Zheng [1,2], Lie Chen [1,2], Xiaojie Guo [1,2], Yuansheng Lin [1,2] and Zhiqiang Qiu [1,2]

1 Key Laboratory of Thermal Power Technology, Wuhan 430064, China; ch.dai@163.com (C.D.); macan1234@sina.com (C.M.); sworkcsic@163.com (K.Z.); zhengwei@mail.ustc.edu.cn (W.Z.); helloworldum@163.com (L.C.); lucky_gxj_0372@163.com (X.G.); keylab_rndl@163.com (Y.L.); jchdlzhjs@126.com (Z.Q.)
2 Wuhan Second Ship Design and Research Institute, Wuhan 430064, China
* Correspondence: pingsong@stu.xjtu.edu.cn

Abstract: With the development of GEN-IV nuclear reactor technology, the supercritical carbon dioxide (SCO_2) Brayton cycle has attracted wide attention for its simple structure and high efficiency. Correspondingly, a series of research has been carried out to study the characteristics of the cycle. The control flexibility of the power generation system has rarely been studied. This paper carried out a dynamic performance of the 20 MW-SCO_2 recompression cycle based on the Simulink software. In the simulation, the response characteristics of the system main parameters under the disturbances of cooling water temperature, split ratio, main compressor inlet temperature and pressure were analyzed. The results show that the turbine inlet temperature is most affected by the disturbances, with a re-stabilization time of 2500–3000 s. According to the response characteristics of the system after being disturbed, this study proposed a stable operation control scheme. The scheme is coordinated with the main compressor inlet temperature and pressure control, recompressor outlet pressure control, turbine inlet temperature control and turbine load control. Finally, the control strategy is verified with the disturbance of reduced split ratio, and the results show that the control effect is good.

Keywords: supercritical carbon dioxide Brayton cycle; Simulink; disturbance; control strategy

Citation: Dai, C.; Song, P.; Ma, C.; Zhang, K.; Zheng, W.; Chen, L.; Guo, X.; Lin, Y.; Qiu, Z. Research on Response Characteristics and Control Strategy of the Supercritical Carbon Dioxide Power Cycle. *Processes* **2021**, 9, 1943. https://doi.org/10.3390/pr9111943

Academic Editor: Pei Liu

Received: 2 September 2021
Accepted: 27 October 2021
Published: 29 October 2021

Publisher's Note: MDPI stays neutral with regard to jurisdictional claims in published maps and institutional affiliations.

Copyright: © 2021 by the authors. Licensee MDPI, Basel, Switzerland. This article is an open access article distributed under the terms and conditions of the Creative Commons Attribution (CC BY) license (https://creativecommons.org/licenses/by/4.0/).

1. Introduction

The supercritical carbon dioxide (SCO_2) Brayton cycle has been widely used in a variety of power generation scenarios, including nuclear energy [1], solar energy [2], and thermal power [3]. The concept of SCO_2 Brayton cycle can be traced back to the 1960s [4]. With the development of the GEN-IV nuclear reactor, the SCO_2 Brayton cycle attracted attention from all over the world for its simple structure and high efficiency. Scholars also proposed the latest advancement of Allam cycle power generation technologies, making the CO_2 available at pipeline specification [5,6] and the corresponding dynamic simulation and control strategies were also studied [7,8], which directly promoted the sequestration and utilization of SCO_2.

Correspondingly, a series of research has been carried out to study the characteristics of the cycle [9–12]. However, most of the existing SCO_2 cycle studies are limited to basic thermodynamics research, parameter optimizations, and system design in different application fields. Considering that the SCO_2 cycle is widely used in the power generation system, attention needs to be paid to the performance of variable load and flexibility. Thus, the research on its dynamic model and control strategy is also essential, but this part of the work is still in infancy.

Hu et al. [13] carried out dynamic modeling on the recuperator of a 2 MW-SCO_2 power generation system, and mainly studied the fluid parameter change law in the recuperator during cold start and variable load. Ma et al. [14] developed the dynamic model of the recuperators for the 1000 MW-SCO_2 coal-fired power unit and analyzed the parameter dynamic response trend with the inlet temperature and mass flow rate change. Then, the neural network was trained to predict the performance of the Printed Circuit Heat Exchanger (PCHE). Due to the difficulty in manufacturing SCO_2 compressors and turbines, especially for high-load power generation systems, some scholars fitted empirical formulas and expanded the scale based on the performance curves of the existing SCO_2 turbomachines, aiming to increase the performance prediction accuracy of CO_2 rotating machinery under non-design conditions, such as Cho et al. [15] and Lee et al. [16].

As for the overall system, Dyreby [17] provided a detailed SCO_2 system modular modeling method and principles under non-design conditions, including the modeling methods of key components and the solution process of the overall system, but there was little analysis on the system dynamic performance. Zhu [18] developed a 350 kW SC recompression closed-loop dynamic model with Panysimu software and analyzed the main parameter response characteristics under the disturbance of the heat source duty, regulating valve opening and cooling water flow, but the research on the control strategy was not conducted.

Luu et al. [19] proposed a combined start-up control scheme of the simple recuperation cycle and recompression cycle with the background of solar energy applications. The simple recuperation cycle was applied to meet the design point at the initial start-up period, and then the system was merged into the recompression branch after the stable operation for 1 h. The total start-up time is about 4 h. As opposed to Luu's combined branch startup program, Liese et al. [20] adopted an inventory control scheme to directly adjust the system flow through the storage tank and realize the startup, shutdown and load tracking condition switch of the recompression system, which can shorten the startup and shutdown time. At the same time, the study pointed out that the model of the cooler should be at least one-dimensional.

In addition to the start-stop control scheme, the control strategy for the stable operation of the system has also been preliminary studied. Olumayegun et al. [21] studied the operation control strategy of the waste heat recovery system. The simulation results pointed out that the turbine inlet temperature should change with the change of waste heat, but the model was developed as an open loop. Sun [22] developed a closed-loop dynamic model of the system against the background coal-fired power generation and compared the effects of the constant flow control strategy and the constant turbine inlet temperature control strategy. The results pointed out that the former had stable control effects, while the latter had a higher cycle efficiency, especially under low-load operating conditions. The study by Du et al. [23] also reached the same conclusion.

Overall, part of the existing research mainly focused on the component level and some studies provided general modeling methods and dynamic responses to parameter disturbances. Moreover, the existing control strategy research mainly pays attention to the system start-stop plan. Thus, there is a lack of research on the dynamic characteristics of SCO_2 cycle and the control strategy for stable operation. As a supplement to the previous research, this paper studies the dynamic performance of the SCO_2 recompression cycle based on a 20 MW sodium-cooled reactor with Simulink software. This study mainly analyzes the dynamic response characteristics of the recompression cycle with the disturbances of cooling water temperature, split ratio, main compressor inlet temperature and pressure, which have been rarely studied previously. According to the response characteristics, a control strategy for the stable operation of the system is proposed and the split ratio disturbance is verified.

2. Simple Description of System

2.1. Comparison of Typical SCO_2 Cycles

Numerous layouts of the SCO_2 Brayton cycle have been researched. Typically, simple cycle, simple recuperated cycle and recompression cycle are included in the general SCO_2

Brayton cycle. Figures 1–3 shows the three-type cycle diagram and T-s curve respectively, where T means temperature and s means entropy. What's more in the T-s curves, the Brayton cycle thermodynamic process are presented and the numbers mean the sequence of the process.

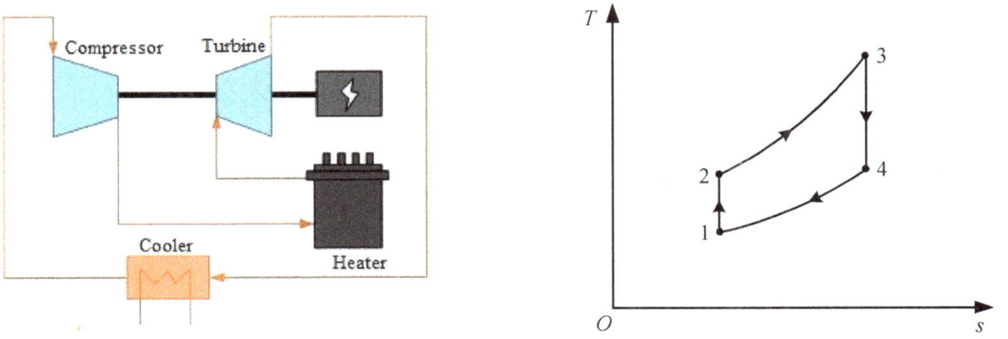

(a) cycle diagram (b) T-s curve

Figure 1. SCO_2 simple cycle.

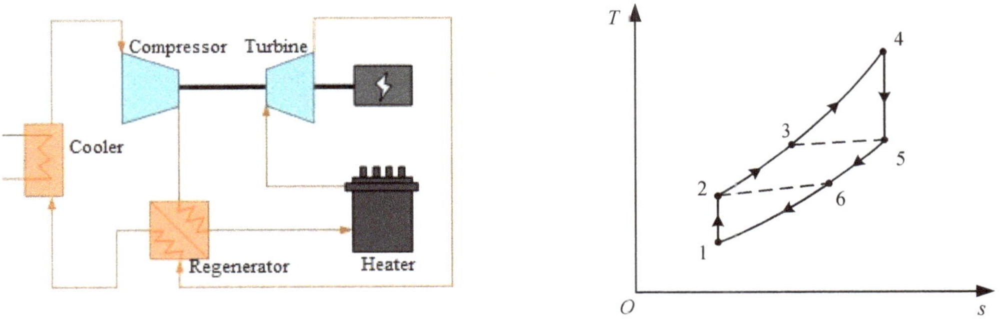

(a) cycle diagram (b) T-s curve

Figure 2. SCO_2 simple recuperated cycle.

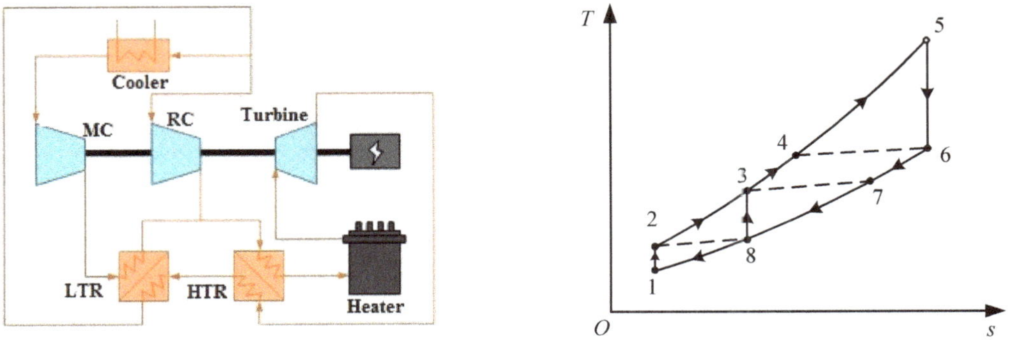

(a) cycle diagram (b) T-s curve

Figure 3. SCO_2 recompression cycle.

The SCO_2 simple cycle is based on the typical Brayton cycle principle. The included compressor, heater, turbine and cooler correspond to the Brayton cycle thermodynamic process of isentropic compression (point1 to point 2 in Figure 1b), constant pressure heat absorption (point 2 to point 3 in Figure 1b), adiabatic expansion (point 3 to point 4 in Figure 1b) and constant pressure release (point 3 to point 4 in Figure 1b), as shown in Figure 1. In the simple SCO_2 cycle system, the low-temperature and low-pressure SCO_2 enters the heater to absorb heat after being boosted by the compressor, and then the high-temperature and high-pressure SCO_2 enters the turbine to perform work. Finally, the exhausted SCO_2 flows into the compressor through the cooler. Thus, a closed cycle is completed. Part of the turbine output power is used to drive the compressor to pressurize the SCO_2, and the rest drives the generator to generate electricity. Given that the physical properties of SCO_2 change drastically near the critical point, the density is relatively large, and the compression coefficient is relatively small, the compressor consumes less power. Moreover, the back pressure of turbine and the exhausted SCO_2 temperature are relatively high, and the direct heat release in the cooler causes a large amount of heat loss, so the efficiency of the SCO_2 simple cycle is extremely low.

The simple recuperated cycle arranges a regenerator based on the simple cycle, as shown in Figure 2. The high-temperature exhausted SCO_2 from the turbine is used to preheat the low-temperature SCO_2 from the compressor outlet. Thus, part of the exhausted heat is recovered and the energy loss of the cooler is reduced. At the same time, the heat absorption temperature of SCO_2 at the inlet of the heater is increased, and the heat absorption of the cycle is reduced, so the cycle efficiency is improved.

Due to the large difference in temperature and pressure of SCO_2 on both sides of the regenerator, the difference in the specific heat capacity is obvious. Meanwhile, the flow rate of the fluid on both sides is equal and the heat capacity is quite different. The above factors result in the pinch point of the regenerator, which will have a great impact on the cycle efficiency.

The SCO_2 recompression cycle solves the pinch point problem of the simple recuperated cycle by adding the thermal process of split flow and recompression, as shown in Figure 3. With the design of split flow, the regenerator is divided into two stages. On the one hand, the temperature change on both sides of the high-temperature regenerator is small. On the other hand, the flow rate of the low-temperature fluid in the low temperature regenerator becomes less after the split, and the heat capacity is reduced. Thus, the high and low temperature regenerators are not prone to the problem of pinch points. Meanwhile, the split circulation sacrifices part of the work to compress the fluid, so that the fluid temperature is increased after being reheated. The heat absorption at the heater is reduced under the same conditions, and the heat loss of the pre-cooling is reduced, which improves the cycle efficiency.

2.2. Description of Recompression System

Considering the problem of the pinch point in the simple recuperated cycle, the power generation system described in this paper adopts the SCO_2 recompression cycle, and the configuration is shown in Figure 4. A small part of the supercritical carbon dioxide is separated and compressed in the RC, and the other is still compressed in the MC. The regenerator is divided into two stages. This can effectively avoid the pinch point of the regenerator. The overall system includes turbine (T), main compressor (MC), recompressor (RC), high-temperature and low-temperature recuperators (HTR and LTR), intermediate heat exchanger (IHE), pre-cooler (PC), surge tank (ST) and heater.

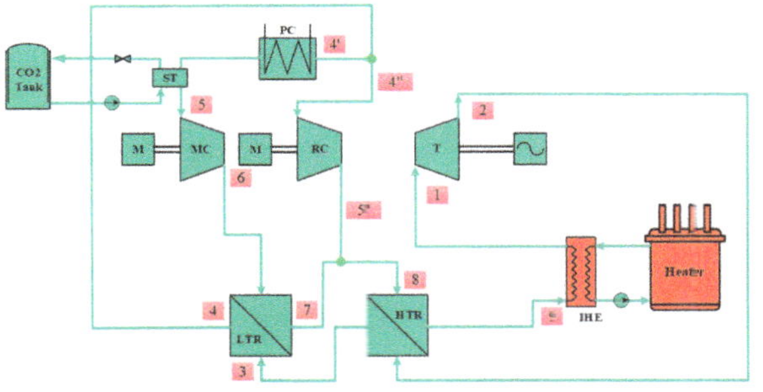

Figure 4. Supercritical carbon dioxide recompression cycle power generation system.

The closed-loop system model was debugged and the initial operating parameters were calculated. The Simulink configuration diagram is shown in Figure 5. The heat duty is inputted to the system to simulate the heater at a value of 64.3 MW. The SCO_2 in the IHE absorbs the heat and flows into the turbine to do work. After the heat release in the HTR and LTR, with a given split ratio (*SR*) at 0.3852, about 38.52% of the SCO_2 flows into the RC to be compressed directly. The rest is cooled in the PC and compressed in the MC. Then, the two streams converge at the outlet of the high-pressure side of LTR and continually flow into the HTR and IHE. Thus, an integral process is completed. The input parameters of the design condition are shown in Table 1.

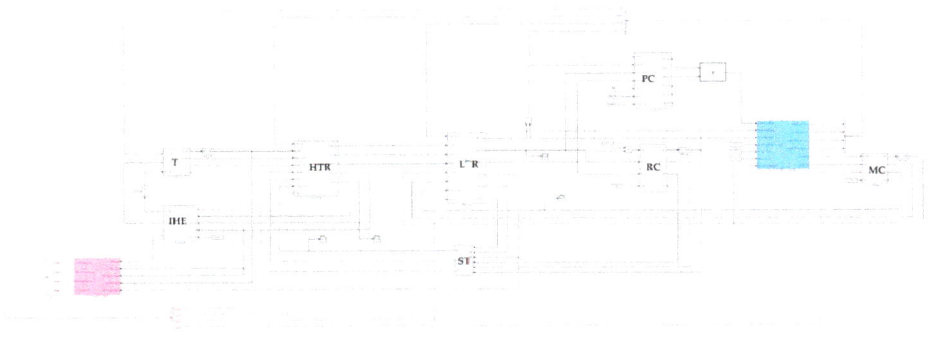

Figure 5. Simulink dynamic model diagram of SCO_2 recompression cycle.

Table 1. Input parameters under design condition of recompression cycle model.

Input Parameters	Values
Rotating speed of MC\% Design rotating speed	100
Rotating speed of RC\% Design rotating speed	100
Spilit ratio	0.3853
Cooling water temperature\K	304.4
Cooling water mass flow rate\kg/s	1400
Cooling water pressure\MPa	0.101
Heater duty\MW	64.3

Based on the developed dynamic model of the 20 MW-SCO_2 recompression cycle with Simulink software, this paper studied the response characteristics and control strategy of the SCO_2 recompression cycle.

3. Disturbance Analysis

In this paper, the response characteristics of the SCO_2 recompression system with the disturbances of cooling water temperature, split ratio (*SR*), MC inlet pressure and temperature were first analyzed.

3.1. Cooling Water Temperature

The cooling water is obtained from outside in the actual operation. On the one hand, if the water temperature is too high, the CO_2 cannot be cooled to the design temperature, which will reduce the cycle efficiency; on the other hand, the water temperature being too low may cause the CO_2 to exit the supercritical state, resulting in the system failing to operate normally. Therefore, the cooling water temperature disturbance is simulated and analyzed in this section. In the simulation, the compressor and turbine run at a constant rotating speed, the cooling water flow rate and the heater duty remain unchanged, while the cooling water temperature steps up by 2 K. The simulation results are shown in Figure 6.

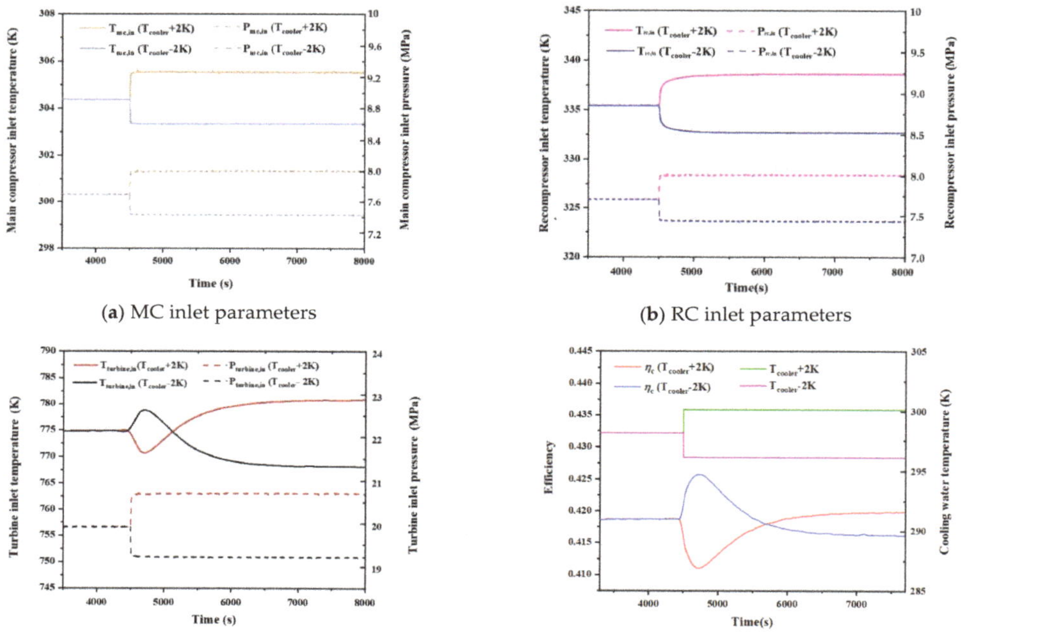

(a) MC inlet parameters (b) RC inlet parameters

(c) Turbine inlet parameters (d) Efficiency

Figure 6. Response curves of main parameters with cooling water temperature disturbance.

As can be seen from Figure 6, the MC inlet temperature is the first to be affected, rising by about 1.1 K and showing an approximate step change with the cooling water temperature stepping up by 2 K. Figure 6b shows that the RC inlet temperature increases by about 3.1 K, and the time for re-stabilization is about 1500 s. In Figure 6c, the inlet temperature of turbine decreases first and then increases, and the variation range is about 6 K. Due to the strong coupling of the closed-loop system and the thermal inertia of the HTR and LTR, the re-stabilization time of the turbine inlet temperature is as long as 3000 s The volume of each component is lumped in the ST and IHE for the integral calculation of the density and enthalpy, and the physical property software is applied to calculate the pressure change of the working fluid. Pressure responds very quickly to the system disturbance, and the pressure at the key points of the system shows an approximate step change. Compared with the design conditions, the MC inlet pressure increased by about 0.3 MPa and the turbine inlet pressure increased by about 0.77 MPa. Under the disturbance of the cooling water temperature, the cycle efficiency change and the turbine inlet temperature change show the same trend as shown in Figure 6d. When the cooling water temperature steps down by 2 K, the parameters change laws are opposite to the above.

3.2. Split Ratio (SR)

The split ratio was used to express the share of flow entering the RC. The low-pressure side outlet fluid of the LTR is divided into two streams: one stream is cooled in the PC and then compressed in the MC, while the other stream is directly compressed in the RC. Under the premise of the design condition, assuming that the split ratio (*SR*) steps up by 0.01 (from 0.3853 to 0.3953) at 4500 s, other input parameters of the system remain unchanged. The system is simulated and the results are shown in Figure 7.

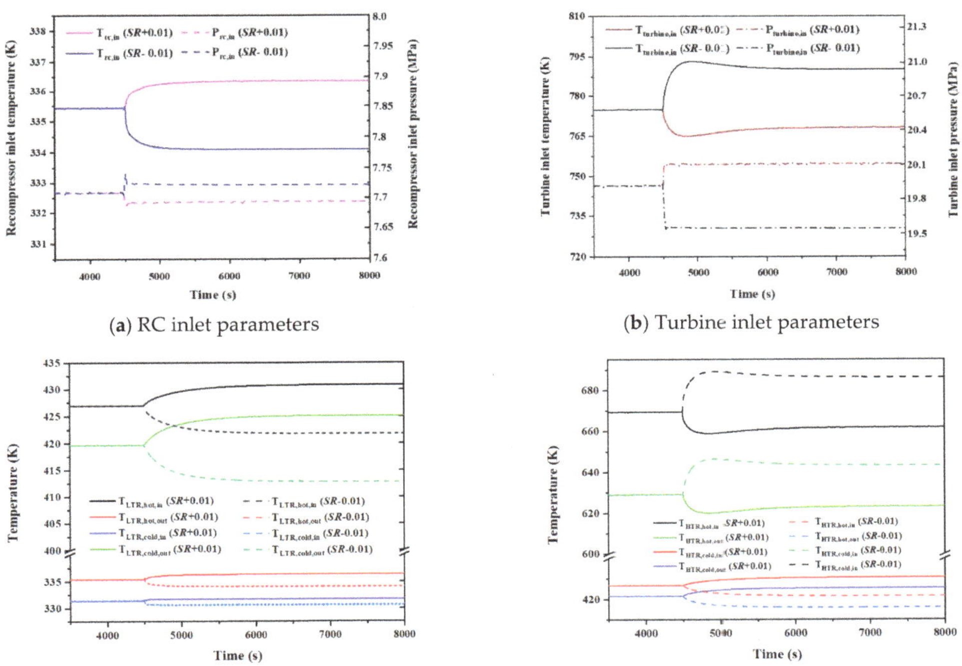

(**a**) RC inlet parameters (**b**) Turbine inlet parameters

(**c**) LTR inlet and out temperatures (**d**) HTR inlet and out temperatures

Figure 7. Response curves of main parameters with *SR* disturbance.

When the split ratio steps up by 0.01, the temperature and pressure at the inlet of MC change slightly. The RC inlet temperature increases by about 1.0 K, and the time for re-stabilization is about 1500 s (Figure 7a). On the contrary, the inlet temperature of the turbine decreases by about 6.4 K, while the re-stabilization time after disturbance is as long as about 2500 s (Figure 7b). These changes are triggered by the change of the mass flow rate in each branch. The mass flow rate of the working fluid at the MC and the cold end of LTR decreases as the split ratio increases, so the temperature of the working fluid at the hot end outlet of the LTR and the inlet of RC increases. At this time, both the inlet and outlet temperatures of the LTR increase (Figure 7c). However, the disturbance of *SR* increases the CO_2 flow rate in the main circuit. When the heater duty is constant, the turbine inlet temperature decreases. Correspondingly, the temperatures at the hot end inlet and cold end outlet of the HTR decrease (Figure 7d). Under the disturbance of the split ratio, the pressures of the system change rapidly, and behave similarly to the step responses. Specifically, the RC inlet pressure decreases and the turbine inlet pressure increases.

When the split ratio steps down by 0.01, the parameter variation trends are opposite to the above and the parameter change ranges are large.

3.3. MC Inlet Pressure

In this simulation, the MC inlet pressure steps up by 0.2 MPa (from 7.7 to 7.9 MPa) at 4500 s with the premise of the design condition, and the other input parameters of the system remain unchanged. The system simulation was carried out and the results are shown in Figure 8.

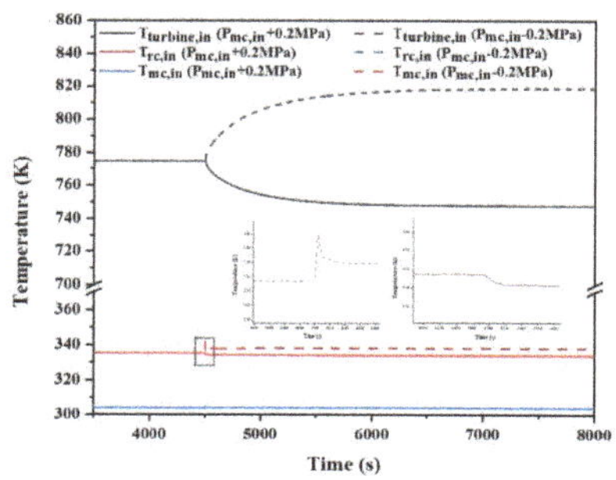

(**a**) T, MC, RC inlet temperatures

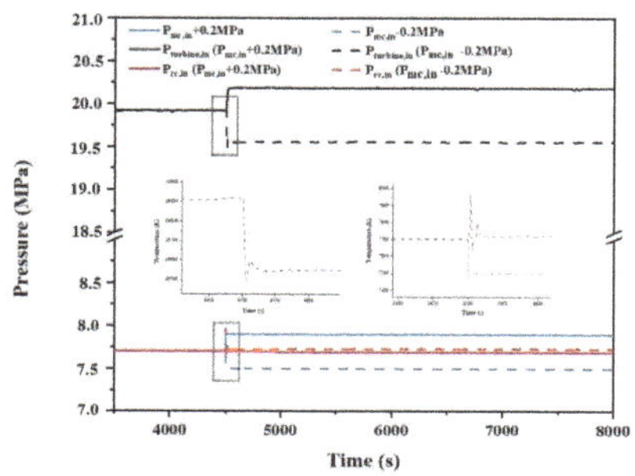

(**b**) T, MC, RC inlet pressures

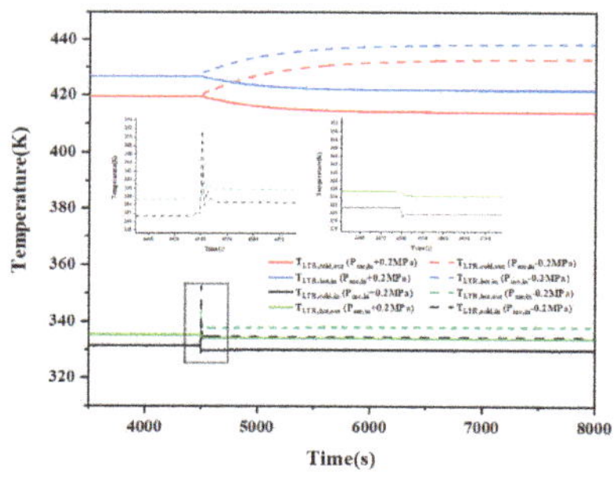

(**c**) LTR inlet and out temperatures

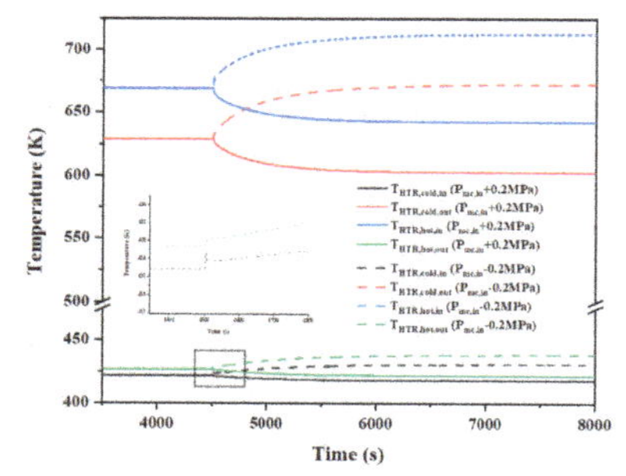

(**d**) HTR inlet and out temperatures

Figure 8. Response curves of main parameters with MC inlet pressure disturbance.

When the MC inlet pressure steps up by 0.2 MPa, the MC inlet temperature changes slightly, the RC inlet temperature decreases by about 1.3 K, and the re-stabilization time is about 80 s. The inlet temperature of the turbine is greatly affected by the disturbance of the MC inlet pressure, which is reduced by about 26.2 K, and the re-stabilization time is as long as 2500 s (Figure 8a). Because of the rapid response of pressure, the turbine inlet pressure increases by 0.25 MPa, which shows an approximate step change. Instead, the RC inlet pressure is almost constant (Figure 8b). As the MC inlet pressure increases, the MC outlet temperature decreases, and thus the inlet and outlet temperatures of LTR and HTR decrease. The temperature re-stabilization time of each point on the recuperator cold side increases with the flow direction of low-temperature CO_2, while the temperature re-stabilization time of each point on the hot side of the recuperators decreases with the flow direction of high-temperature CO_2 (Figure 8c,d).

When the MC inlet pressure steps down by 0.2 MPa, the parameters change trend are opposite to the above and the parameters change ranges are larger. It is worth noting that when the MC inlet pressure is reduced from 7.7 to 7.5 MPa, the pressure at this point is closer to the critical pressure. The pressures and temperatures of some key points of the system will fluctuate for a short time, such as the RC inlet temperature and pressure responses. In addition, the LTR cold end inlet temperature and hot end outlet temperature response may even cause the deterioration of LTR heat transfer, as shown in Figure 8c.

3.4. MC Inlet Temperature

In the SCO_2 recompression system, the MC inlet temperature and pressure need to stay above the critical point. However, when the system is disturbed by the split ratio, cooling water temperature and mass flow rate, the MC inlet temperature may drop below the critical point. Therefore, this section will discuss the dynamic characteristics of the system main parameters when the MC inlet temperature drops below the critical point.

Assuming that the MC inlet temperature is stepped down by 1.0 K (from 304.4 to 303.4 K) at 4500 s with the premise of the design condition, and other input parameters of the system remain unchanged. The system was simulated and the results are shown in Figure 9. When the MC inlet temperature steps down by 1.0 K, the RC inlet temperature decreases by about 2.7 K and the time for re-stabilization is about 500 s. As shown in Figure 9a, the inlet temperature of the turbine is greatly affected by the disturbance of the MC inlet temperature, which is reduced by about 15 K, and the re-stabilization time is as long as 3000 s. The temperature of each point of the recuperators decreases accordingly as the disturbance occurs. The temperature re-stabilization time of each point on the recuperator cold side increases with the flow direction of low-temperature CO_2; the temperature re-stabilization time of each point on the hot side of the recuperators decreases with the flow direction of high-temperature CO_2, as shown in Figure 9b,c.

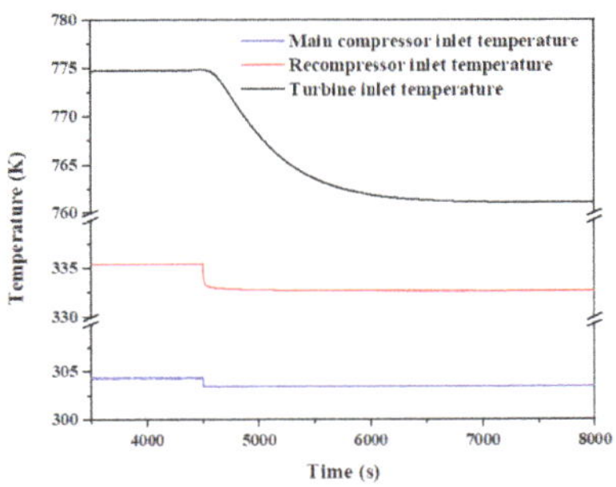

(**a**) T, MC, RC inlet temperatures

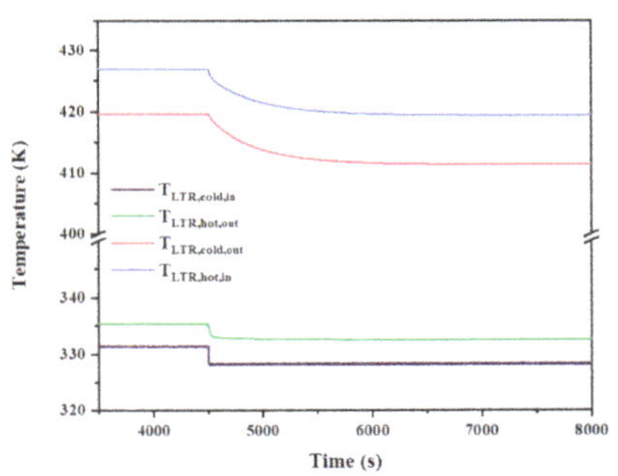

(**b**) LTR inlet and out temperatures

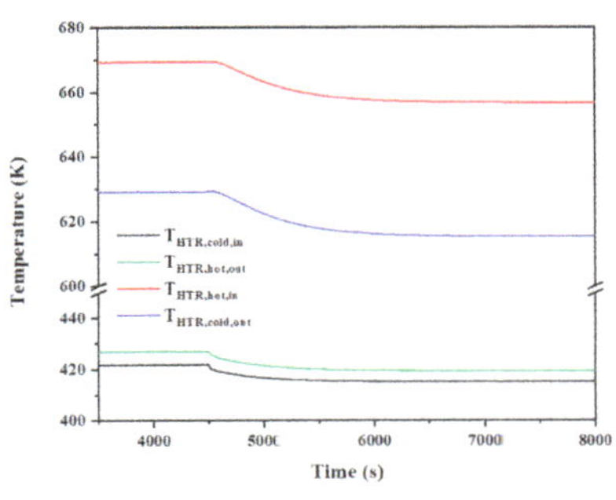

(**c**) HTR inlet and out temperatures

Figure 9. Response curves of main parameters with MC inlet temperature disturbance.

4. Control Strategy of Stable Operation

The control strategy for the stable operation was proposed based on the response characteristics. Finally, the system control effects with the disturbance of the split ratio were verified.

4.1. Control Strategy

From the above disturbances simulation results, it can be seen that the turbine inlet temperature is the most sensitive parameter to all disturbances, with large fluctuation range and long re-stabilization time. Therefore, this study adopts the stable operation control strategy and maintains the turbine inlet temperature constant. The overall control system is divided into the following five parts, as shown in Figure 10:

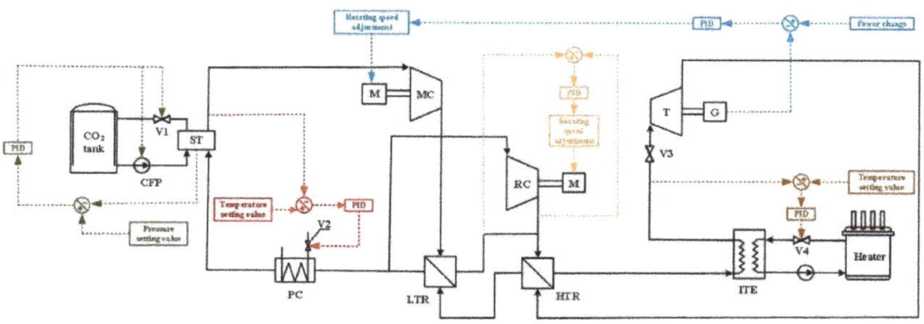

Figure 10. The operation control strategy of the SCO_2 recompression system.

The operation control strategy also consists of five parts: MC inlet pressure control module, MC inlet temperature control module, turbine inlet temperature control module, turbine output power control module and RC outlet pressure control module.

1. MC inlet pressure control module

The ST is connected to the front pipeline of the MC and the CO_2 storage tank, and its function is to maintain the MC inlet pressure at the set value, so the ST pressure is used as the controlled parameter. When the ST pressure is higher than the set value, the PRV (V1) is opened to draw a part of the working fluid from the ST, thereby reducing the pressure in the surge tank. When the ST pressure is lower than the set value, the CFP is ran to fill the CO_2 with specific parameters into the ST, thereby increasing the pressure in the ST.

2. MC inlet temperature control module

The MC inlet temperature was controlled to the set value by changing the cooling water flow rate. When the MC inlet temperature is higher than the set value, the difference signal between the MC inlet temperature and the set value is converted into a signal of the opening of the cooling water valve (V2) by the PID controller. The cooling water mass flow rate is increased by increasing the valve opening, thereby reducing the MC inlet temperature to achieve control effect. When the MC inlet temperature is lower than the set value, the control command is opposite.

3. Turbine inlet temperature control module

In order to reduce the influence of disturbance on the system, it is necessary to control the turbine inlet temperature. The turbine inlet temperature is controlled to the set value by changing the hot fluid mass flow rate of the ITE. When the turbine inlet temperature is lower than the set value, the difference signal between the turbine inlet temperature and the set value is converted into a signal of opening the valve (V4) by the PID controller. The hot fluid mass flow rate of ITE increases with the increase of the valve opening, and thus the turbine inlet temperature will decrease. When the turbine inlet temperature is higher than the set value, the control command is opposite.

4. Turbine output power control module

The system flow is charged by adjusting the rotating speed of the MC to realize the turbine load tracking in this study. When the turbine load demand changes, the difference signal between the demand value and the current turbine load is transmitted to the PID controller, and then converted into a variable frequency signal of the MC rotating speed by the PID controller. The MC speed increases as the turbine load demand increases, and vice versa. The system flow rate changes with the MC rotating speed, and finally the turbine load is adjusted to meet the demand.

5. RC outlet pressure control module

In order to ensure that the pressure of the RC outlet working fluid is consistent with the LTR cold end outlet working fluid when they merge, the RC outlet pressure is taken as the controlled parameter. The RC flow rate is determined by the main circuit flow rate and the split ratio, so the outlet pressure can be adjusted by the rotating speed. The difference signal between the LTR cold end outlet pressure and the RC outlet pressure is transmitted to the PID controller and then converted into a frequency conversion signal of the RC rotating speed, which can realize the control function.

4.2. Control Effect Analysis

This section discusses the control effect with the disturbance of the split ratio. The split ratio is stepped down by 0.01 (decreased from 0.3853 to 0.3753) at 4500 s when operating under design condition. The PID parameter tuner of Simulink is used to tune the controller parameters.

In the SCO_2 system, considering that the MC inlet parameters must be kept above the critical point, the MC inlet temperature control has the highest priority in the coordinated control of the overall system. Because the turbine inlet temperature has a greater influence on the turbine power, the control command for the turbine inlet temperature is also faster than the mass flow rate control command using the power difference as a signal. The above is achieved by tuning the PID controller parameters.

The main control parameters and the response of the controlled quantity are shown in Figure 11. When the split ratio is stepped down, the recompressor flow rate decreases accordingly. The MC mass flow rate is controlled by the rotating speed and exhibits hysteresis, so the working fluid of the main circuit is reduced, and the output power of the turbine is also decreased. In the control system, the turbine output power demand remains unchanged, while the actual output is reduced. Thus, the signal of the difference is transmitted to the MC through the PID controller, and the mass flow rate control system starts to work then. The MC flow rate is increased with the increase of the MC rotating speed. The flow rate of the main circuit returns to the initial state in about 130 s, and the flow rate of each branch reaches the distribution balance under the new split ratio. During this period, the heater load and the cooling water mass flow rate are adjusted to maintain the turbine inlet temperature and MC inlet temperature at the set value. The MC inlet temperature is the first to return to the set value, and the fluctuating range of the turbine inlet temperature is ± 5 K, as shown in Figure 11c,d. With the cooperation of all control modules, the turbine power is almost restored to the demand value synchronously with the flow rate of the main circuit, and the control effects are good.

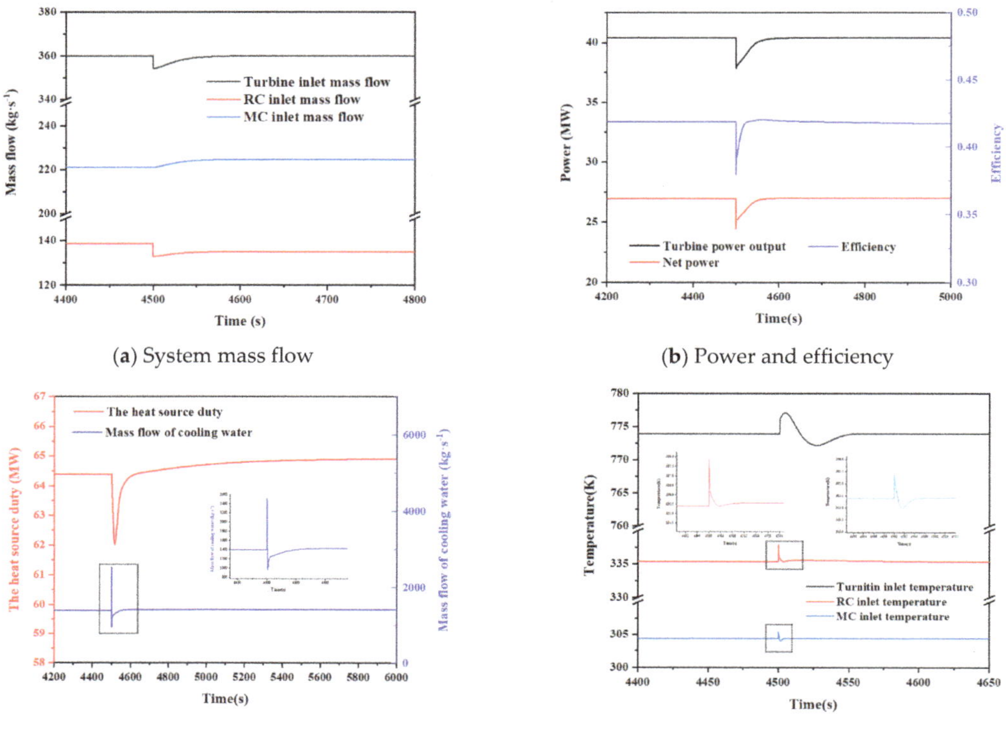

(**a**) System mass flow

(**b**) Power and efficiency

(**c**) The heater duty and flow of cooling water

(**d**) T, MC, RC inlet temperatures

Figure 11. Response curves of controlled parameter with *SR* disturbance.

5. Conclusions

In this study, the response characteristics of the SCO_2 recompression cycle system were analyzed on the disturbances of cooling water temperature, split ratio, main compressor inlet temperature and pressure. Subsequently, according to the response characteristics of the system, a control strategy of stable operation was proposed. The verification effect of the split ratio disturbance was carried out to provide a reference for future SCO_2 dynamic research. The main conclusions of the study are as follows:

- When the system is subjected to a step disturbance, the pressure shows a similar step response rapidly, while the temperature change is relatively slow, especially for the turbine inlet temperature. Due to the closed-loop system features and the recuperator thermal inertia, the turbine inlet temperature fluctuates in a wide range and the re-stabilization time is the longest, about 2500–3000 s. The inlet temperature of the MC is easily reduced to below the critical point in the cooling water parameters disturbance, but adjusting the cooling water flow is an effective way to control the MC inlet temperature to be constant.
- For the split ratio disturbance and the MC inlet pressure disturbance, the reduction of the value has a greater impact on the system parameters than the increase of the value. When the MC inlet pressure is reduced from 7.7 to 7.5 MPa, the pressure at this point is closer to the critical pressure. The pressures and temperatures of the system key points will fluctuate for a short time, especially the LTR cold end inlet temperature and hot end outlet temperature responses, which may even cause the deterioration of LTR heat transfer.

- According to the response characteristics of the system after being disturbed, this study proposed a stable operation control scheme to coordinate the MC inlet temperature and pressure control, PC outlet pressure control, turbine inlet temperature control and turbine load control. When the split ratio decreases by 0.01, the parameters deviate from the set values and the turbine output power decreases. After the control system acts on for about 130 s, the flow rate of each branch reaches the distribution balance under the new split ratio, and meanwhile the turbine load returns to the set demand. The MC inlet temperature is the first to return to the set value, and the fluctuating range of the turbine inlet temperature is ±5 K. In general, the overall control effect is good.

Author Contributions: Conceptualization, K.Z.; Formal analysis, L.C.; Investigation, P.S.; Methodology, C.M.; Project administration, Y.L. and Z.Q.; Software, W.Z. and X.G.; Writing—original draft, P.S.; Writing—review and editing, C.D. All authors have read and agreed to the published version of the manuscript.

Funding: This research was funded by National Key R&D Program of China, grant number 2019YFB1901201.

Conflicts of Interest: The authors declare no conflict of interest.

References

1. Li, H.; Fan, G.; Cao, L.; Yang, Y.; Yan, X.; Dai, Y.; Zhang, G.; Wang, J. A comprehensive investigation on the design and off-design performance of supercritical carbon dioxide power system based on the small-scale lead-cooled fast reactor. *J. Clean. Prod.* **2020**, *256*, 120720. [CrossRef]
2. Qiao, Z.; Cao, Y.; Li, P.; Wang, X.; Romero, C.E.; Pan, L. Thermoeconomic analysis of a CO_2 plume geothermal and supercritical CO_2 Brayton combined cycle using solar energy as auxiliary heat source. *J. Clean. Prod.* **2020**, *256*, 120374.
3. Guo, J.; Li, M.; Xu, J.; Yan, J.; Ma, T. Energy, exergy and economic (3E) evaluation and conceptual design of the 1000 MW coal-fired power plants integrated with S-CO_2 Brayton cycles. *Energy Convers. Manag.* **2020**, *211*, 112713. [CrossRef]
4. Liu, M.; Yang, K.; Zhang, X.; Yan, J. Design and optimization of waste heat recovery system for supercritical carbon dioxide coal-fired power plant to enhance the dust collection efficiency. *J. Clean. Prod.* **2020**, *275*, 122523. [CrossRef]
5. Allamab, R.; Martin, S.; Forrest, B.; Fetvedt, J.; Lu, X.; Freed, D.; Brown, G.W., Jr.; Sasaki, T.; Itoh, M.; Manning, J. Demonstration of the Allam cycle: An update on the development status of a high efficiency supercritical carbon dioxide power process employing full carbon capture. *Energy Procedia* **2017**, *114*, 5948–5966. [CrossRef]
6. Fernandes, S.; Wang, Q.; Xu, B.R.; Chen, D. Process and carbon footprint analyses of the Allam cycle power plant integrated with an air separation unit. *Clean Technol.* **2019**, *1*, 325–340. [CrossRef]
7. Fernandes, D.; Wang, S.; Xu, Q.; Chen, D. Dynamic Simulations of the Allam Cycle Power Plant Integrated with an Air Separation Unit. *Int. J. Chem. Eng.* **2019**, *10*, 6035856 . [CrossRef]
8. Xiong, J.; Zhao, H.; Chen, H.; Zheng, C. Dynamic simulation of a conceptual 600 MWe oxy-fuel combustion power plant. In Proceedings of the 3rd Oxyfuel Combustion Conference, IEAGHG, Wuhan, China, 9–13 September 2013.
9. Sarkar, J. Second law analysis of supercritical CO_2 recompression Brayton cycle. *Energy* **2009**, *34*, 1172–1178. [CrossRef]
10. Ahn, Y.; Bae, S.J.; Kim, M.; Cho, S.K.; Baik, S.; Lee, J.I.; Cha, J.E. Review of supercritical CO_2 power cycle technology and current status of research and development. *Nucl. Eng. Technol.* **2015**, *47*, 647–661. [CrossRef]
11. Persichilli, M.; Kacludis, A.; Zdankiewicz, E.; Held, T. Supercritical CO_2 Power Cycle Developments and Commercialization: Why sCO_2 can Displace Steam Ste. *Power-Gen India Cent. Asia* **2012**, *2012*, 19–21.
12. Ma, Y.; Liu, M.; Yan, J.; Liu, J. Thermodynamic study of main compression intercooling effects on supercritical CO_2 recompression Brayton cycle. *Energy* **2017**, *140*, 746–756. [CrossRef]
13. Hu, H.; Guo, C.; Cai, H.; Jiang, Y.; Liang, S.; Guo, Y. Dynamic characteristics of the recuperator thermal performance in a S–CO_2 Brayton cycle. *Energy* **2021**, *214*, 119017. [CrossRef]
14. Ma, T.; Li, M.; Xu, J.; Cao, F. Thermodynamic analysis and performance prediction on dynamic response characteristic of PCHE in 1000 MW S-CO_2 coal fired power plant. *Energy* **2019**, *175*, 123–138. [CrossRef]
15. Cho, S.K.; Lee, J.; Lee, J.I. Off design performance map similarity study of radial type turbomachinery in supercritical co brayton cycle. In Proceedings of the Asme-jsme-ksme Joint Fluids Engineering Conference, Seoul, Korea, 26–31 July 2015.
16. Lee, J.; Ahn, Y.; Jung, H.Y. Off design performance of compressors of 150MWe S-CO_2 Cycle for Sodium Cooled Fast Reactor. In Proceedings of the Transactions of the Korean Nuclear Society Spring Meeting, Jeju, Korea, 29–30 May 2014.
17. Dyreby, J.J. *Modeling the Supercritical Carbon Dioxide Brayton Cycle with Recompression*; ProQuest Dissertations Publishing: Morrisville, NC, USA, 2014.
18. Zhu, Y. Research on Dynamic Characteristics of Supercritical Carbon Dioxide Brayton Cycle System. Master's Thesis, North China Electric Power University, Beijing, China, 2018.

19. Luu, M.T.; Milani, D.; McNaughton, R.; Abbas, A. Dynamic modelling and start-up operation of a solar-assisted recompression supercritical CO_2 Brayton power cycle. *Appl. Energy* **2017**, *199*, 247–263. [CrossRef]
20. Liese, E.; Albright, J.; Zitney, S.A. Startup, shutdown, and load-following simulations of a 10 MWe supercritical CO_2 recompression closed Brayton cycle. *Appl. Energy* **2020**, *277*, 115628. [CrossRef]
21. Olumayegun, O.; Wang, M. Dynamic modelling and control of supercritical CO_2 power cycle using waste heat from industrial processes. *Fuel* **2019**, *249*, 89–102. [CrossRef]
22. Jia, S. Characteristic Simulation and Control of Supercritical Carbon Dioxide Cycle Power Generation System. Master's Thesis, Harbin Institute of Technology, Harbin, China, 2018.
23. Du, X.; Hu, J.; Xia, G. Operation characteristic of supercritical carbon dioxide-cooled reactor system under coordination control scheme. *Int. J. Adv. Robot. Syst.* **2020**, *17*, 172988142093383. [CrossRef]

processes

MDPI

Article

Solid Oxide Fuel Cell-Based Polygeneration Systems in Residential Applications: A Review of Technology, Energy Planning and Guidelines for Optimizing the Design

Farah Ramadhani [1,*], M. A. Hussain [2,*], Hazlie Mokhlis [3,*] and Oon Erixno [4]

1. Physics Department, Faculty of Science, University of Malaya, Kuala Lumpur 50603, Malaysia
2. Chemical Engineering Department, Faculty of Engineering, University of Malaya, Kuala Lumpur 50603, Malaysia
3. Electrical Engineering Department, Faculty of Engineering, University of Malaya, Kuala Lumpur 50603, Malaysia
4. Higher Institution Centre of Excellence (HICoE), UM Power Energy Dedicated Advanced Centre (UMPEDAC), Level 4, Wisma R&D Universiti Malaya, Jalan Pantai Baharu, Kuala Lumpur 59990, Malaysia

* Correspondence: farahr@um.edu.my (F.R.); mohd_azlan@um.edu.my (M.A.H.); hazli@um.edu.my (H.M.)

Abstract: Solid oxide fuel cells are an emerging energy conversion technology suitable for high-temperature power generation with proper auxiliary heat. Combining SOFCs and polygeneration has produced practical applications for modern energy system designs. Even though many researchers have reviewed these systems' technologies, opportunities and challenges, reviews regarding the optimal strategy for designing and operating the systems are limited. Polygeneration is more complicated than any other energy generation type due to its ability to generate many types of energy from various prime movers. Moreover, integration with other applications, such as vehicle charging and fueling stations, increases the complication in making the system optimally serve the loads. This study elaborates on the energy planning and guidelines for designing a polygeneration system, especially for residential applications. The review of polygeneration technologies also aligns with the current research trend of developing green technology for modern and smart homes in residential areas. The proposed guideline is expected to solve the complication in other applications and technologies and design the polygeneration system optimally.

Keywords: polygeneration; SOFC; optimal design; residential; electric vehicles; hydrogen vehicle

Citation: Ramadhani, F.; Hussain, M.A.; Mokhlis, H.; Erixno, O. Solid Oxide Fuel Cell-Based Polygeneration Systems in Residential Applications: A Review of Technology, Energy Planning and Guidelines for Optimizing the Design. *Processes* **2022**, *10*, 2126. https://doi.org/10.3390/pr10102126

Academic Editors: Pei Liu, Ming Liu and Xiao Wu

Received: 22 September 2022
Accepted: 17 October 2022
Published: 19 October 2022

Publisher's Note: MDPI stays neutral with regard to jurisdictional claims in published maps and institutional affiliations.

Copyright: © 2022 by the authors. Licensee MDPI, Basel, Switzerland. This article is an open access article distributed under the terms and conditions of the Creative Commons Attribution (CC BY) license (https://creativecommons.org/licenses/by/4.0/).

1. Introduction

The increase in energy demand has triggered the development and exploration of various types of energy supply. One of the recent developments in energy supply is polygeneration. Polygeneration is an energy generation system that can provide three or more energy types by integrating one or two prime movers. The kinds of energy generation and prime mover sometimes relate to applying polygeneration and available sources. Rapid developments of renewable energy sources have also affected the excitement surrounding the use of polygeneration to tackle climate change and neutralize carbon emissions from non-renewable-based energy generation [1].

The increased attention regarding polygeneration systems resonates with the fact that people need various energy types in one integrated system. The rapidness of technology is a significant reason for the variety of energy required by people. Currently, the needs are increasing and are not limited to electric consumption but also include the need for heating, water cooling and clean water, clean air and green fuels. The ineffectiveness of separated systems related to their efficiency and fuel consumption is another issue that increases the implementation of the polygeneration system [2]. In the separation concept, only one primary source can generate energy with single conversion efficiency and much waste. However, a polygeneration system works with the waste and converts it to become a useful side product that increases the conversion efficiency. In line with that, a split system is much more costly than polygeneration in terms of energy cost. The increase in fuel price must be a consideration when increasing the effectiveness of an energy generation system to have lower energy costs and affordable operation costs.

Solid oxide fuel cells (SOFCs) are attractive as fuel cells in stationary applications such as polygeneration systems. Their high-temperature characteristic is valuable for heating and cooling generation. When a SOFC is combined with a solid oxide electrolyzer cell (SOEC) or works with hydrogen fuel, it becomes a carbon-neutralized system that can produce various energy types and fuels. SOFC-based polygeneration systems have become widely used for industrial, public and private applications such as those in residential areas [3,4].

Despite the effectiveness of polygeneration, the drawbacks of implementing the system are still challenging. Due to the various sources and products generated, the biggest challenge is designing an effective polygeneration system. Creating a polygeneration system does not mean combining two or more primary sources to have three or more products. With ineffective design, a polygeneration system might have less efficiency and a higher energy cost than a separated system. Moreover, when optimizing the system design, more than one aim might need to be considered, making the system more complicated to develop. Therefore, optimization and management are needed when designing a polygeneration system starting with synthesizing the sources and components, sizing the capacity and operating the system depending on the schedule, use or achievement objectives.

In the literature, polygeneration systems and fuel cells are topics that have been receiving attention since the beginning of this decade. Many fuel-cell-based polygeneration system concepts have been reviewed for various applications, such as desalination [3,5–7], industries [8], public services [9] and community-based power generation [10]. Among the reviewed literature, the prospect of polygeneration and its challenges are discussed quite prevalently [10–12]. Other studies about the state of the art and technology reviews of polygeneration systems were discussed in refs. [13–15]. However, none of the review papers focused on the optimal development and preparation of polygeneration system design. Even though the design is the first step in developing a system, there is an urgent need to discuss the topic and give insight into how to design a polygeneration system, especially when using fuel cells as the prime mover.

This research aims to review polygeneration technologies and propose a guideline for performing energy planning and optimizing the system. This work has continuity with the previous review paper [16], which concerned cogeneration systems. However, polygeneration systems have advanced purposes and complicated designs compared to cogeneration systems, making the procedure more challenging. Therefore, with the energy planning and guideline proposed, it is expected that researchers can gain insight when developing a polygeneration system. On top of that, this work could be a guideline for various applications of a polygeneration system with various primary energy sources, products and objectives.

This paper consists of a brief description of energy design in residential applications in Section 2, continuing with a general presentation on energy system designs in Section 3. Section 4 explains the overview of polygeneration systems, while Section 5 discusses SOFC-based polygeneration system applications. Section 6 presents the energy planning for designing a polygeneration system, while Section 7 explains the guideline for optimizing polygeneration system design.

2. The Evolution of Residential Energy Supply Designs

Residential areas are among those with high energy use, which raises global greenhouse gas (GHG) emissions. The production of residential buildings accounts for more than 40% of power and 21% of GHG emissions [17]. According to projections, energy use will climb by more than 67% by 2050 due to an increase in the global population, the number of houses and new energy services [18]. These issues increase the energy cost of the building paid by resident users.

In Malaysia, the national energy consumption increased by 210.7% from 1990 to 2009 [19]. Because of a lack of efficiency techniques in the Malaysian building sector, carbon emissions have increased by 235.6%. Policies should be put in place to lower greenhouse gas (GHG) emissions from buildings and boost residential buildings' energy supply efficiency through integrated supply techniques.

2.1. Conventional Energy System Designs

A traditional, stationary supply is a non-renewable energy system that generates power and delivers it to a residential area. The electric efficiency from grid generation is about 0.4–0.48 [20]; hence, enormous electricity usage can increase energy loss and cost. As depicted in Figure 1, the separated system in a conventional, stationary supply for residential applications includes the usage of gasoline-fueled vehicles as well as electricity, heat and cooling energy. The classic car uses gasoline as fuel, which has high energy prices and high GHG emissions. The efficiency of the gasoline vehicle is 236.8 MJ/100 km, or 20 km/L [20]. Moreover, vehicles with gasoline fuel contribute to the increase in GHG emissions more than natural gas vehicles. There are 2.344 kg of carbon emissions in one liter of gasoline [21]. Due to the issue of fossil fuel depletion, gasoline utilization must be reduced or it must be substituted by another form of fuel, for example, electricity or hydrogen.

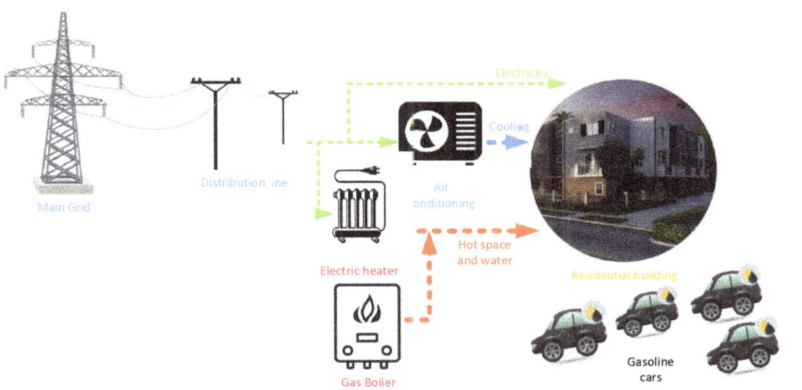

Figure 1. Conventional energy system design for residential application.

2.2. Integrated Residential Energy System Designs

Integration of one or several energy sources to overcome the low efficiency of grid distribution has been made into a standalone system. For small-scale applications, one house can build 1 kW to 2 kW of standalone power supply, which can be separated or connected to the primary grid. This integration is more efficient, avoiding power losses in the distribution line and optimizing electricity usage, as depicted in Figure 2. A microgrid configuration can be developed in multi-family homes or apartments from one or combined supply sources of non-renewable or renewable energies.

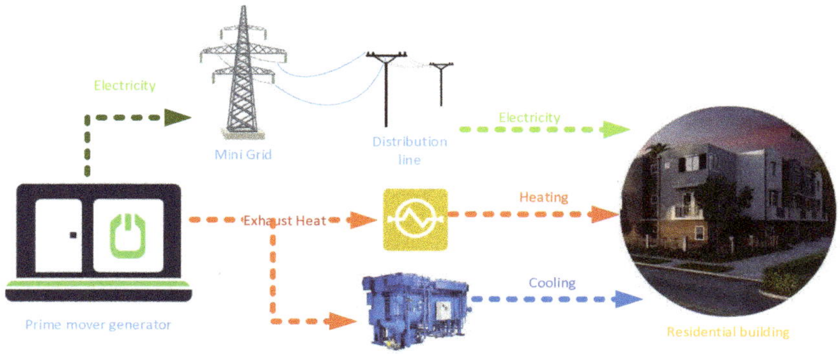

Figure 2. Integrated energy supply design for residential application.

The microgrid energy supply's development includes giving residents access to heating and cooling energy and electricity. Various energy sources can simultaneously work as a prime mover to produce heat and electricity. Combined heat and power (CHP) or cogeneration systems employ the prime mover's exhaust heat, which is typically squandered, to generate both electricity and heat for the consumer. The CHP system has been developed into a combined cooling, heating and power (CCHP), or trigeneration, system for further applications, which offers more forms of energy, including electricity, hot water and space and cold water and space. Trigeneration, sometimes known as polygeneration, is a system that may create three or more energy sources from a single or integrated supply system. A polygeneration system can produce fresh water, oxygen, hydrogen and other forms of energy in addition to electricity, heat and cooling.

2.3. *Advanced Residential Energy System Design*

Advanced residential energy systems combine stationary and mobile power supplies, which are attractive for future developments. The idea comes from the trend of vehicle technologies increasing the use of electric and hydrogen as fuels instead of gasoline in modern vehicle types, as depicted in Figure 3. With the development of hybrid or pure electric and hydrogen cars, the opportunity for building a personal recharging/fueling station is not a difficult thing. Personal vehicle chargers can be implemented for single- or multi-family usage, for private use or be shared with the community, while, for a fueling station, an electrolyzer can produce hydrogen from electric sources such as photovoltaic, wind and fuel cells.

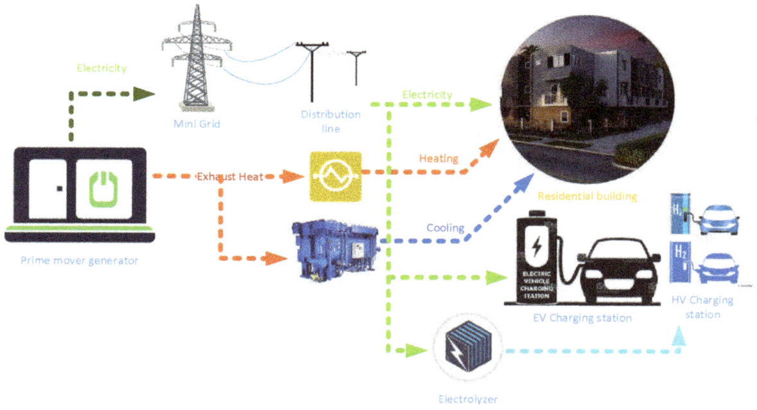

Figure 3. Advanced energy system design with a vehicle charging/fueling station in a residential area.

Combining stationary and vehicle energy supplies can lower the amount of non-renewable energy used for primary energy needs, improve energy utilization effectiveness and prevent power loss during distribution. Additionally, an integrated system can dramatically lower airborne GHG emissions by using RE as the prime mover. A short payback period and lower energy costs are further economic benefits of an integrated system [22]. Therefore, the design of an energy system that integrates stationary and vehicle energy supply should be well planned based on the users' requirements.

3. Energy Generation Systems

Energy systems can be categorized into three types based on their design: single generation, cogeneration and polygeneration/trigeneration systems. Depending on customers' needs, a single power generator generates a single supply consisting of electricity, district heating or cooling. A cogeneration system can produce electricity and hot water or space, electricity and cold water or space or electricity and hydrogen from a single or integrated system. From a single or combined energy source, polygeneration systems can produce more than three energy types simultaneously; while known as trigeneration, the system may have more than three different types of energy, i.e., CCHP and quad-generations.

3.1. Single Power Generation Systems

Single power generators are commonly used to generate electricity through standalone or grid-connected configurations. To enhance the performance of the prime mover, many forms of renewable or non-renewable energy sources can be blended into a hybrid system. A system that combined SOFCs and a gas turbine was implemented in [23], using a system with a size of 500 kW. Power generation systems using fuel cells as the prime mover have been designed and operated efficiently [24]. A study of a hybrid system of SOFCs and PEMFCs was conducted in ref. [25] to increase system efficiency.

The addition of thermoelectric material to the prime mover's exhaust heat is another way to raise efficiency. According to the research, adding a thermoelectric generator to the burner, air preheater or fuel preheater boosts the prime mover's ability to produce electricity [26–29]. Thermoelectric devices can be coupled with most prime mover types to provide additional electricity. Due to their low-temperature operation, thermoelectric materials cannot be associated with all renewable energy sources. Some practical, renewable energy sources that combine well with thermoelectricity include photovoltaic and fuel cells. Fuel cells outperform photovoltaic cells in heat quality and acceptable temperature range for thermoelectric materials [4].

A single power system can produce cooling and/or heat in addition to electric power by using a separate system. For multi-family residential settings, some district heating or cooling systems have been created [30–33]. A single or hybrid prime mover produces the district heating or cooling system and commonly uses a combusting engine to produce more heat. For residential dwellings, a complex analysis was also performed for power sharing design in a district heating and combined heat and power system [34].

3.2. Cogeneration System

Electrical power is not the sole energy required for several power generation applications, especially those in office and residential buildings. Other sources such as air and water heating and cooling systems are also needed continuously. Based on ref. [35], for community residential complexes, unconventional sources are needed, including energy, hydrogen for vehicles, recycled fuel, fresh water and oxygen. To address such needs, the majority of commercial and residential buildings, however, employ separate systems (SP) for power, heating and cooling, which results in inefficient energy use and drives up the cost of energy. An integrated system that can produce electricity, heating, cooling and other supplemental outputs such as fresh water, hydrogen and oxygen is needed to increase system efficiency, energy consumption and energy cost.

The combined heat and power system (CHP) can be regarded as the most straightforward system when building an integrated system. It comprises a prime mover that produces power and heat simultaneously, either as a single prime mover or as a hybrid, and a heating system that recovers the heat energy from the prime mover. When the heat demand exceeds the heat supply, several applications of CHP systems may include an auxiliary boiler. Additionally, it has a heat storage system to keep the extra energy for use when the temperature is low. The CHP is advantageous because it can create heat power as if it were a separate system without additional fuel. This system, which has been developed by a number of businesses, including a collaboration between Ceres Power in the U.K., Hexis in Switzerland and British Gas and Ceramic Fuel Cells Ltd., is anticipated to become the first commercialized, energy-generating system deployed for residential purposes (Australia) [36].

3.3. *Polygeneration System*

Despite the success in developing the cogeneration system, the system cannot fulfil the cooling demand for office and residential buildings. Therefore, when the heat requirement is low, the extra heat from the prime mover is useless, reducing the total system efficiency. Further technology in energy generation has triggered the development of polygeneration, which is equipped with a cooling system and other types of energy, i.e., fresh water, hydrogen and oxygen, to optimize the utilization of heat and fuel and satisfy the requirement of users. In a polygeneration system, heating and cooling demand can be met and electrical power can be generated as the primary output with the same amount of fuel utilization. In several scenarios, cooling devices can be improved using an electric chiller and refrigerator as auxiliary cooling devices when the heat power produced is less than the requirement [37–41].

Polygeneration is also called trigeneration or a combined cooling, heating and electricity (CCHP) system in other applications. A trigeneration or CCHP system serves three types of energy from one or more prime movers in an integrated system. When the system can provide more than three, it can be called quad-generation; thus, commonly, all these systems are named polygeneration systems. Polygeneration systems can produce more energy types than cogeneration and trigeneration systems. For instance, they are more energy efficient and have superior energy waste recovery in addition to providing hydrogen for automobiles and fresh water, energy and heating and cooling for water and space for residential purposes [6,42–44]. Therefore, implementing a polygeneration system integrated with a vehicle station is promising for future residential buildings.

4. Overview of Polygeneration Systems

In most applications, a polygeneration system consists of several device types: prime mover, heating, cooling, storage, hydrogen generator and other purifying devices.

4.1. *Prime Mover*

Based on the literature, the development of polygeneration systems has resulted in different types of prime movers being used to serve the system. In [45], combustion-based prime mover technologies included gas turbines, Stirling engines and Rankine cycle units, while electrochemical technologies included fuel cells and electrolyzers. Technologies such as gas turbines and reciprocating engines are commercially viable, while Stirling engines, Rankine cycle units and fuel are still in their early stages. A comparison of the prime mover technologies used in polygeneration systems is shown in Table 1. The comparison includes the efficiency of electrical and system generations and the emissions (CO_2 and NOx) from the system generation, including electric and thermal generation.

Table 1. Evaluation and comparison of several polygenerational prime movers [45–47].

Prime Movers	Reciprocating Engines	Gas Turbine	Stirling Engines	Steam Turbine	Rankine Engines	Fuel Cells
Advantages	Effective part load performance, very flexible and quick start-up time necessary.	Low maintenance requirements, flexible and compact design and moderate output heat temperature.	Low noise and emission levels, appropriate for household use and possible operation using renewable energy.	Flexibility in fuel, including renewable energy, long life cycle, high system efficiency and flexibility in fuel-to-heat ratio.	Simple design, increased flexibility, low operating temperature and pressure and a wide fuel range are necessary.	High output temperature, low operation emission and noise and high electrical efficiency.
Disadvantages	A substantial number of moving parts, mechanical solid noise and vibration and a high rate of emission.	Inefficient part load performance, unsuitable for start/stop application.	Long start-up time, high investment cost, limited adaptability component.	Bulk construction High initial cost, slow response at partial load, slow start up.	Low efficiency, infrequent research and commercialism in the ORC field.	High capital and investment cost, the design is complex.
Size capacity	Up to 75 MW	Up to 250 MW	Up to 55 kW	50 kW to 100 MW	Up to 250 MW	Up to 2 MW
Electric efficiency	25–45%	18–36%	15–35%	10–37%	15–38%	37–60%
System efficiency	65–80%	65–75%	60–80%	60–80%	80%	55–80%
Life time (h)	20,000–50,000	5000–40,000	10,000–30,000	50,000–100,000	30,000–50,000	10,000–65,000
Fuels used	Diesel, natural gas, propane.	Natural gas, biogas, propane, distillate oil.	Any fuel (natural gas and bio-fuels).	Any fuel (natural gas, bio-fuels and renewable energies).	Any fuel (natural gas and bio-fuels).	Hydrogen, natural gas, propane, methane.
Electrical-to-thermal ratio	0.5–1	0.4–0.7	0.15–0.4	0.9–1.13	0.15–0.4	0.5–2
Waste heat temperature (°C)	80–200	120–350	Up to 85	Up to 85	Up to 100	Up to 1000
Thermal output (kJ/kWh)	3376–5908	3376–7174	-	-	1065–52,753	1900–4431
Part load efficiency	High	Low	Moderate	Low	Moderate	Very high
Start-up time	>10 s	>10 min	-	>5 h	>1 h	>3 h
Noise level	High	Moderate	Moderate	Moderate	Low	Moderate
Investment cost ($/kWe)	340–1600	450–1500	1300–2000	2000–32,000	1000–2000	2500–3500
NOx emissions (Kg/MWh)	Up to 10	0.1–0.5	0.23	Fuel dependent	Fuel dependent	0.005–0.01
CO_2 emissions (Kg/MWh)	Up to 650	580–720	672	Fuel dependent	Fuel dependent	430–490

Similar to batteries, which transform electrochemical energy into electrical energy, fuel cells are devices that do the same thing. Heat and water are by-products of a chemical reaction between hydrogen and oxygen or other oxidizing agents, but electrical power is the primary result. Since it is seen as a nicer primary mover than traditional generators, which produce poisonous gas as a by-product, this technology has raised environmental concerns. Proton exchange membrane fuel cells (PEMFC), alkaline fuel cells (AFC), molten carbonate fuel cells (MCFC), microbial fuel cells, phosphoric acid fuel cells (PAFC), direct methanol fuel cells (DMFC) and solid oxide fuel cells (SOFCs) are the six varieties of FC based on membrane characteristics. The benefits and downsides of each type are listed in Table 2.

SOFCs are attractive for power and heat generation, especially for polygeneration systems. Three main components govern the operation of SOFCs: the reformer, SOFC stack and inverter. The reformer extracts hydrogen from the injected gases, and the electrolyte of the SOFC stack converts the injected hydrogen; then, the DC output from the stack is converted to AC power using an inverter. SOFC-based trigeneration systems can utilize natural gas and reach significant system efficiency of up to 55% [45]. A micro-scaled cogeneration system was successfully developed in ref. [48], including a 4.6 kW Vaillant PEMFC unit and a 1 kW Sulzer Hexis SOFC unit.

In recent applications, renewable energy sources such as sunlight, wind and geothermal sources have been actively developed as the sources are replenishable, constantly replenished and will never run out. Few researchers have studied the implementation of renewable energy as a single or combined prime mover. An integrated photovoltaic (PV) source with a cogeneration system was reviewed by ref. [49]. Three modes were studied: combined PV-SOSE system, standalone SOFC and PV-SOFC system. The study improved the system efficiency for standalone SOFC, PV-SOSE and PV-SOFC systems by around 83.6%, 20% and 23%, respectively. Recent studies, such as [1,4], have proposed a combination of SOFCs and renewable energy for a polygeneration system in residential applications, as depicted in Figure 4.

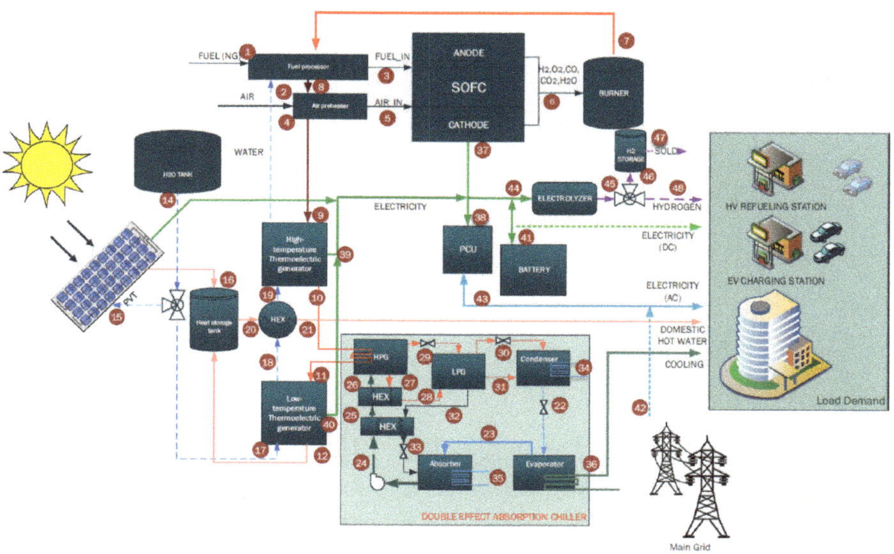

Figure 4. A polygeneration system with integrating SOFCs and renewable energy for residential use [2,4]. PCU: power conditioning unit, H_2O: water, H_2: hydrogen, HEX: heat exchanger, LPG: low pressure gradient (pump), HPG: high pressure gradient (pump), EV: electric vehicle, HV: hydrogen vehicle.

Table 2. Comparison between different types of fuel cells [47,50–52].

Fuel Cell Type	PEMFC	AFC	DMFC	PAFC	MCFC	SOFC
Operating temp (°C)	30–100	90–100	50–100	160–220	600–700	500–1000
Electrical efficiency (power) (%)	30–40	60	20–25	40–42	43–47	50–60
System efficiency (heat and power) (%)	85–90	85	85	85–90	85	Up to 90
Typical stack size	<1 kW–100 kW	10 kW–100 kW	Up to 1.5 kW	50 kW–1 MW (250 kW typical module)	<1 kW–1 MW (250 kW typical module)	5 kW–3 MW
Electrolyte	Solid polymeric membrane	An aqueous solution of potassium hydroxide soaked in a matrix	Solid organic polymer poly-perfluorosulfonic acid	100% phosphoric acid stabilized in a SiC-based matrix	Li_2CO_3/K_2CO_3 materials that are stabilized in an alumina-based matrix	Solid and stabilized zirconia ceramic matrix with free oxide ions
Fuels	Methanols or hydrocarbons	Pure hydrogen	Methanol	Natural gas	Natural gas, biogas, others	Natural gas or propane, hydrocarbons or methanol
Applications	Transportation, DG, backup power, portable electricity and speciality vehicles.	Military, space, backup power, transportation	Consumer goods, laptops and mobile phones	Distributed generations (DG)	Electric utilities, distributed generations	Electric utilities, distributed generations, and auxiliary powers
Advantages	Advanced start-up time, varies output quickly, compact, no corrosive fluid used	Lower cost components due to a broader range of stable materials, low temperature, quick start-up	A high density of energy storage, no reforming required, easy storage and transport	High cogeneration efficiency	High efficiency, scalable and flexibility of fuel	Internal reformation is made possible by high temperature, which eliminates the need for liquid electrolytes, makes high-temperature, valuable output for various applications and does not require pure H2 (low-price fuel)
Disadvantages	Expensive platinum catalysts used, high-purity H2 required	Sensitive to CO_2 in both air and fuel, electrolyte conductivity, aqueous electrolyte management (polymer)	Low power output, methanol is toxic and flammable	High-cost catalysts, long start-up time, sulphur sensitivity	High-temperature corrosion and cell component failure, low power density and a lengthy start-up time	Lengthy start-up times, problems with temperature stress during load after the operation and the need for pricey, heat-resistant materials

In addition to utilizing sunlight as the combined prime mover, a few studies have also concerned geothermal sources using ground source heat pumps (GSHPs). A GSHP works with electrical energy to extract heat from the ground and transfer the heat into the building. It utilizes geothermal energy to meet heating demand instead of burning fuel as boilers do. A study developed a combined GSHP-SOFC prime mover for a trigeneration system in office and residential houses [53]. The analysis considered three types of trigeneration combinations utilizing a boiler, chiller, GSHP and fuel cell based on their capacity. The result revealed that a variety of SOFC-GSHPs achieved significant energy savings.

4.2. Heat Recovery Units

Heat recovery equipment works to recover the thermal energy from prime mover exhaust gas streams to meet heating demands [46]. The units can increase system efficiency without any increase in fuel utilization. Heat exchangers are the most accurate heat recovery equipment and can be made of a simple tube and shell, plate, fin or even micro-channel design. They are marginally different to a HRSG, which can generate heat and steam water from the rejected exhaust gas. The components are frequently used in heat recovery mufflers, which are utilized with reciprocating engines to recover exhaust heat. They produce less noise and pressurized steam and hot water than HRSG devices in the 370 °C to 540 °C range [54]. However, the kind of exhaust gas has an impact on the muffler's stack exit temperature and the quantity of heat that can be recovered [55]. Since the high-back-pressure issue can decrease muffler capacity, regular maintenance and cleaning are required to maintain the engine's performance.

Thermoelectric generators (TEGs) are yet another choice of heat recovery device. An appealing method for converting heat to electricity is by TEG. By converting exhaust heat into electric power, a TEG serves to raise the primary mover's efficiency. The electrons from n to p semiconductors travel through the TEG material when there is a temperature gradient, creating the voltage needed to drive the external load [56]. The TEG performance is different based on the semiconductor type, p or n, and the operating temperature range.

TEGs have been widely applied for recovering heat in both stationary and mobile machinery. Heat recovery for a high-temperature prime mover is optimally used for fuel-cell-based power generation systems. The system efficiency of a heat recovery system can be increased by up to 5% from the base case by using a TEG in several components, such as in the burner, exhaust heat and the air and fuel preheater [57]. As a result, using a TEG to produce power backup for the prime mover in both single and multigeneration systems has been commercialized.

4.3. Cooling Units

Polygeneration systems generate cooling from the rejected heat exhaust of the prime mover, eliminating the need for additional fuel to meet cooling and air conditioning demands. Hot water, steam and exhaust gas range in temperature and can all be wasted heat sources [46]. In order to continually satisfy the need for power, heating and cooling, the optimum design of a polygeneration system must consider the best combination of prime mover and heating and cooling unit based on their specifications.

Instead of the mechanical compression used by refrigerators, sorption cooling methods (such as absorption and adsorption) use thermal compression to produce cooling. While an adsorption cooler creates cold air or water through a material that adsorbs the refrigerant vapors, an absorption cooler operates with a pair of working fluids: refrigerant and absorber [58]—the basic absorption cooling cycle includes four parts: generator, absorber, evaporator and condenser. Lithium bromide–water ($LiBr/H_2O$) and water–ammonia (H_2O/NH_3) are commonly the functional absorbent/refrigerant pairings for absorption cooling.

4.4. Hydrogen Generation Units

Some diversifications have been created to manufacture a new fuel to decrease the consumption of non-renewable primary energy sources, fossil fuels and coals. Hydrogen combustion is cleaner and better for the environment than oil and natural gas combustion. However, hydrogen costs more than natural gas and oil because it is a relatively new technology. As a result, there has been an attractive increase in the development of hydrogen production, particularly for household fueling applications [59].

The development of diverse hydrogen production systems employing non-renewable and renewable energy began when hydrogen was established as the future fuel. Steam methane reforming, coal gasification, biomass burning, nuclear power and water hydrolysis are a few methods for creating hydrogen [60]. The method for producing hydrogen through water electrolysis is significantly more environmentally benign than other methods. The technique for creating ultra-pure hydrogen through hydrolysis was discovered 2000 years ago [61]. However, only small-scale applications are covered by the production. Of the hydrogen produced worldwide, 4% is produced via electrolysis of water. Hydrogen production on a big scale is used in home fueling stations and the marine, rocket, spacecraft, electrical and food industries [62]. Polygeneration can be explicitly applied to produce fuels such as hydrogen and oxygen through an integrated system, as depicted in Figure 5. In the research, a system was designed to supply 500 MW of electricity, hot water and hot air, while the excess electricity was used for hydrogen production [63].

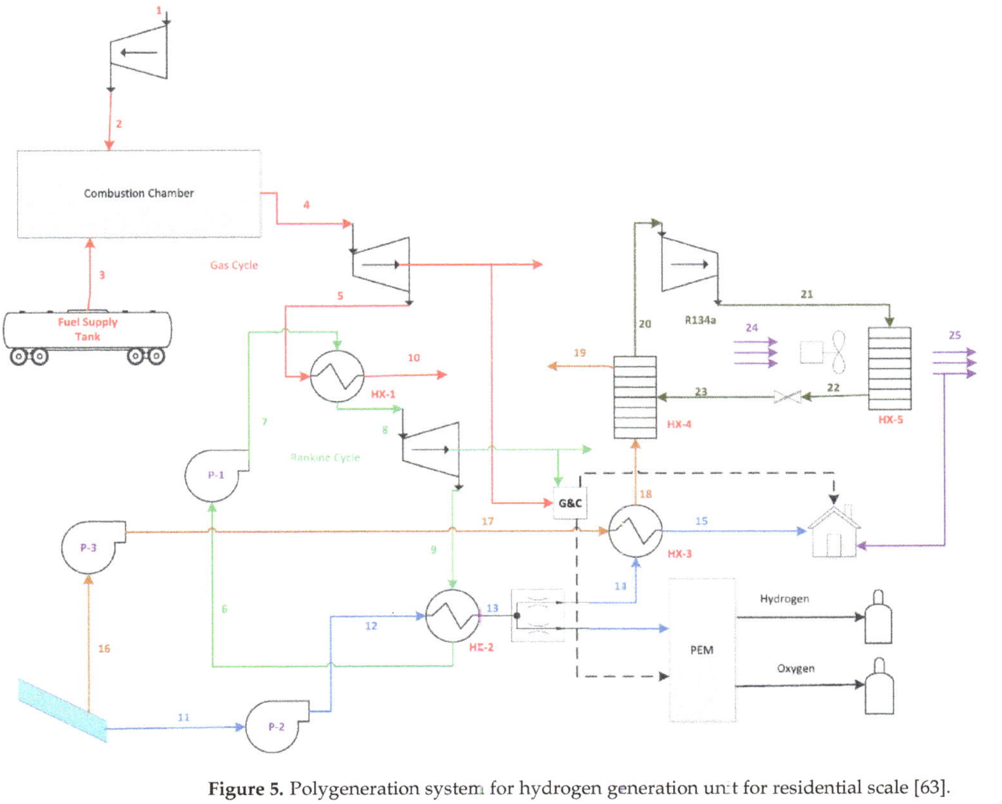

Figure 5. Polygeneration system for hydrogen generation unit for residential scale [63].

4.5. Storage Units

In the production of heat and power, erratic and disjointed power, heating and cooling reduce system efficiency, raising energy costs. Storage units are required to store excess energy during times of low demand and use the excess during periods of high demand. Utilizing storage facilities improves operational hours, fuel efficiency and system effectiveness while decreasing capacity. When applying electric tariffs, using storage can also lower energy costs when peak load hours are considered.

Utilizing electrical energy storage becomes more crucial in order to reduce the import of grid electricity. Additionally, it extends the system's operational hours during high grid demand and when energy prices are high. Based on the chemical components of BES, certain contemporary technologies have been adopted: lead–acid (LA), nickel (Ni), sodium (Na), lithium-ion (Li) and metal–air batteries. LA batteries are the oldest technology, are rechargeable, have low cost and simplicity when being manufactured and have a quick chemical reaction and good life cycle [64]. Despite the advantages that make this technology the largest in the market, LA batteries also have some drawbacks: heavy metal components which generate toxic and hazardous material and a lot of water production from the discharging process [65].

In addition to electrical demand, the fluctuating characteristics of heat and cooling demand also significantly affect system efficiency. The occupant's behaviors, particularly for residential applications, can occasionally necessitate more electrical energy than is required for heating and cooling. Due to the increased electrical demand, the primary mover's exhaust heat is temporarily rendered useless. In order to improve the system efficiency, thermal storage is essential as electrical storage.

The most popular type of heat storage utilized for combined heat and power generation is sensible heat storage (SHS). It functions by supplying the substance with energy and raising the temperature without altering the material's phase [66]. The temperature of the substance used to transfer heat energy (often solid or liquid materials), the fluctuating inside temperature of the storage and the quantity of material stored all have an impact on SHS capacity. Water, sands, oil and rock beds are used to transport heat for any application. Water has more benefits than other materials since it is inexpensive and readily available, but it also uses solid resources such as metals, rocks, concrete and bricks. To reduce upfront and ongoing expenditures, the appropriate materials for sensible heat storage are chosen [67]. The multi-objective optimization of this study also took into account accessibility and environmental consequences such as carbon footprint.

Hydrogen is typically created using an electrolyzer and stored in a hydrogen tank for some purposes, mainly when using fuel cells as the primary mover. Hydrogen-based energy storage (HES), which is ecologically friendly, has various benefits for both fixed and mobile applications. Additionally, it provides a level of electrical generation continuity that is not possible with intermittent, renewable energy sources such as photovoltaic and wind generators. This storage type, however, lacks design and evaluation studies due to its infrequent application. According to research by the Ministry of Energy, Green Technology and Water [67], hydrogen-based storage is frequently paired with a hybrid renewable energy system (HRES), which mainly look at design factors, including security, operability, final use, energy and efficiency performance.

5. Applications of SOFC-Based Energy Generation Systems in Residential Applications

In the most recent advancements in renewable energy, fuel cells (FCs) are recognized as the new and most promising kind of source for the future [68]. PEMFCs, DMFCs and PAFCs are designated for low-temperature operations in the fuel cell application, whereas MCFCs and SOFCs are designated for high-temperature activities. Low-temperature fuel cells can be used in mobile applications such as vehicle transportation and in military and governmental buildings [69–71]. On the other hand, high-temperature fuel cells are more suitable for stationary applications such as power generation systems due to their higher-temperature functioning and higher efficiency [72,73].

Because of its benefits relating to the flexibility of the electrolyte material and adaptability in using hydrocarbon fuel, the SOFC is the fuel cell type designed for stationary power production [74]. The SOFC can potentially be applied for hybrid power generation to boost electrical output and for combined heat and power generation to deliver more than electricity due to its high efficiency and proportionate heat-to-power ratio.

The number of papers describing the SOFC as a power producer has been rising yearly, as can be seen in Figure 6. SOFC application is split into single power generation, cogeneration and trigeneration/polygeneration systems based on the type of power generated. Early in 1992, research into SOFCs as a prime mover for single power generation began. The majority of research from 1992 to 1993 focused on optimizing the stack SOFC to produce more electricity and boost power efficiency. In 1994 and beyond, research on the hybrid system consisting of SOFCs and gas turbine grew since the hybrid configuration significantly affected system efficiency. Since 1995, there has been a significant increase in studies on using SOFCs' exhaust heat. The figure shows that the cogeneration system's application increased from 2014 to 2015. The number of papers on cogeneration SOFCs in residential settings declined from 2016 to 2017. The trend for cogeneration SOFCs increased from 2018 to 2021 due to larger applications that implement the fuel cell. The increase in attention paid to cogeneration systems cannot be separated from the fact that the output is not limited to heat and power but also includes heat and cooling, power and cooling, syngas and hydrogen or clean water as a side production.

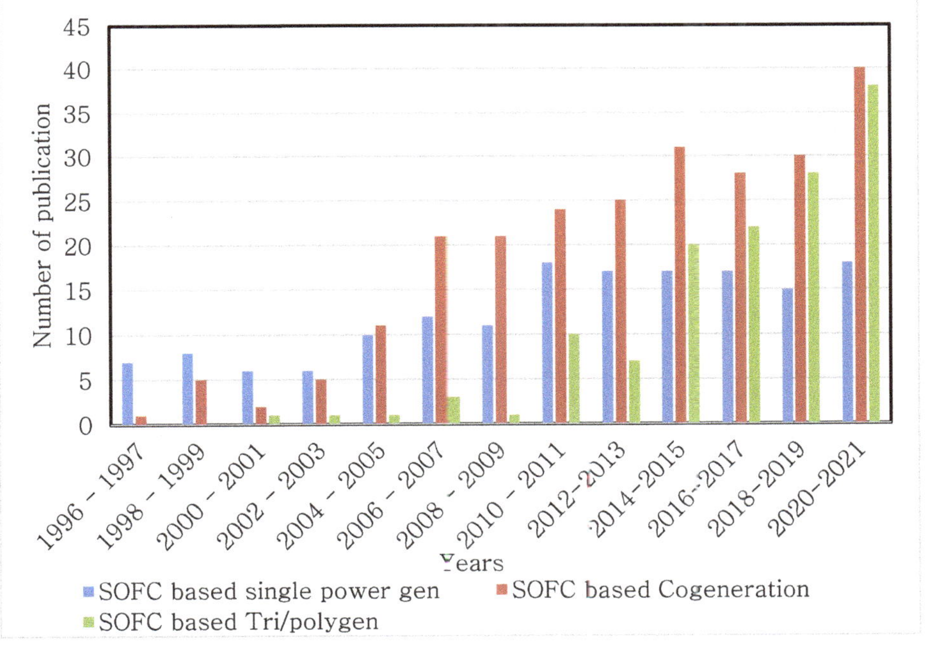

Figure 6. The trend of publication numbers for SOFC power generation in residential applications.

On the other hand, even though it is not as substantial as the research into the cogeneration system, research into trigeneration or polygeneration SOFCs is growing. This research began in early 2000 and was designated as a novel topic. Therefore, the trend of research into trigeneration/polygeneration systems was still attractive for the next five to ten years. It was analyzed by a number of publications on SOFC trigeneration/polygeneration systems for residential applications. The increase in interest in utilizing SOFCs in polygeneration systems was triggered by the prediction of the price reduction of SOFCs for the next five to ten years [75–77].

Moreover, the increase in polygeneration for residential applications cannot be separated from the environmental issue. Based on the energy plan of the government, Malaysia has a target to reduce carbon emissions by up to 45% by 2030 [78]. Therefore, applying SOFC-based polygeneration systems is attractive for saving primary energy and reducing carbon emissions in Malaysia.

The use of SOFCs as a primary mover is divided into two configurations based on the architecture of household energy planning, namely integrated stationary and vehicle supply. Integrated supply covers the application of single power generation, cogeneration and a trigeneration/polygeneration system. Meanwhile, integrated energy supply and vehicles combine stationary and mobile applications, for example, single-family or multi-family homes with electric vehicle charging stations. Another application affixes a small-scale fueling system to a residential structure. A recent study focused on the design of a power-to-hydrogen station to provide fuel for vehicles using polygeneration systems. The combinations of energy supply to generate hydrogen were assessed based on real-time demand profiles [79]. It was found that the modular, hybrid energy station that was proposed had a significant impact on future refueling station design and residential housing in the neighboring area.

In its application related to fueling stations, consideration must be given to designing a transport system for hydrogen production. An integrated hydrogen production and transportation system was designed to support the government in developing the early stage of a fuel cell electric vehicle (FCEV) [80]. The research intended to develop a transportable 700 bar hydrogen fueling station with an off-the-shelf component cost of less than 23% of the capital cost of the current stations in the U.S., as depicted in Figure 7.

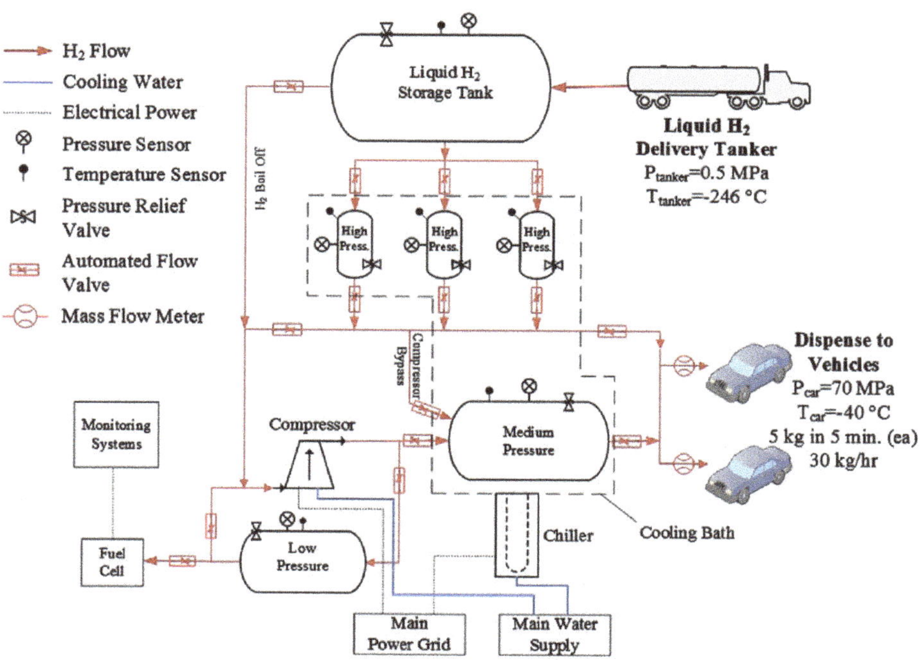

Figure 7. The design of a transportable vehicle fueling station [80].

6. Energy Planning for Designing a Polygeneration System

Due to the high investment cost of SOFC-based energy generation systems, energy planning should be conducted to design a highly efficient polygeneration system. The literature on energy planning for residential applications lists three strategies: configuration review, operating strategy design and capacity and operating point optimization. The detail of the strategy is presented in Sections 6.1–6.3.

6.1. Evaluation of Configuration Design

Structure evaluation, topology assessment or system synthesizing is essential for designing a polygeneration system. Starting with the choice of the prime mover and its topology, polygeneration structure optimization is undertaken. A SOFC can be employed as the prime mover in a standalone or hybrid form in SOFC-based polygeneration. In a freestanding arrangement, the SOFC's electrical output and waste heat are used; in a hybrid configuration, the SOFC can be coupled with a turbine-driven system to produce more electricity. The organic Rankine cycle (ORC), biomass and solar system have all been used to study a combined prime mover integrating SOFCs and renewable energy.

Further, selecting one or more prime movers allows for the examination of polygeneration [81]. Studies have shown polygeneration structure optimization based on the chosen primary mover from a gas engine, PEMFC or SOFC. Another study focused on LNG-based SOFC polygeneration system assessment in desalination water application [82]. The research assessed two combinations of SOFCs with heat recovery systems to analyze the system performance through system efficiency. A polygeneration system employing SOFCs as a prime mover was investigated for water desalination and hydrogen production [83]. Based on the evaluation, the system produced electricity of 4.4 MW, a refrigeration capacity of 0.16 MW and 0.96 kg/s of desalinated water.

Structure evaluation is not concerned with choosing the right prime mover but with designing a suitable heating, cooling and storage system and its connection to the grid. Guidance for selecting the configuration of a polygeneration system based on energy and economic and environment factors was presented in the review in [46]. The heat-to-power ratio and its initial cost can be used to select the appropriate heating and cooling units, particularly for SOFC-integrated cooling, heating and power systems. Burners and heat exchangers are the most widely utilized components for recovering SOFC exhaust heat. Some research used an HRSG unit to provide space heating from the exhaust gas of SOFCs. A double- or triple-effect absorption chiller is the best cooling system for maximizing system efficiency due to the high exhaust temperature from SOFCs. However, additional heating in the form of an auxiliary boiler and additional cooling in the form of an electric chiller can be employed for periods of peak heating and cooling demand.

Since the environmental aspect has become an urgent parameter when creating a sustainable system, several emerging polygeneration systems have been considered, reducing carbon emissions through some technologies such as carbon capture and carbon converters. A study investigated a polygeneration system that captures and converts carbon into methanol/power [84]. The carbon footprint assessment showed the remarkable ability of the proposed system to reduce carbon emissions.

Due to their linear power-to-capacity ratio, batteries and sensible heat storage are frequently utilized for electrical and thermal storage. While sensible heat storage capacity is determined by the storage temperature and pressure, batteries measure capacity, charge rate and discharge. Both electrical and heat storage significantly affect the continuity of CCHP operation and reduce the appropriateness of the grid-connected system. However, capacity and structure evaluations of these storage units should be conducted to optimize their cost and efficiency.

6.2. Operational Strategy Design

Operating strategy optimization has been carried out through various management techniques to decrease energy consumption and enhance energy savings. In contrast to electricity generation, polygeneration management takes into account electric, cooling, heating and other energy generation requirements. Optimizing energy flow while balancing each component to meet all demands is the operational strategy's most important goal. The operating procedures now in use for polygeneration systems are classified into three categories based on their approaches: conventional, novel and intelligent.

According to the required load, conventional operating techniques manage the energy supply from the prime mover and cooling and heating units. Electrical and thermal loads are used to control the energy source in CCHP or polygeneration systems. Conventional operating strategies consider three techniques: following thermal load (FTL), following electrical load (FEL) and combined FEL and FTL strategies.

The FEL uses an operational method known as electric demand management (EDM) to control energy sources in order to satisfy electrical load without taking the thermal load into account. The prime mover for the application is created and run optimally based on the electrical load profile. The thermal requirement in FEL is now the second criterion that must be met by prime mover generation. The auxiliary boiler creates an additional heat source if the primary heating and cooling systems cannot keep up with demand. The following electrical load is superior as it increases energy savings for some countries with hot, humid climates or tropical environments since it reduces the need for heating and cooling equipment [85]. The polygeneration system, which is connected to an electrical heater and air conditioning and refrigeration units, known as the heating, ventilation and air conditioning system (HVAC), also uses the FEL method.

On the other hand, while developing and running a polygeneration system, FTL or thermal demand management (TDM) uses an operation to meet thermal demand. The primary mover's capacity is intended to meet thermal demand; however, if electrical need cannot be met, further grid imports of electricity are necessary. The FTL approach is typically used in four-season countries and regions with significant thermal demand. To meet cooling demand in summer, the prime mover produces more heat energy, whereas the exhaust heat is needed for space heating and the household hot water system throughout the winter season. A heat storage unit should be utilized to boost the polygeneration system's continuity and reduce energy consumption during the peak load period.

Utilizing conventional strategies to operate polygeneration systems results in concerns relating to running costs, system efficiency and energy conservation. Energy waste occurs when an electrical or thermal source produces more energy than can be used. As a result, some research explored combining the FEL and FTL strategies in an approach used for the system [86]. In other studies [87,88], in order to cut down on energy usage, operating expenses and carbon dioxide emissions, a combined FEL/FTL working approach was investigated. In ref. [89], to maximize the excess from both the FEL and FTL, researchers developed a novel operation called minimum distance (MD) and matching performance (MP) assessment.

Numerous unique techniques, including the seasonal operation strategy, time-based, emission-based strategy and economic-based and energy-storage-based strategy, were brought about by the more intense development of optimization strategies for polygeneration systems. The seasonal operation approach operates the heating and cooling systems to meet both demands while taking into account climate change throughout the year. During winter, a heating unit is utilized heavily while a cooling unit is utilized to the fullest. The design should take into account a few possibilities for using auxiliary heating or cooling units during the period of peak load. In order to protect the primary heating and cooling system from the exhaust heat of the prime mover, an auxiliary boiler and electric chiller are used as an alternative. In ref. [90], a trigeneration system was optimized by the proposal of two season-based strategies: summer operation and winter operation. The research took into account energy tariff parameters to decrease energy costs and increase energy saving Another study proposed a variable electric-to-cooling ratio following climate profiles to enhance the operating strategy of a polygeneration system [91]. Considering three climate temperatures, i.e., hot, cool and moderate, the study defined a variable electric cooling ratio compared to a constant electric cooling ratio (ECR). In ref. [92], three operation strategies were investigated for optimizing the design of a SOFC-based trigeneration system, i.e., based load operation, thermal match operation and electrical match operation.

Energy storage is considered to enhance energy operation and increase energy saving. Some studies considered employing heat energy storage to increase heat energy saving [93], while others preferred to use a series of batteries as electrical energy storage to improve electrical energy saving.

A time-based operation, or schedule, is applied during a specific time and period to fill the load. For a polygeneration system, the prime mover and heating and cooling units are operated during a particular time and turned off if the period is already finished. This strategy was applied in ref. [94] to improve energy saving and heat coverage in a polygeneration system. The scheduling operation strategy was also used in [95] to increase economic profit from a combined heat and power system. The study proposed four schedules for selling excess energy from the system: selling electricity to the neighbor, selling cooling energy for the neighbor, selling heat energy to the neighbor and selling electric power to the grid. A time-based operational strategy was also proposed in [96] to design a SOFC-based CHP generation system. The study investigated three schedules: winter summer day, peak day and combined summer–winter day for two days.

In economic-based strategies, energy tariffs are primarily applied to minimize energy costs. Interestingly, application of energy tariffs has been proposed for four different load management types: following electrical load, base load, continuous operation and peak saving load [93]. A study considered two types of tariffs for buying or selling via the grid connection. Cost optimization was also implemented to reduce the system's cost and emission of carbon dioxide [97,98]. It compared cost optimization with FEL and FTL as the existing operation strategies.

Management of energy dispatch is essential for improving the effectiveness of the energy flow between the prime mover and the heating and cooling units. The objectives are to satisfy all the load demands and enhance the system's profit.

An adaptive neural network (ANN) for energy management was implemented to design the model of an energy source system and predict the future load requirement. In ref. [99], an optimum design based on an expected model using the Lavenberg–Marquardt ANN method was developed. The research performed the technique to design a CCHP system and to analyze the energy destruction and efficiency of the optimum model. In ref. [100], a CCHP design was optimized using a back propagation technique taking into account energy and exergy analyses. The research claimed that the result of the predicted model had a precision of approximately 0.1% compared to the actual model.

Fuzzy logic (FL) is a simple approach that can be implemented for several types of power and heat generation studies. The input and output are determined in the fuzzy range from 0 to 1, and several rules are applied between the input and output. FL works with multiple inputs and single/multiple outputs. For controlling energy output, the temperature from a storage tank and the temperature from the room are employed as the input for fuzzy logic control [101]. Another study used fuzzy selection to optimize the size and operating strategy of a power generation unit (PGU) in a polygeneration system. [102]. It considered seven models of a prime mover with their operating strategy and three sub-models, i.e., energy, economic and environment. The fuzzy selection technique was applied to analyze the optimum design based on energy, economic and environmental considerations. Another study conducted optimal, FL-based energy management [103], considering thermal and electrical loads and storage temperature as the inputs to generate electricity from the prime mover to reduce energy costs and CO_2 emissions. A recent study proposed two-stage fuzzy management of electricity, cooling, heating and hydrogen supplies in a polygeneration system [2].

6.3. Optimal Design of Polygeneration Systems

The design of a polygeneration system must be adequately developed for various reasons, including fluctuating load pattern, climatic circumstances, governmental restrictions and other factors. An inefficient design approach can result in a system's emissions rate increasing, significant, high investment costs and a decline in system efficiency. Moreover, an overdesigned polygeneration system leads to energy waste, while a less designed system leads to an intermittent supply and decreases its reliability. Two types of factors influence the design of polygeneration capacity, namely internal and external factors. Regarding internal factors, some parameters cause the design difference: the components' specification and capacity. The prime mover capacity is the most crucial consideration when developing a polygeneration system. Increasing system efficiency also depends on the effectiveness of the heating and cooling equipment. The lifetime of the prime mover, system continuity and increased system dependability during peak hours can all be improved by adding or not attaching the storage units.

External design considerations for polygeneration systems include application type, load behavior, climate, regional conditions and governmental laws. Polygeneration is divided into four varieties based on its capacity: large-size, medium-size, small-size and micro-size capacities. Large- and medium-sized structures are frequently employed in industrial applications, whilst medium-sized and small-sized structures are commonly utilized in office and academic buildings. However, for residential applications, a micro-sized polygeneration system that can work with microgrid power generation is preferred.

The heating and cooling unit requirements differ in different climates and regional conditions. In four-season countries, heating units are more dominant during winter and fall than cooling units are during the summer and spring seasons. In contrast, hot and tropical countries need more cooling units than heating for space and water. Some regulations mean that exporting excess electricity to the grid is not allowed in some countries, i.e., China and Japan [104,105], while others in the U.K. and U.S. allow exporting electricity to the grid through feed-in tariffs regulation. For ASEAN countries, feed-in tariffs for renewable energy sources, i.e., solar and geothermal sources, have been regulated, as in Vietnam, Thailand, Singapore, Indonesia and Malaysia. However, the regulation for exporting electricity back to the grid has not been established yet for other renewable energy types such as fuel cells. The government regulation allowing or not allowing the export of the excess influences the operating strategy and design of polygeneration (grid-connected or standalone application), affecting operating cost and profit from feed-in surplus energy. For several applications of CCHP independent from the grid, battery utilization as energy storage is preferred to save excess energy and increase system efficiency [106–110].

In several studies, optimizing the design included the system size and operating strategy at two optimization stages. The complexity of the optimization procedures can be tackled by implementing robust techniques and simply modeling the components. A study implemented the two-stage optimization to design a polygeneration system in desalination water application [111].

(a) Size Optimization

After reaching the optimum topology of the system, it needs to size the component capacity to avoid oversized or lack of source during operation. For a polygeneration system, the optimization of size capacity should be conducted with various power generators, including electricity, heat and cooling generators. Although the prime mover is the main component that influences heat and power generation, heating and cooling units' capacity must be determined correctly in line with the load requirement. For the system attached to storage units, such as storage units for heat, electrical energy or both, accurate sizing influences the initial cost reduction and increases the system's continuity to satisfy the load demand. Adequate sizing is also required for auxiliary devices, i.e., power electronic units (PCU) such as inverters and converters, BOP units such as reformers, the preheater and combustor and the boiler and electric chiller for additional heating and cooling units.

In applications combining SOFC-based polygeneration with renewable energy such as geothermal, wind and photovoltaic energy, component sizing is required to design a continuous energy generator that can meet the electric load.

Optimization in sizing the polygeneration system has been attractive in recent years. This increase aligns with the future objective to design an efficient and compact trigeneration system for research and commercialism. A study optimized a system for the Mediterranean, considering the size of the prime mover and absorption chiller using partial loads [112]. Moreover, a study performed size optimization of SOFC capacity and SOFC shares in a CCHP system [113]. Some parameters, such as electric-to-heating and electric parameters, were considered to determine the optimum combination of SOFCs and other prime movers, such as steam turbines and gas turbines, with HRSG as a heat recovery unit. In addition to conducting topology evaluation, the study also optimized the cogeneration capacities, including that of cogenerated thermal and electrical power [114]. The capacity of the auxiliary boiler and the capacity of sold/bought electricity to/from the grid were also considered. In his thesis, Elmer [115] designed a multigeneration system consisting of SOFCs, a dehumidifier and cooling units to provide a heat and power supply for a residential area. Sizing of the SOFC capacity, dehumidifier and cooling capacity was conducted to increase efficiency and decrease system cost, primary energy demand and carbon emission. In addition to performing topology evaluation, the authors in [81] also performed sizing of the prime mover, hydrogen tank and battery.

For polygeneration generation, the storage is crucial, affecting system efficiency and energy savings. Some studies were concerned with optimizing and conducting parameter analysis of storage utilization. In ref. [116], the research proposed a novel technique to design optimum heat storage. The study considered the volume and operation period to estimate the contribution of heat storage and increase primary energy saving, the coverage ratio of the heat load and the estimated economic benefits. In ref. [98], the authors optimized SOFC capacity and heat storage to optimally meet a single family's residential power and heat requirements. They analyzed the effect of utilizing batteries for heat and power generation energy saving, considering four strategies and combinations of prime movers and heating units. Another study designed hot water storage and its operating temperature to optimize heat and power generation [117]. The study revealed that the operating temperature of heat storage affects the primary energy saving of the system. In ref. [118], research estimated a model of optimized heat storage capacity based on the input temperature from the power and heat generation system to optimize heat accumulation throughout night hours (8 h).

However, most previous studies used a grid connection as power backup for when the polygeneration system cannot fulfil the electric load demand [93,105,119–123]. Prime mover sizing is conducted to satisfy load base requirements, while electric energy is imported from the grid during peak load duration. The tariffs for importing and exporting electricity fluctuate based on the load requirement. Especially for residential applications, the peak hours rate usually occurs at night and is higher than the daily consumption. In some countries, regulation for grid-connected and feed-in tariffs is far from government regulation. Some studies analyzed battery utilization as a replacement for a grid connection [106,108,124–126]. However, further investigation is required to model the design, analyze and manage an integrated polygeneration system that is fully independent of grid connection.

(b) Operating parameters optimization

In addition to the proper topology design of the polygeneration system, optimum operating conditions can affect the system efficiency and reduce the operation cost. Especially for SOFC-based polygeneration systems, running the prime mover in the part load operation benefits energy efficiency and economic aspects [127,128]. Part load operation in SOFCs can be conducted by controlling current density, fuel utilization, flow rate and temperature. Performance analysis is performed based on these parameters by implementing two operating point cases to minimize renewable energy consumption and carbon dioxide equivalent [129,130]. The result of multiple studies revealed that a system with lower fuel utilization also has lower electrical output power and efficiency yet higher system efficiency. In ref. [131], the researchers conducted a performance analysis of a SOFC-based polygeneration system considering cathode and anode flow input and temperature.

An optimum operating condition is not only applied for the prime mover but also heating, cooling and storage units. Optimizing working conditions for heating units include defining the optimum temperature for the heat exchanger, heater unit and domestic heat water output. Furthermore, the optimum power ratio for heating also affects the heating unit's performance. For cooling unit optimization, defining the cooling temperature for chiller or refrigerator units and the power-to-cooling ratio are also beneficial for increasing cooling performance. Cooling performance is defined as the cooling power at the evaporator divided by the cooling at the generator. The cooling ratio for a single-effect absorption chiller is about 0.7, while, for a double- and triple-effect absorption chiller, it can reach up to 1.2 and 1.5, respectively [132]. For storage unit operation, an optimum operating condition value can be defined from the working temperature and pressure of the heat storage. For electric storage, operating conditions are characterized by working voltage and charge/discharge rate. Working parameters for hydrogen storage can be defined from its operating pressure and the charge/discharge rate of hydrogen.

7. Guidelines for Optimizing the Design of Polygeneration Systems

Designing a polygeneration system is the first step before executing and developing an entire system. General guidance for planning a polygeneration system is presented in Figure 8. Defining the prime mover is the primary step in determining whether the polygeneration system will use a single energy conversion device or integrated devices. The type of prime mover is also essential for designing it for single use or in a hybrid configuration. If the prime mover is expected to have emissions that can be used, it is better to combine it with another conversion device that makes it worthwhile to generate additional power.

The second step in designing an optimal polygeneration system is determining the type of primary source. The primary source, called fuels, is mainly related to the kind of prime mover chosen. However, since several prime movers have become more flexible in using renewable or non-renewable sources, the decision is from the perspective of cost or the environment. Renewable energy is the best choice to make a system with a low carbon footprint, yet a mixture of renewable and non-renewable sources sometimes happens when considering fuel costs [133–135].

In polygeneration system design, auxiliary energy production must be determined depending on its application. When using polygeneration for residential areas, basic needs such as heating water and space, cold water and space, fresh air and water are essential. Thanks to the development of construction and urban layout technology, it is possible to integrate residential areas and refueling stations in the community [1,21,136–138]. For an integrated residential area with a vehicle fueling station, extra electricity and/or hydrogen is an excellent choice to be considered. After deciding on the polygeneration components, a criterion must be determined for assessment. It can be one or more to assess the possibility of the design through some synthesis analyses [139]. When the first assessment has been performed, the third stage can continue to define input parameters for the optimization process. When using renewable energy sources, some parameters usually used for residential applications are electricity usage, hot or cold water, hot or cold space and climate profiles such as sunlight, wind speed or rain precipitation. The tenants' driving distance and behavior must be considered when integrating the residential area with a vehicle fueling station to know the system's load profile. The second assessment can be performed by running the simulated system with the load profiles based on its application.

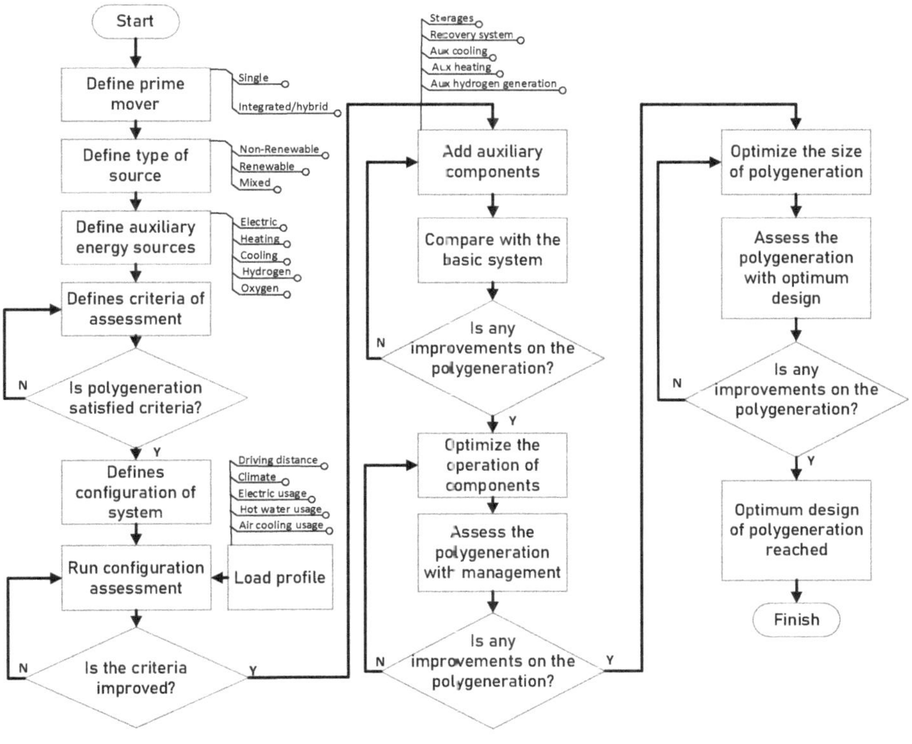

Figure 8. A flow chart of guidelines for optimizing a polygeneration system.

When running the second assessment, the same criteria must be used to analyze whether the configuration can cope with the load profiles. If imbalanced situations exist between the supply and demand in the system, some auxiliary storage could be an option. If there is a huge gap that is a waste of the supply, storage can be added for electricity, heating, cooling or even for hydrogen if the prime mover needs hydrogen as a fuel or the vehicles are based on fuel cell engines. A comparison between a separated basic system and the proposed polygeneration system should be carried out to analyze the improvement in terms of prior criteria.

An additional operating system can be added to improve the operation of the polygeneration system. Conventionally, the process of a separated system follows one criterion to satisfy either based on electricity or heat profile. However, since more energy types are produced, polygeneration needs more complicated operating procedures. More measures can be considered to set a non-dominated solution for the system. Some advanced management approaches have been proposed to overcome the complexity of polygeneration, such as artificial intelligence (AI) and metaheuristic methods [2,43,44].

After assessing the polygeneration system for better operational management, the last stage is to optimize the capacity of the polygeneration system. Even though additional storage and/or auxiliary devices have been attached, the management improvement affects the system's energy flow. The size can be optimized after all the components and management are set up. Since many elements, parameters and criteria must be considered, the optimization process can be challenging. The conventional numeric method might not be suitable for the polygeneration case, yet, thanks to the development of metaheuristic methods and AI, complex problems such as polygeneration can be solved by those methods without hassle.

By using this guideline for designing a polygeneration system, researchers can save more time and find an optimum solution for their design before executing it as a prototype or an entire system. Even though the parameters mentioned in this guideline are specified for residential applications, other applications can use a similar procedure. The flexibility in designing a complex system such as a polygeneration system could facilitate a brighter future for the system to be more applicable and widely used for other applications.

8. Summary

The evolution of energy systems has been a game changer for researchers designing a more effective, efficient and sustainable approach. Polygeneration is the answer to the problem of the complexity in a separate system. This study presented a technical review of SOFCs-based polygeneration systems implemented for residential applications. By reviewing polygeneration technology, this paper presented related technologies used to determine the prime mover, heat recovery systems, cooling units, hydrogen generation units and storage. The application and research trends of SOFC-based polygeneration systems were presented based on the number of publications since 1996. From the publication trend, the increased attention paid to polygeneration system cannot be denied and has become an essential key to implementing the system in modern energy planning. Potentialpro research topics could be synthesizing the system configurations and components and designing the capacity, operating conditions and the operating strategies of the polygeneration system. Moreover, with further consideration of sustainability purposes, the technology of the polygeneration could be advantageous for generating cleaner and better energy for the environment and society. A guideline for designing a polygeneration system was presented with specific applications for residential areas to prepare for the future. However, the procedure can be applied for any other application and configuration that will become broader and widely used.

Author Contributions: F.R.: Conceptualization, writing; M.A.H.: Supervising, revising; H.M.: Supervising, editing; O.E.: revising, visualization. All authors have read and agreed to the published version of the manuscript.

Funding: This research is supported by Universiti Malaya under UM International Collaboration Grant (ST027-2022).

Institutional Review Board Statement: Not applicable.

Informed Consent Statement: Not applicable.

Data Availability Statement: Not applicable.

Conflicts of Interest: The authors declare no conflict of interest.

References

1. Ramadhani, F.; Hussain, M.; Mokhlis, H.; Fazly, M.; Ali, J.M. Evaluation of solid oxide fuel cell based polygeneration system in residential areas integrating with electric charging and hydrogen fueling stations for vehicles. *Appl. Energy* **2019**, *238*, 1373–1388. [CrossRef]
2. Ramadhani, F.; Hussain, M.A.; Mokhlis, H.; Illias, H.A. Two-stage fuzzy-logic-based for optimal energy management strategy for SOFC/PV/TEG hybrid polygeneration system with electric charging and hydrogen fueling stations. *J. Renew. Sustain. Energy* **2021**, *13*, 024301. [CrossRef]
3. Luqman, M.; Ghiat, I.; Maroof, M.; Lahlou, F.-Z.; Bicer, Y.; Al-Ansari, T. Application of the concept of a renewable energy based-polygeneration system for sustainable thermal desalination process—A thermodynamics' perspective. *Int. J. Energy Res.* **2020**, *44*, 12344–12362. [CrossRef]
4. Ramadhani, F.; Hussain, M.; Mokhlis, H.; Illias, H.A. Optimal heat recovery using photovoltaic thermal and thermoelectric generator for solid oxide fuel cell-based polygeneration system: Techno-economic and environmental assessments. *Appl. Therm. Eng.* **2020**, *181*, 116015. [CrossRef]
5. Giwa, A.; Yusuf, A.; Dindi, A.; Balogun, H.A. Polygeneration in desalination by photovoltaic thermal systems: A comprehensive review. *Renew. Sustain. Energy Rev.* **2020**, *130*, 109946. [CrossRef]
6. Rokni, M. Analysis of a polygeneration plant based on solar energy, dual mode solid oxide cells and desalination. *Int. J. Hydrogen Energy* **2019**, *44*, 19224–19243. [CrossRef]

7. Sahoo, U.; Kumar, R.; Pant, P.; Chaudhary, R. Development of an innovative polygeneration process in hybrid solar-biomass system for combined power, cooling and desalination. *Appl Therm. Eng.* **2017**, *120*, 560–567. [CrossRef]
8. Villarruel-Jaramillo, A.; Pérez-García, M.; Cardemil, J.M.; Escobar, R.A. Review of Polygeneration Schemes with Solar Cooling Technologies and Potential Industrial Applications. *Energies* **2021**, *14*, 6450. [CrossRef]
9. Linde, L. Implementation of BECCS in a Polygeneration System—A Techno-Economic Feasibility Study in the District Heating Network of Stockholm. Master Thesis, KTH Industrial Engineering and Management, Stockholm, Sweden, 2016.
10. Lee, I.; Tester, J.W.; You, F. Systems analysis, design, and optimization of geothermal energy systems for power production and polygeneration: State-of-the-art and future challenges. *Renew. Sustain. Energy Rev.* **2019**, *109*, 551–577. [CrossRef]
11. Parraga, J.; Khalilpour, R.; Vassallo, A. Polygeneration with biomass-integrated gasification combined cycle process: Review and prospective. *Renew. Sustain. Energy Rev.* **2018**, *92*, 219–234. [CrossRef]
12. Rong, A.; Lahdelma, R. Role of polygeneration in sustainable energy system development challenges and opportunities from optimization viewpoints. *Renew. Sustain. Energy Rev.* **2016**, *53*, 363–372. [CrossRef]
13. Kwan, T.H.; Katsushi, F.; Shen, Y.; Yin, S.; Zhang, Y.; Kase, K.; Yao, Q. Comprehensive review of integrating fuel cells to other energy systems for enhanced performance and enabling polygeneration. *Renew. Sustain. Energy Rev.* **2020**, *128*, 109897. [CrossRef]
14. Cigolotti, V.; Genovese, M.; Fragiacomo, P. Comprehensive Review on Fuel Cell Technology for Stationary Applications as Sustainable and Efficient Poly-Generation Energy Systems. *Energies* **2021**, *14*, 4963. [CrossRef]
15. Baroutaji, A.; Arjunan, A.; Robinson, J.; Wilberforce, T.; Abdelkareem, M.A.; Olabi, A.G. PEMFC Poly-Generation Systems: Developments, Merits, and Challenges. *Sustainability* **2021**, *13*, 11696. [CrossRef]
16. Department of Energy, USA. *Chapter 5: Increasing Efficiency of Building Systems and Technologies*; Department of Energy (DOE): Washington, DC, USA, 2015.
17. International Energy Agency, I. *Energy-Efficient Buildings: Heating and Cooling Equipment, in Technology Roadmap*; International Energy Agency (IEA): Paris, France, 2011.
18. Zaid, S.; Graham, P. The Need for Energy Efficiency Legislation in the Malaysian Building Sector. In Proceedings of the 3rd International Building Control Conference, Kuala Lumpur, Malaysia, 21 November 2013.
19. Wakui, T.; Yokoyama, R. Optimal sizing of residential SOFC cogeneration system for power interchange operation in housing complex from energy-saving viewpoint. *Energy* **2012**, *41*, 65–74. [CrossRef]
20. Wakui, T.; Wada, N.; Yokoyama, R. Feasibility study on combined use of residential SOFC cogeneration system and plug-in hybrid electric vehicle from energy-saving viewpoint. *Energy Convers. Manag* **2012**, *60*, 170–179. [CrossRef]
21. Jana, K.; Ray, A.; Majoumerd, M.M.; Assadi, M.; De, S. Polygeneration as a future sustainable energy solution—A comprehensive review. *Appl. Energy* **2017**, *202*, 88–111. [CrossRef]
22. Palsson, J.; Selimovic, A.; Sjunnesson, L. Combined solid oxide fuel cell and gas turbine systems for efficient power and heat generation. *J. Power Sources* **2000**, *86*, 442–448. [CrossRef]
23. Chachuat, B.; Srinivasan, B.; Bonvin, D. Adaptation strategies for real-time optimization. *Comput. Chem. Eng.* **2009**, *33*, 1557–1567. [CrossRef]
24. Dicks, A.L.; Fellows, R.G.; Mescal, C.M.; Seymour, C. A study of SOFC–PEM hybrid systems. *J. Power Sources* **2000**, *86*, 501–506. [CrossRef]
25. Islam, S.; Dincer, I.; Yilbas, B.S. Development, analysis and assessment of solar energy-based multigeneration system with thermoelectric generator. *Energy Convers. Manag.* **2017**, *156*, 746–756. [CrossRef]
26. Zhang, H.; Kong, W.; Dong, F.; Xu, H.; Chen, B.; Ni, M. Application of cascading thermoelectric generator and cooler for waste heat recovery from solid oxide fuel cells. *Energy Convers. Manag.* **2017**, *148*, 1382–1390. [CrossRef]
27. Zhang, H.; Xu, H.; Chen, B.; Dong, F.; Ni, M. Two-stage thermoelectric generators for waste heat recovery from solid oxide fuel cells. *Energy* **2017**, *132*, 280–288. [CrossRef]
28. Islam, S.; Dincer, I.; Yilbas, B.S. Analysis and Assessment of A Biomass Energy-Based Multigeneration System with Thermoelectric Generators. *Energy Fuels* **2017**, *31*, 10901–10915. [CrossRef]
29. Li, Y.; Rezgui, Y.; Zhu, H. District heating and cooling optimization and enhancement—Towards integration of renewables, storage and smart grid. *Renew. Sustain. Energy Rev.* **2017**, *72*, 281–294. [CrossRef]
30. Franco, A.; Versace, M. Multi-objective optimization for the maximization of the operating share of cogeneration system in District Heating Network. *Energy Convers. Manag.* **2017**, *139*, 33–44. [CrossRef]
31. Ameri, M.; Besharati, Z. Optimal design and operation of district heating and cooling networks with CCHP systems in a residential complex. *Energy Build.* **2016**, *110*, 135–148. [CrossRef]
32. Verrilli, F.; Parisio, A.; Glielmo, L. Stochastic model predictive control for optimal energy management of district heating power plants, in Decision and Control (CDC). In Proceedings of the 2016 IEEE 55th Conference, Las Vegas, NV, USA, 12–14 December 2016; pp. 807–812.
33. Gładysz, P.; Ziębik, A. Complex analysis of the optimal coefficient of the share of cogeneration in district heating systems. *Energy* **2013**, *62*, 12–22. [CrossRef]
34. Ahmadi, P. Modeling, Analysis and Optimization of Integrated Energy Systems for Multigeneration Purposes. In *Mechanical Engineering*; University of Ontario Institute of Technology: Oshawa, ON, Canada, 2013.
35. Irvine, J.T.S.; Connor, P. *Solid Oxide Fuels Cells: Facts and Figures: Past Present and Future Perspectives for SOFC Technologies*; Springer: London, UK, 2012.

36. Szega, M.; Żymełka, P. Thermodynamic and Economic Analysis of the Production of Electricity, Heat And Cold in the CHP Unit with the Absorption Chillers. *J. Energy Resour. Technol.* **2017**, *140*, 052002. [CrossRef]
37. Ciampi, G.; Rosato, A.; Scorpio, M.; Sibilio, S. Experimental analysis of a micro-trigeneration system composed of a micro-cogenerator coupled with an electric chiller. *Appl. Therm. Eng.* **2014**, *73*, 1309–1322. [CrossRef]
38. Palmero-Marrero, A.I.; Oliveira, A.C. Performance simulation of a solar-assisted micro-tri-generation system: Hotel case study. *Int. J. Low Carbon Technol.* **2011**, *6*, 309–317. [CrossRef]
39. Malico, I.; Carvalhinho, A.P.; Tenreiro, J. Design of a trigeneration system using a high-temperature fuel cell. *Int. J. Energy Res.* **2008**, *33*, 144–151. [CrossRef]
40. Liu, Y.; Leong, K. Numerical study of an internal-reforming solid oxide fuel cell and adsorption chiller co-generation system. *J. Power Sources* **2006**, *159*, 501–508. [CrossRef]
41. Sigurjonsson, H.; Clausen, L.R. Solution for the future smart energy system: A polygeneration plant based on reversible solid oxide cells and biomass gasification producing either electrofuel or power. *Appl. Energy* **2018**, *216*, 323–337. [CrossRef]
42. Sigarchian, S.G.; Malmquist, A.; Martin, V. The choice of operating strategy for a complex polygeneration system: A case study for a residential building in Italy. *Energy Convers. Manag.* **2018**, *163*, 278–291. [CrossRef]
43. Calise, F.; Figaj, R.D.; Vanoli, L. A novel polygeneration system integrating photovoltaic/thermal collectors, solar assisted heat pump, adsorption chiller and electrical energy storage: Dynamic and energy-economic analysis. *Energy Convers. Manag.* **2017**, *149*, 798–814. [CrossRef]
44. Jradi, M.; Riffat, S. Tri-generation systems: Energy policies, prime movers, cooling technologies, configurations and operation strategies. *Renew. Sustain. Energy Rev.* **2014**, *32*, 396–415. [CrossRef]
45. Al Moussawi, H.; Fardoun, F.; Louahlia-Gualous, H. Review of tri-generation technologies: Design evaluation, optimization, decision-making, and selection approach. *Energy Convers. Manag.* **2016**, *120*, 157–196. [CrossRef]
46. Bernard, A. Micro Combined Heat and Power Operating on Renewable Energy for Residential Building. Ph.D. Thesis, CEP-Centre Energétique et Procédés, Paris, France, 2008.
47. Akikur, R.; Saidur, R.; Ping, H.; Ullah, K. Performance analysis of a co-generation system using solar energy and SOFC technology. *Energy Convers. Manag.* **2014**, *79*, 415–430. [CrossRef]
48. Elmer, T.; Worall, M.; Wu, S.; Riffat, S.B. Fuel cell technology for domestic built environment applications: State of-the-art review. *Renew. Sustain. Energy Rev.* **2015**, *42*, 913–931. [CrossRef]
49. Ellamla, H.R.; Staffell, I.; Bujlo, P.; Pollet, B.G.; Pasupathi, S. Current status of fuel cell based combined heat and power systems for residential sector. *J. Power Sources* **2015**, *293*, 312–328. [CrossRef]
50. US DOE. *Energy Efficiency and Renewable Energy*; US DOE: Washington, DC, USA, 2007.
51. Entchev, E.; Yang, L.; Ghorab, M.; Lee, E. Simulation of hybrid renewable microgeneration systems in load sharing applications. *Energy* **2013**, *50*, 252–261. [CrossRef]
52. Oland, C.B. *Guide to Low-Emission Boiler and Combustion Equipment Selection*; The Laboratory: Oak Ridge, TN, USA, 2002.
53. Oland, C.B. *Guide to Combined Heat and Power Systems for Boiler Owners and Operators*; United States Department of Energy: Washington, DC, USA, 2004.
54. Bensaid, S.; Brignone, M.; Ziggiotti, A.; Specchia, S. High efficiency Thermo-Electric power generator. *Int. J. Hydrogen Energy* **2012**, *37*, 1385–1398. [CrossRef]
55. Terayama, T.; Nagata, S.; Tanaka, Y.; Momma, A.; Kato, T.; Yamamoto, A. Improvement of SOFC System Efficiency by Incorporating Thermoelectric Power Generation Heat Exchanger. *ECS Trans.* **2013**, *57*, 2627–2636. [CrossRef]
56. Sonar, D.; Soni, S.; Sharma, D. Micro-trigeneration for energy sustainability: Technologies, tools and trends. *Appl. Therm. Eng.* **2014**, *71*, 790–796. [CrossRef]
57. Siyal, S.H.; Mentis, D.; Howells, M. Economic analysis of standalone wind-powered hydrogen refueling stations for road transport at selected sites in Sweden. *Int. J. Hydrogen Energy* **2015**, *40*, 9855–9865. [CrossRef]
58. Sharma, S.; Ghoshal, S.K. Hydrogen the future transportation fuel: From production to applications. *Renew. Sustain. Energy Rev.* **2015**, *43*, 1151–1158. [CrossRef]
59. Stojić, D.L.; Marčeta, M.P.; Sovilj, S.P.; Miljanić, Š.S. Hydrogen generation fromwater electrolysis—Possibilities of energy saving. *J. Power Sources* **2003**, *118*, 315. [CrossRef]
60. Dunn, S. Hydrogen futures: Toward a sustainable energy system. *Int. J. Hydrogen Energy* **2002**, *27*, 235–264. [CrossRef]
61. Rabbani, H.; Dincer, I.; Rahnamayan, S. A novel multigenerational hydrogen production system: Performance evaluation. *Int. J. Hydrogen Energy* **2017**, *42*, 12888–12895. [CrossRef]
62. Sarhaddi, F.; Farahat, S.; Ajam, H.; Behzadmehr, A.; Adeli, M.M. An improved thermal and electrical model for a solar photovoltaic thermal (PV/T) air collector. *Appl. Energy* **2010**, *87*, 2328–2339. [CrossRef]
63. Suberu, M.Y.; Mustafa, M.W.; Bashir, N. Energy storage systems for renewable energy power sector integration and mitigation of intermittency. *Renew. Sustain. Energy Rev.* **2014**, *35*, 499–514. [CrossRef]
64. Mahlia, T.; Saktisahdan, T.; Jannifar, A.; Hasan, M.; Matseelar, H. A review of available methods and development on energy storage; technology update. *Renew. Sustain. Energy Rev.* **2014**, *33*, 532–545. [CrossRef]
65. New Mexico Environment Department. *Air-Water Temperature Correlation*; Surface Water Quality Bureau: Santa Fe, NM, USA, 2011.
66. Ang, S.M.C.; Fraga, E.S.; Brandon, N.P.; Samsatli, N.J.; Brett, D.J. Fuel cell systems optimisation—Methods and strategies. *Int. J. Hydrogen Energy* **2011**, *36*, 14678–14703. [CrossRef]

67. Isa, Z.M.; Rahim, N.A. PEM fuel cell model parameters optimization using modified particle swarm optimization algorithm. In Proceedings of the IEEE Conference on Clean Energy and Technology (CEAT), Langkawi, Malaysia, 18–20 November 2013.
68. Ramadhani, F.; Abu Bakar, K.; Shafer, M.G. Optimization of standalone street light system with consideration of lighting control. In Proceedings of the 2013 International Conference on Technological Advances in Electrical, Electronics and Computer Engineering (TAEECE), Konya, Turkey, 9–11 May 2013.
69. Ramadhani, F.; Abu Bakar, K.; Hussain, M.; Erixno, O.; Nazir, R. Optimization with traffic-based control for designing standalone streetlight system: A case study. *Renew. Energy* **2016**, *105*, 149–159. [CrossRef]
70. Choudhury, A.; Chandra, H.; Arora, A. Application of solid oxide fuel cell technology for power generation—A review. *Renew. Sustain. Energy Rev.* **2013**, *20*, 430–442. [CrossRef]
71. Mohamad, H.; Abu Bakar, A.; Ping, H.; Mokhlis, H. An adaptive controller of hydro generators for smart grid application in Malaysia. In Proceedings of the 2010 International Conference on Power System Technology, Hangzhou, China, 24–28 October 2010; pp. 1–6.
72. Burer, M.; Tanaka, K.; Favrat, D.; Yamada, K. Multi-criteria optimization of a district cogeneration plant integrating a solid oxide fuel cell–gas turbine combined cycle, heat pumps and chillers. *Energy* **2003**, *28*, 497–518. [CrossRef]
73. Akikur, R.K.; Saidur, R.; Ullah, K.R.; Hajimolana, S.A.; Ping, H.W.; Hussain, M.A. Economic feasibility analysis of a solar energy and solid oxide fuel cell-based cogeneration system in Malaysia. *Clean Technol. Environ. Policy* **2015**, *18*, 669–687. [CrossRef]
74. Mahlia, T.M.I.; Chan, P. Life cycle cost analysis of fuel cell based cogeneration system for residential application in Malaysia. *Renew. Sustain. Energy Rev.* **2011**, *15*, 416–426. [CrossRef]
75. Al Moussawi, H.; Fardoun, F.; Louahlia, H. 4-E based optimal management of a SOFC-CCHP system model for residential applications. *Energy Convers. Manag.* **2017**, *151*, 607–629. [CrossRef]
76. *Eleventh Malaysia Plan, Strategy Paper 17: Sustainable Usage of Energy to Support Growth*; Federal Government Administrative Centre: Putrajaya, Malaysia, 2017.
77. Genovese, M.; Fragiacomo, P. Hydrogen station evolution towards a poly-generation energy system. *Int. J. Hydrogen Energy* **2021**, *47*, 12264–12280. [CrossRef]
78. Richardson, I.A.; Fisher, J.T.; Frome, P.E.; Smith, B.O.; Guo, S.; Chanda, S.; McFeely, M.S.; Miller, A.M.; Leachman, J.W. Low-cost, transportable hydrogen fueling station for early market adoption of fuel cell electric vehicles. *Int. J. Hydrogen Energy* **2015**, *40*, 8122–8127. [CrossRef]
79. Wakui, T.; Kawayoshi, H.; Yokoyama, R. Optimal structural design of residential power and heat supply devices in consideration of operational and capital recovery constraints. *Appl. Energy* **2016**, *163*, 118–133. [CrossRef]
80. Atsonios, K.; Samlis, C.; Manou, K.; Nikolopoulos, A.; Sfetsioris, K.; Mitsotakis, A.; Grammelis, P. Technical assessment of LNG based polygeneration systems for non-interconnected island cases using SOFC. *Int. J. Hydrogen Energy* **2020**, *46*, 4827–4843. [CrossRef]
81. Holagh, S.G.; Haghghi, M.A.; Mohammadi, Z.; Chitsaz, A. Exergoeconomic and environmental investigation of an innovative poly-generation plant driven by a solid oxide fuel cell for production of electricity, cooling, desalinated water, and hydrogen. *Int. J. Energy Res.* **2020**, *44*, 10126–10154. [CrossRef]
82. Li, M.; Zhuang, Y.; Song, M.; Li, W.; Du, J. Techno-economic and carbon footprint feasibility assessment for polygeneration process of carbon-capture coal-to-methanol/power and molten carbonate fuel cell. *Energy Convers. Manag.* **2021**, *235*, 114015. [CrossRef]
83. Wang, J.; Yang, Y.; Mao, T.; Sui, J.; Jin, H. Life cycle assessment (LCA) optimization of solar-assisted hybrid CCHP system. *Appl. Energy* **2015**, *146*, 38–52. [CrossRef]
84. Mago, P.J.; Fumo, N.; Chamra, L.M. Performance analysis of CCHP and CHP systems operating following the thermal and electric load. *Int. J. Energy Res.* **2009**, *33*, 852–864. [CrossRef]
85. Mago, P.; Chamra, L.; Ramsay, J. Micro-combined cooling, heating and power systems hybrid electric-thermal load following operation. *Appl. Therm. Eng.* **2010**, *30*, 800–806. [CrossRef]
86. Smith, A.; Luck, R.; Mago, P.J. Analysis of a combined cooling, heating, and power system model under different operating strategies with input and model data uncertainty. *Energy Build.* **2010**, *42*, 2231–2240. [CrossRef]
87. Zheng, C.; Wu, J.; Zhai, X. A novel operation strategy for CCHP systems based on minimum distance. *Appl. Energy* **2014**, *128*, 325–335. [CrossRef]
88. Kavvadias, K.; Maroulis, Z. Multi-objective optimization of a trigeneration plant. *Energy Policy* **2010**, *38*, 945–954. [CrossRef]
89. Hajabdollahi, H.; Ganjehkaviri, A.; Jaafar, M.N.M. Assessment of new operational strategy in optimization of CCHP plant for different climates using evolutionary algorithms. *Appl. Therm. Eng.* **2015**, *75*, 468–480. [CrossRef]
90. Basrawi, F.; Yamada, T.; Obara, S. Economic and environmental based operation strategies of a hybrid photovoltaic–microgas turbine trigeneration system. *Appl. Energy* **2014**, *121*, 174–183. [CrossRef]
91. Kavvadias, K.; Tosios, A.; Maroulis, Z. Design of a combined heating, cooling and power system: Sizing, operation strategy selection and parametric analysis. *Energy Convers. Manag.* **2010**, *51*, 833–845. [CrossRef]
92. Martínez-Lera, S.; Ballester, J. Analysis and sizing of thermal energy storage in combined heating, cooling and power plants for buildings. *Appl. Energy* **2013**, *106*, 127–142. [CrossRef]
93. Ondeck, A.D.; Edgar, T.F.; Baldea, M. Optimal operation of a residential district-level combined photovoltaic/natural gas power and cooling system. *Appl. Energy* **2015**, *156*, 593–606. [CrossRef]

94. Adam, A.; Fraga, E.S.; Brett, D.J. Modelling and Optimisation in Terms of CO2 Emissions of a Solid Oxide Fuel Cell based Micro-CHP System in a Four Bedroom House in London. *Energy Procedia* **2013**, *42*, 201–209. [CrossRef]
95. Hawkes, A.; Leach, M. Cost-effective operating strategy for residential micro-combined heat and power. *Energy* **2007**, *32*, 711–723. [CrossRef]
96. Hawkes, A.; Aguiar, P.; Croxford, B.; Leach, M.; Adjiman, C.; Brandon, N. Solid oxide fuel cell micro combined heat and power system operating strategy: Options for provision of residential space and water heating. *J. Power Sources* **2007**, *164*, 260–271. [CrossRef]
97. Taghavifar, H.; Anvari, S.; Saray, R.K.; Khalilarya, S.; Jafarmadar, S. Towards modeling of combined cooling, heating and power system with artificial neural network for exergy destruction and exergy efficiency prognostication of tri-generation components. *Appl. Therm. Eng.* **2015**, *89*, 156–168. [CrossRef]
98. Anvari, S.; Taghavifar, H.; Saray, R.K.; Khalilarya, S.; Jafarmadar, S. Implementation of ANN on CCHP system to predict trigeneration performance with consideration of various operative factors. *Energy Convers. Manag.* **2015**, *101*, 503–514. [CrossRef]
99. Entchev, E. Residential fuel cell energy systems performance optimization using "soft computing" techniques. *J. Power Sources* **2003**, *118*, 212–217. [CrossRef]
100. Li, M.; Mu, H.; Li, N.; Ma, B. Optimal design and operation strategy for integrated evaluation of CCHP (combined cooling heating and power) system. *Energy* **2016**, *99*, 202–220. [CrossRef]
101. Shaneb, O.A.; Taylor, P.C.; Coates, G. Real time operation of μCHP systems using fuzzy logic. *Energy Build.* **2012**, *55*, 141–150. [CrossRef]
102. Chen, J.M.P.; Ni, M. Economic analysis of a solid oxide fuel cell cogeneration/trigeneration system for hotels in Hong Kong. *Energy Build.* **2014**, *75*, 160–169. [CrossRef]
103. Weber, C.; Marechal, F.; Favrat, D.; Kraines, S. Optimization of an SOFC-based decentralized polygeneration system for providing energy services in an office-building in Tōkyō. *Appl. Therm. Eng.* **2006**, *26*, 1409–1419. [CrossRef]
104. Antonucci, V.; Brunaccini, G.; De Pascale, A.; Ferraro, M.; Melino, F.; Orlandini, V.; Sergi, F. Integration of μ-SOFC Generator and ZEBRA Batteries for Domestic Application and Comparison with other μ-CHP Technologies. *Energy Procedia* **2015**, *75*, 999–1004. [CrossRef]
105. Nataf, K.; Bradley, T.H. An economic comparison of battery energy storage to conventional energy efficiency technologies in Colorado manufacturing facilities. *Appl. Energy* **2016**, *164*, 133–139. [CrossRef]
106. Jabari, F.; Nojavan, S.; Ivatloo, B.M.; Sharifian, M.B. Optimal short-term scheduling of a novel tri-generation system in the presence of demand response programs and battery storage system. *Energy Convers. Manag.* **2016**, *122*, 95–108. [CrossRef]
107. Isa, N.M.; Das, H.S.; Tan, C.W.; Yatim, A.; Lau, K.Y. A techno-economic assessment of a combined heat and power photovoltaic/fuel cell/battery energy system in Malaysia hospital. *Energy* **2016**, *112*, 75–90. [CrossRef]
108. Misra, S.; Satyaprasad, G.R.K.D.; Mahapatra, S.S.; Mohanty, A.; Biswal, S.R.; Dora, A. Design of Fuel cell based Cogeneration systems: An approach for battery less Solar PV system. *Int. Res. J. Eng. Technol.* **2016**, *3*, 47–50.
109. Fragiacomo, P.; Lucarelli, G.; Genovese, M.; Florio, G. Multi-objective optimization model for fuel cell-based poly-generation energy systems. *Energy* **2021**, *237*, 121823. [CrossRef]
110. Cardona, E.; Piacentino, A. A methodology for sizing a trigeneration plant in mediterranean areas. *Appl. Therm. Eng.* **2003**, *23*, 1665–1680. [CrossRef]
111. Baghernejad, A.; Yaghoubi, M.; Jafarpur, K. Optimum power performance of a new integrated SOFC-trigeneration system by multi-objective exergoeconomic optimization. *Int. J. Electr. Power Energy Syst.* **2015**, *73*, 899–912. [CrossRef]
112. Arcuri, P.; Beraldi, P.; Florio, G.; Fragiacomo, P. Optimal design of a small size trigeneration plant in civil users: A MINLP (Mixed Integer Non Linear Programming Model). *Energy* **2015**, *80*, 628–641. [CrossRef]
113. Elmer, T. *A Novel SOFC Tri-Generation System for Building Applications*; University of Nottingham: Nottingham, UK, 2015.
114. Martínez-Lera, S.; Ballester, J. A novel method for the design of CHCP (combined heat, cooling and power) systems for buildings *Energy* **2010**, *35*, 2972–2984. [CrossRef]
115. Windeknecht, M.; Tzscheutschler, P. Optimization of the Heat Output of High Temperature Fuel Cell Micro-CHP in Single Family Homes. *Energy Procedia* **2015**, *78*, 2160–2165. [CrossRef]
116. Liso, V.; Zhao, Y.; Yang, W.; Nielsen, M.P. Modelling of a Solid Oxide Fuel Cell CHP System Coupled with a Hot Water Storage Tank for a Single Household. *Energies* **2015**, *8*, 2211–2229. [CrossRef]
117. Facci, A.L.; Cigolotti, V.; Jannelli, E.; Ubertini, S. Technical and economic assessment of a SOFC-based energy system for combined cooling, heating and power. *Appl. Energy* **2017**, *192*, 563–574. [CrossRef]
118. Wang, J.-J.; Jing, Y.-Y.; Zhang, C.-F. Optimization of capacity and operation for CCHP system by genetic algorithm. *Appl. Energy* **2010**, *87*, 1325–1335. [CrossRef]
119. Guo, L.; Liu, W.J.; Cai, J.J.; Hong, B.W.; Wang, C.S. A two-stage optimal planning and design method for combined cooling, heat and power microgrid system. *Energy Convers. Manag.* **2013**, *74*, 433–445. [CrossRef]
120. Wang, J.; Zhai, Z.; Jing, Y.; Zhang, C. Optimization design of BCHP system to maximize to save energy and reduce environmental impact. *Energy* **2010**, *35*, 3388–3398. [CrossRef]
121. Wang, J.; Zhai, Z.; Jing, Y.; Zhang, C. Particle swarm optimization for redundant building cooling heating and power system. *Appl. Energy* **2010**, *87*, 3668–3679. [CrossRef]

122. Wakui, T.; Yokoyama, R. Impact analysis of sampling time interval and battery installation on optimal operational planning of residential cogeneration systems without electric power export. *Energy* **2015**, *81*, 120–136. [CrossRef]
123. Chen, X.; Wu, Q.; Li, Z.; Wang, M. Investigation of a CHP-EES based on Dymola. In Proceedings of the 2015 IEEE International Conference on Signal Processing, Communications and Computing (ICSPCC), Ningbo, China, 19–22 September 2015.
124. Antonucci, V.; Branchini, L.; Brunaccini, G.; De Pascale, A.; Ferraro, M.; Melino, F.; Orlandini, V.; Sergi, F. Thermal integration of a SOFC power generator and a Na–NiCl2 battery for CHP domestic application. *Appl. Energy* **2017**, *185*, 1256–1267. [CrossRef]
125. Hajabdollahi, Z.; Fu, P.-F. Multi-objective based configuration optimization of SOFC-GT cogeneration plant. *Appl. Therm. Eng.* **2016**, *112*, 549–559. [CrossRef]
126. Wang, J.; Wu, J.; Zheng, C. Analysis of tri-generation system in combined cooling and heating mode. *Energy Build.* **2014**, *72*, 353–360. [CrossRef]
127. Kazempoor, P.; Dorer, V.; Ommi, F. Evaluation of hydrogen and methane-fuelled solid oxide fuel cell systems for residential applications: System design alternative and parameter study. *Int. J. Hydrogen Energy* **2009**, *34*, 8630–8644. [CrossRef]
128. Kazempoor, P.; Dorer, V.; Weber, A. Modelling and evaluation of building integrated SOFC systems. *Int. J. Hydrogen Energy* **2011**, *36*, 13241–13249. [CrossRef]
129. Perdikaris, N.; Panopoulos, K.; Hofmann, P.; Spyrakis, S.; Kakaras, E. Design and exergetic analysis of a novel carbon free tri-generation system for hydrogen, power and heat production from natural gas, based on combined solid oxide fuel and electrolyser cells. *Int. J. Hydrogen Energy* **2010**, *35*, 2446–2456. [CrossRef]
130. Braun, R.; Kazempoor, P. *Application of SOFCs in Combined Heat, Cooling, and Power Systems. Solid Oxide Fuel Cells: From Materials to System Modeling*; Ni, M., Zhao, T.S., Eds.; Energy and Environment Series; Royal Society of Chemistry: London, UK, 2013.
131. Adams, S.; Klobodu, E.K.M.; Apio, A. Renewable and non-renewable energy, regime type and economic growth. *Renew. Energy* **2018**, *125*, 755–767. [CrossRef]
132. Baul, T.; Datta, D.; Alam, A. A comparative study on household level energy consumption and related emissions from renewable (biomass) and non-renewable energy sources in Bangladesh. *Energy Policy* **2018**, *114*, 598–608. [CrossRef]
133. Narayan, S.; Doytch, N. An investigation of renewable and non-renewable energy consumption and economic growth nexus using industrial and residential energy consumption. *Energy Econ.* **2017**, *68*, 160–176. [CrossRef]
134. Rosato, A.; Sibilio, S.; Ciampi, G.; Entchev, E.; Ribberink, H. Energy, Environmental and Economic Effects of Electric Vehicle Charging on the Performance of a Residential Building-integrated Micro-trigeneration System. *Energy Procedia* **2017**, *111*, 699–709. [CrossRef]
135. Tie, C.H.; Gan, C.K.; Ibrahim, K.A. The impact of electric vehicle charging on a residential low voltage distribution network in Malaysia. In Proceedings of the 2014 IEEE Innovative Smart Grid Technologies-Asia (ISGT Asia), Kuala Lumpur, Malaysia, 20 May 2014.
136. Wakui, T.; Wada, N.; Yokoyama, R. Energy-saving effect of a residential polymer electrolyte fuel cell cogeneration system combined with a plug-in hybrid electric vehicle. *Energy Convers. Manag.* **2014**, *77*, 40–51. [CrossRef]
137. Aviso, K.B.; Tan, R.R. Fuzzy P-graph for optimal synthesis of cogeneration and trigeneration systems. *Energy* **2018**, *154*, 258–268. [CrossRef]
138. Andiappan, V.; Ng, D.K. Synthesis of tri-generation systems: Technology selection, sizing and redundancy allocation based on operational strategy. *Comput. Chem. Eng.* **2016**, *91*, 380–391. [CrossRef]
139. Tan, L.; Yang, C.; Zhou, N. Synthesis/design optimization of SOFC–PEM hybrid system under uncertainty. *Chin. J. Chem. Eng.* **2015**, *23*, 128–137. [CrossRef]

processes MDPI

Article

Optimization of Heat Recovery Networks for Energy Savings in Industrial Processes

Jui-Yuan Lee * and Po-Yu Chen

Department of Chemical Engineering and Biotechnology, National Taipei University of Technology, 1, Sec. 3, Zhongxiao E. Rd., Taipei 10608, Taiwan
* Correspondence: juiyuan@ntut.edu.tw; Tel.: +886-2-2771-2171 (ext. 2524)

Abstract: Among the pillars of decarbonization of the global energy system, energy efficiency plays a key role in reducing energy consumption across end-use (industry, transport and buildings) sectors. In industrial processes, energy efficiency can be improved by exploiting heat recovery via heat exchange between process streams. This paper develops a stage-wise superstructure-based mathematical programming model for the optimization of heat exchanger networks. The model incorporates rigorous formulation to handle process streams with phase change (condensation or evaporation), and is applied to a case study of an ethylene glycol production plant in Taiwan for minimizing utility consumption. The results show a compromise between steam savings and process feasibility, as well as how the model is modified to reflect practical considerations. In the preliminary analysis, with a substantial potential steam saving of 15,476 kW (28%), the solution involves forbidden matches that pose a hazard to the process and cannot be implemented. In the further analysis without process streams that cause forbidden matches, although the space limitation in the plant renders the best solution infeasible, the compromise solution can achieve a considerable steam saving of up to 8448 kW (91%) and is being evaluated by the plant managers and operators.

Keywords: energy conservation; heat integration; heat exchanger network synthesis; retrofit; mathematical programming; superstructure

Citation: Lee, J.-Y.; Chen, P.-Y. Optimization of Heat Recovery Networks for Energy Savings in Industrial Processes. *Processes* **2023**, *11*, 321. https://doi.org/10.3390/pr11020321

Academic Editors: Pei Liu, Ming Liu and Xiao Wu

Received: 28 November 2022
Revised: 10 January 2023
Accepted: 14 January 2023
Published: 18 January 2023

Copyright: © 2023 by the authors. Licensee MDPI, Basel, Switzerland. This article is an open access article distributed under the terms and conditions of the Creative Commons Attribution (CC BY) license (https://creativecommons.org/licenses/by/4.0/).

1. Introduction

To tackle the climate crisis and avoid the worst effects of climate change, many countries in the world have pledged to reduce global CO_2 emissions to net zero by 2050, while limiting the long-term increase in average global temperatures to 1.5 °C. Achieving the rapid reduction in CO_2 emissions in the net-zero emissions scenario (NZE) requires a range of technologies. Energy efficiency plays a key role in reducing energy consumption across the industry, buildings and transport sectors, and is among the key pillars of decarbonization of the global energy system, delivering around 50% of emissions reductions to 2030 in the NZE together with wind and solar [1].

In the industry sector, energy efficiency can be improved by exploiting process integration options such as waste heat recovery via heat exchange. The idea is to reduce the need for external heating and cooling using utilities (e.g., steam and cooling water) by transferring heat from process streams that need cooling to those that need heating in heat exchangers. However, synthesizing heat exchanger networks (HENs) using heuristics could lead to suboptimal solutions as a result of simply matching process streams near each other and losing sight of the big picture. Improving and optimizing the heat exchange between process streams involves systematic design procedures. Many techniques for heat integration and HEN synthesis have been developed since the 1970s [2], following two main approaches: pinch analysis [3] and mathematical programming [4]. Although pinch analysis allows a visualization of the process and provides insights into the energy use, mathematical programming would be preferred when addressing process constraints (e.g., forbidden matches) and cost trade-offs. Arguably, all the problems that can be solved by pinch analysis can equally well be solved by mathematical programming, but not vice versa.

Recent works on HEN synthesis using mathematical programming have improved the quality of solutions [5–8] and considered retrofits [9,10], multiple plants [11–13] and multiperiod operations [14,15]. Huang and Karimi [5] proposed a multistage, match-centric superstructure and a stage-less, exchanger-centric superstructure for HEN synthesis; the corresponding mixed integer nonlinear programming (MINLP) models consider a wider range of network configurations and allow solutions with lower total annualized costs (TACs) to be obtained. Beck and Hofmann [6] proposed the use of convex linear approximations to reformulate the original MINLP problem for HEN synthesis into a mixed integer linear programming (MILP) problem. Ziyatdinov et al. [7] introduced an iterative sequential method for optimal HEN synthesis. Liu et al. [8] presented an extended stage-wise superstructure with intermediate placement of utilities; the resulting complex mathematical model was solved using a genetic algorithm-based solution approach. For retrofit design, Čuček et al. [9] presented approaches developed for the retrofit of existing HENs within individual processes and industrial sites to achieve energy savings, cost savings and emissions reductions. Pavão et al. [10] presented a superstructure-based model and a meta-heuristic solution approach for the retrofit of HENs. For multi-plant and multiperiod heat integration, Cheng et al. [11] developed a sequential design procedure to generate the inter-plant heat integration scheme in industrial parks. Nair et al. [12] proposed an MINLP model for configuring an eco-industrial park-wide multi-enterprise HEN with a practical and rational strategy. Kachacha et al. [13] used a MILP model for the synthesis of heat exchange and transport networks between processes considering the variability of streams. Pavão et al. [14] used a meta-heuristic two-level method to handle multiperiod HEN optimization with a post-optimization strategy to improve the results. Hofmann et al. [15] carried out the optimization of HENs and storage integration considering operational constraints of existing steam generation units.

Further extensions of HEN models include heat-integrated water integration [16,17], work integration [18–22] and process synthesis [23]. Ibrić et al. [16] proposed a superstructure for the synthesis of heat-integrated water-using and wastewater treatment networks, and a two-step solution strategy to minimize TAC with upper bounds to freshwater and utility consumption. Ibrić et al. [17] later addressed the synthesis of non-isothermal water networks thermally integrated with process streams. Huang and Karimi [18] presented an efficient superstructure-based MINLP formulation to synthesize work-heat exchanger networks. Li et al. [19] proposed a building block-based superstructure and formulated an MINLP optimization model for the synthesis of work and heat exchanger networks for minimum TAC. Viske et al. [20] proposed the use of a nonsmooth extension to handle unclassified process streams in the synthesis of work and heat exchange networks. Yang et al. [21] formulated a thermo-economic multi-objective MINLP model to synthesize sub- and above-ambient HENs of constant-pressure and pressure-changing streams with multi-stream compression. Hamedi et al. [22] presented a simulation-based method for work-integrated HEN synthesis without correlations or simplifying assumptions. Li et al. [23] proposed a surrogate-based optimization framework for the simultaneous synthesis of a chemical process and its HEN.

Most previous studies of HEN synthesis assume a global minimum temperature difference for non-isothermal process streams with constant heat capacities and do not consider isothermal process streams with phase change. In addition, most of the recently developed methods and models were applied only to examples and case studies in the literature, with limited practical applications. In this paper, a superstructure-based mathematical programming model for the optimization of HENs is developed and applied to a new industrial case study for minimizing utility consumption. The model handles process streams with phase change rigorously, as in the formulations of Ponce-Ortega et al. [24] and Watanapanich et al. [25], and can be modified to accommodate various practical considerations by adjusting model parameters and adding appropriate constraints. The rest of the paper is organized as follows A brief problem statement is given in Section 2. The model formulation is presented in Section 3, followed by the industrial case study in Section 4 to demonstrate its application. Conclusions and prospects for future work are given in Section 5.

2. Problem Statement

Given a process with a set of hot streams $i \in \mathbf{I}$ to be cooled and a set of cold streams $j \in \mathbf{J}$ to be heated, some of the process streams involve only sensible heat (temperature change), some involve only latent heat (phase change), and some involve both. The data (e.g., flowrates, inlet and outlet temperatures, heat capacities, latent heats and phase change fractions) for these process streams are known. The objective is to synthesize an optimal HEN of the process for minimum utility consumption, while satisfying the heating and cooling demands as well as practical constraints on HEN design.

3. Model Formulation

Figure 1 shows a condensed stage-wise HEN superstructure, which is divided into a number K of stages by $K + 1$ temperature locations. A hot process stream may be split and matched with cold process streams for heat exchange in each stage. If its cooling demand is not satisfied by heat exchange through the stages, a cold utility will be used at the outlet of the superstructure to provide the unsatisfied cooling. Likewise, a cold process stream may be split and matched with hot process streams in each stage. If its heating demand is not satisfied through the stages, a hot utility will be used at the outlet of the superstructure to provide the unsatisfied heating. Although only a match between one hot and one cold process stream is shown in Figure 1, the condensed superstructure can be expanded for the number of process streams. For split streams, isothermal mixing [4] is assumed to simplify the model formulation. It is also assumed that the heat losses from the system are small (with exchangers properly lagged) and can be neglected. Based on the superstructure in Figure 1, the HEN model is formulated and presented in the following subsections. Notation used is given in the Nomenclature in Appendix A.

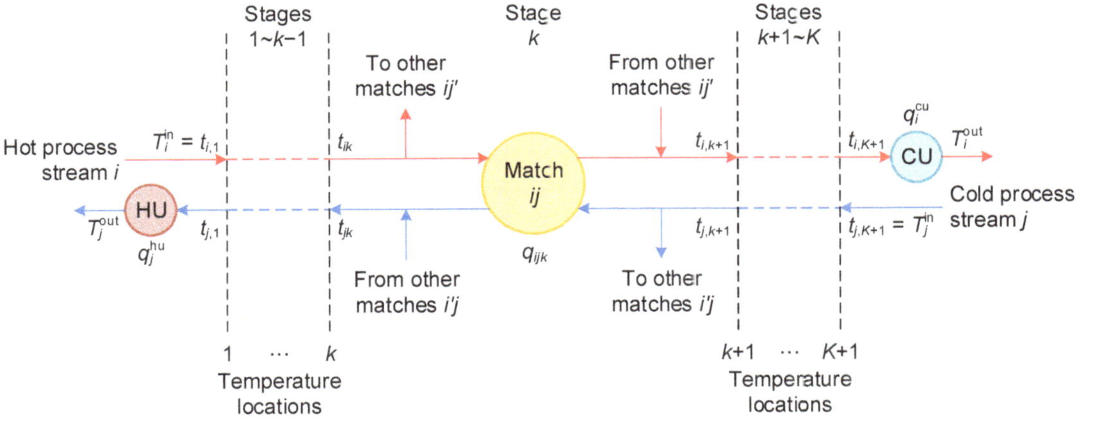

Figure 1. Stage-wise superstructure for a HEN.

3.1. Overall Energy Balance for Process Streams

Equations (1) and (2) describe the overall energy balances for hot and cold process streams, respectively. Hot process stream i can be cooled by transferring heat to cold process streams j in stages k (q_{ijk}) and the cold utility (q_i^{cu}) at the outlet of the superstructure. Cold process stream j can be heated by transferring heat from hot process streams i in stages k (q_{ijk}) and the hot utility (q_j^{hu}) at the outlet of the superstructure.

$$F_i\left(Cp_i\left(T_i^{\text{in}} - T_i^{\text{out}}\right) + \lambda_i\alpha_i\right) = \sum_{j\in\mathbf{J}}\sum_{k\in\mathbf{ST}} q_{ijk} + q_i^{\text{cu}} \quad \forall i \in \mathbf{I} \tag{1}$$

$$F_j\left(Cp_j\left(T_j^{\text{out}} - T_j^{\text{in}}\right) + \lambda_j\alpha_j\right) = \sum_{i\in\mathbf{I}}\sum_{k\in\mathbf{ST}} q_{ijk} + q_j^{\text{hu}} \quad \forall j \in \mathbf{J} \tag{2}$$

where F_i and F_j are the flowrates, Cp_i and Cp_j the heat capacities, λ_i and λ_j the latent heats, and α_i and α_j the phase change fractions of hot process stream i and cold process stream j, respectively. For streams involving only sensible heat, the latent heat and phase change fraction are set to zero; for streams involving only latent heat, the heat capacity is set to zero. Thus, Equations (1) and (2) apply to three types of process streams, including those involving both sensible and latent heat.

3.2. Energy Balance for Process Streams in Each Stage

Equations (3) and (4) describe the energy balances in stage k for hot ($i \in \mathbf{I}^1 \cup \mathbf{I}^3$) and cold process streams involving sensible heat ($j \in \mathbf{J}^1 \cup \mathbf{J}^3$), respectively. Process streams involving only latent heat ($i \in \mathbf{I}^2$ and $j \in \mathbf{J}^2$) are isothermal and can be modelled simply by Equations (1) and (2).

$$F_iCp_i(t_{ik} - t_{i,k+1}) + q_{ik}^{\Lambda} = \sum_{j\in\mathbf{J}} q_{ijk} \quad \forall i \in \mathbf{I}^1 \cup \mathbf{I}^3, k \in \mathbf{ST} \tag{3}$$

$$F_jCp_j\left(t_{jk} - t_{j,k+1}\right) + q_{jk}^{\Lambda} = \sum_{i\in\mathbf{I}} q_{ijk} \quad \forall j \in \mathbf{J}^1 \cup \mathbf{J}^3, k \in \mathbf{ST} \tag{4}$$

where t_{ik} is the temperature of hot process stream i at temperature location k, t_{jk} the temperature of cold process stream j at temperature location k, q_{ik}^{Λ} the condensation heat load of hot process stream i in stage k, and q_{jk}^{Λ} the evaporation heat load of cold process stream j in stage k. For streams involving only sensible heat ($i \in \mathbf{I}^1$ and $j \in \mathbf{J}^1$), the latent heat loads (q_{ik}^{Λ} and q_{jk}^{Λ}) are set to zero.

3.3. Temperature Assignment and Feasibility Constraints

As shown in Figure 1, hot process streams enter the superstructure from the left-hand side, while cold process streams enter from the right-hand side. Therefore, the inlet temperature of hot process stream i (T_i^{in}) is equal to its temperature at the first temperature location ($t_{i,1}$), and the inlet temperature of cold process stream j (T_j^{in}) is equal to its temperature at the last temperature location ($t_{j,K+1}$), as given in Equations (5) and (6). Equations (7)–(10) specify the trends in temperature. For non-isothermal process streams, a monotonic decrease in temperature throughout the superstructure is expected; for isothermal process streams, the temperature remains constant at the inlet temperature. In addition, the outlet temperature of hot process stream i (T_i^{out}) is used as a lower limit for its temperatures in the superstructure, while the outlet temperature of cold process stream j (T_j^{out}) is used as an upper limit, as given in Equations (11) and (12).

$$t_{i,1} = T_i^{\text{in}} \quad \forall i \in \mathbf{I} \tag{5}$$

$$t_{j,K+1} = T_j^{\text{in}} \quad \forall j \in \mathbf{J} \tag{6}$$

$$t_{ik} \geq t_{i,k+1} \quad \forall i \in \mathbf{I}^1 \cup \mathbf{I}^3, k \in \mathbf{ST} \tag{7}$$

$$t_{ik} = T_i^{\text{in}} \quad \forall i \in \mathbf{I}^2, k \in \mathbf{ST} \tag{8}$$

$$t_{jk} \geq t_{j,k+1} \quad \forall j \in \mathbf{J}^1 \cup \mathbf{J}^3, k \in \mathbf{ST} \tag{9}$$

$$t_{jk} = T_j^{\text{in}} \quad \forall j \in \mathbf{J}^2, k \in \mathbf{ST} \tag{10}$$

$$t_{ik} \geq T_i^{\text{out}} \quad \forall i \in \mathbf{I}, k \in \mathbf{ST} \tag{11}$$

$$t_{jk} \leq T_j^{\text{out}} \quad \forall j \in \mathbf{J}, k \in \mathbf{ST} \tag{12}$$

3.4. Utility Requirements and Latent Heat Balance

Equations (13) and (14) are used to calculate the hot and cold utility requirements for non-isothermal process streams.

$$q_i^{\text{cu}} = F_i Cp_i \left(t_{i,K+1} - T_i^{\text{out}}\right) + q_i^{\Lambda,\text{cu}} \quad \forall i \in \mathbf{I}^1 \cup \mathbf{I}^3 \tag{13}$$

$$q_j^{\text{hu}} = F_j Cp_j \left(T_j^{\text{out}} - t_{j,1}\right) + q_j^{\Lambda,\text{hu}} \quad \forall j \in \mathbf{J}^1 \cup \mathbf{J}^3 \tag{14}$$

where $q_i^{\Lambda,\text{cu}}$ is the condensation heat load of hot process stream i in the utility cooler and $q_j^{\Lambda,\text{hu}}$ is the evaporation heat load of cold process stream j in the utility heater. Again, for streams involving only sensible heat, the latent heat loads are set to zero.

Equations (15) and (16) describe the latent heat balances for process streams involving both sensible and latent heat ($i \in \mathbf{I}^3$ and $j \in \mathbf{J}^3$).

$$F_i \lambda_i \alpha_i = \sum_{k \in \mathbf{ST}} q_{ik}^{\Lambda} + q_i^{\Lambda,\text{cu}} \quad \forall i \in \mathbf{I}^3 \tag{15}$$

$$F_j \lambda_j \alpha_j = \sum_{k \in \mathbf{ST}} q_{jk}^{\Lambda} + q_j^{\Lambda,\text{hu}} \quad \forall j \in \mathbf{J}^3 \tag{16}$$

3.5. Feasibility of Latent Heat Exchange

Equations (17)–(20) ensure feasible latent heat exchange for type-3 process streams. Equations (17) and (18) are used for hot process streams that are condensed at the inlet (supply) temperature and cooled to the outlet (target) temperature. If hot process stream i enters stage k at a temperature lower than the condensation temperature ($t_{ik} < T_i^{\text{in}}$), binary variable y_{ik} has to be zero, and q_{ik}^{Λ} is forced to zero, meaning no condensation of hot process stream i in stage k. Equations (19) and (20) are used for cold process streams that are heated from the inlet temperature to the outlet temperature and evaporated. If cold process stream j leaves stage k at a temperature lower than the evaporation temperature ($t_{jk} < T_j^{\text{out}}$), binary variable y_{jk} has to be zero, and q_{jk}^{Λ} is forced to zero, meaning no evaporation of cold process stream j in stage k.

$$\varepsilon - \Gamma y_{ik} \leq T_i^{\text{in}} - t_{ik} \leq \Gamma(1 - y_{ik}) \quad \forall i \in \mathbf{I}^3, k \in \mathbf{ST} \tag{17}$$

$$q_{ik}^{\Lambda} \leq Q_i^{\text{U}} y_{ik} \quad \forall i \in \mathbf{I}^3, k \in \mathbf{ST} \tag{18}$$

$$\varepsilon - \Gamma y_{jk} \leq T_j^{\text{out}} - t_{jk} \leq \Gamma\left(1 - y_{jk}\right) \quad \forall j \in \mathbf{J}^3, k \in \mathbf{ST} \tag{19}$$

$$q_{jk}^{\Lambda} \leq Q_j^{\text{U}} y_{jk} \quad \forall j \in \mathbf{J}^3, k \in \mathbf{ST} \tag{20}$$

where ε is a sufficiently small value, Γ a sufficiently large value, y_{ik} indicates if $t_{ik} \geq T_i^{\text{in}}$, and y_{jk} indicates if $t_{jk} \geq T_j^{\text{out}}$.

3.6. Logical Constraints and Temperature Differences

Equations (21)–(23) set upper limits on the heat loads of exchangers (Q_{ij}^{U}), coolers (Q_i^{U}) and heaters (Q_j^{U}), and force the heat load to zero if the match does not exist.

$$q_{ijk} \leq Q_{ij}^{\text{U}} z_{ijk} \quad \forall i \in \mathbf{I}, j \in \mathbf{J}, k \in \mathbf{ST} \tag{21}$$

$$q_i^{\text{cu}} \leq Q_i^{\text{U}} z_i^{\text{cu}} \quad \forall i \in \mathbf{I} \tag{22}$$

$$q_j^{\text{hu}} \leq Q_j^{\text{U}} z_j^{\text{hu}} \quad \forall i \in \mathbf{I} \tag{23}$$

where z_{ijk}, z_i^{cu} and z_j^{hu} are binary variables indicating the existence of matches between hot process stream i and cold process stream j in stage k, between hot process stream i and a cold utility, and between a hot utility and cold process stream j, respectively.

Equations (24)–(27) ensure feasible driving forces for existing matches. Different minimum temperature differences can be assigned to different matches.

$$\Delta T_{ij}^{\min} - \Gamma\left(1 - z_{ijk}\right) \leq t_{ik} - t_{jk} \quad \forall i \in \mathbf{I}, j \in \mathbf{J}, k \in \mathbf{ST} \tag{24}$$

$$\Delta T_{ij}^{\min} - \Gamma\left(1 - z_{ijk}\right) \leq t_{i,k+1} - t_{j,k+1} \quad \forall i \in \mathbf{I}, j \in \mathbf{J}, k \in \mathbf{ST} \tag{25}$$

$$\Delta T_{i,\text{cu}}^{\min} - \Gamma(1 - z_i^{\text{cu}}) \leq t_{i,K+1} - T_{\text{cu}}^{\text{out}} \quad \forall i \in \mathbf{I} \tag{26}$$

$$\Delta T_{\text{hu},j}^{\min} - \Gamma\left(1 - z_j^{\text{hu}}\right) \leq T_{\text{hu}}^{\text{out}} - t_{j,1} \quad \forall j \in \mathbf{J} \tag{27}$$

where $\Delta T_{ij}^{\min}$, $\Delta T_{i,\text{cu}}^{\min}$ and $\Delta T_{\text{hu},j}^{\min}$ are the minimum temperature differences for matches between hot process stream i and cold process stream j, between hot process stream i and the cold utility, and between the hot utility and cold process stream j, respectively; $T_{\text{cu}}^{\text{out}}$ and $T_{\text{hu}}^{\text{out}}$ are the outlet temperatures of the cold and hot utilities, respectively. Equations (26) and (27) may not be needed assuming the hot utility to be sufficiently hot and the cold utility sufficiently cold. To prevent temperature inversion in heat exchangers involving condensation or evaporation of process streams, additional constraints can be used [25]. Alternatively, the minimum temperature difference can be manipulated (i.e., increased) for specific matches.

3.7. Objective Function

For maximum heat recovery in the process, the objective function is to minimize the hot utility requirement (f_{HU}), as given in Equation (28).

$$\min f_{\text{HU}} = \sum_{j \in \mathbf{J}} q_j^{\text{hu}} \tag{28}$$

A further step may be taken to obtain simpler HEN designs and keep down the capital cost by minimizing the number of matches or units (f_{NU}), as given in Equation (29), subject to the minimum hot utility requirement (f_{HU}^*) in Equation (30).

$$\min f_{\text{NU}} = \sum_{i \in \mathbf{I}} \sum_{j \in \mathbf{J}} \sum_{k \in \mathbf{ST}} z_{ijk} + \sum_{i \in \mathbf{I}} z_i^{\text{cu}} + \sum_{j \in \mathbf{J}} z_j^{\text{hu}} \tag{29}$$

$$\sum_{j \in \mathbf{J}} q_j^{\text{hu}} \leq f_{\text{HU}}^* \tag{30}$$

Equations (1)–(30) constitute a MILP model, for which global optimality can be guaranteed. The model is solved to determine the optimal HEN. The main results are stream temperatures and heat duties of the matches between process and utility streams. These details can then be used in the economic assessment involving the selection of heat transfer equipment and the calculation of heat transfer area and exchanger cost.

In the next section, an industrial case study is presented to illustrate the application of the proposed MILP model, which is solved in GAMS using solver CPLEX. Solutions were found with reasonable processing time. It is worth noting that numerical uncertainty comes from numerical errors and approximations in the implementation of the mathematical model In this work, the accuracy of the proposed model was confirmed by comparing the base-case results with the actual conditions. Numerical errors are small enough to be neglected.

4. Industrial Case Study

This case study considers an ethylene glycol (EG) production plant located in southern Taiwan. The EG process involves a reactor, two absorbers, two strippers, an eight-stage evaporator, a drying tower and a distillation column followed by three separation columns [26]. Tables 1 and 2 show the stream data extracted from the material and energy balance. The EG process flowsheet is complicated and is not shown for confidentiality reasons. Two levels of saturated steam at 213 °C and 180 °C are available as hot utilities. Cooling water as the cold utility is available at 29 °C with a return temperature of 39 °C.

Table 1. Hot process stream data.

Stream	Description	Flowrate (t/h)	Inlet Temperature (°C)	Outlet Temperature (°C)	Heat Capacity (J/g/°C)	Latent Heat (J/g)	Phase Change Fraction
H1	Outlet recycle gas	981	220	60	2.11	-	0
H2	Recycled water	1150	107.4	31	4.118	-	0
H3	Recycled water	37.1	116.1	35	4.118	-	0
H4	Process condensate	347	106.3	52	4.226	-	0
H5	Process vapor	17.4	124	124	-	2194.508 (at 124 °C)	1
H6	Process vapor	7.8	171	139	1.925	2092 (at 171 °C)	1
H7	Absorbent	354.1	45	33.1	4.163	-	0
H8	MEG product	48.5	138	60	2.761	-	0
H9	DEG product	5.1	123	38	2.552	-	0
H10	Water vapor	13.5	61	40	4.184	2397.432 (at 61 °C)	1
H11	Glycol vapor	1.2	130	43	2.741	1125.496 (at 130 °C)	1
H12	Glycol vapor	5.1	92	60	2.678	962.32 (at 92 °C)	1
H13	Glycol vapor	10.9	147	120.5	2.72	627.6 (at 147 °C)	1
H14	Glycol vapor	2.2	152	139	2.761	589.944 (at 152 °C)	1

MEG, monoethylene glycol; DEG, diethylene glycol.

Table 2. Cold process stream data.

Stream	Description	Flowrate (t/h)	Inlet Temperature (°C)	Outlet Temperature (°C)	Heat Capacity (J/g/°C)	Latent Heat (J/g)	Phase Change Fraction
C1	Inlet recycle gas	981	45.5	192.5	1.979	-	0
C2	Recycled water	1189.2	45.8	102.4	4.118	-	0
C3	EO solution	400	44.8	146.3	4.079	-	0
C4	Recycled water	35.44	32	100	4.226	-	0
C5	Process condensate	30	30.7	90.1	4.226	-	0
C6	Process condensate	12	40	60	4.226	-	0
C7	EG solution	40.03	186.4	186.4	-	1966.48 (at 186.4 °C)	1
C8	Glycol mixture	25.03	134	134	-	874.456 (at 134 °C)	1
C9	Glycol mixture	341.2	163.3	170	2.954	782.408 (at 170 °C)	0.2
C10	Glycol mixture	151.02	171.5	179.2	2.889	-	0
C11	Glycol mixture	283.221	167.5	172.1	2.933	-	0
C12	Glycol mixture	53.64	161.8	163.9	2.964	-	0

EO, ethylene oxide.

It should be noted that the HEN model presented in Section 3 does not consider multiple hot or cold utilities. For two levels of steam, this model is solved to determine the minimum total steam demand first. The loads on the two steam levels are then determined according to the outlet temperatures of cold streams that require steam.

4.1. Preliminary Analysis

Currently, the EG process has a HEN with nine matches between H1–H9 and C1–C6, as shown in Table 3, where the heat loads were determined by solving the HEN model considering only the existing matches. More details (e.g., intermediate temperatures of the process streams) are given in Table S1. These heat loads and stream temperatures agree with the operating conditions in the plant, thus validating the proposed model. The condensers

(H10–H14), evaporator (C7) and reboilers (C8-C12) were not included for heat integration. With the current HEN, the process requires 40,567 kW of 213 °C steam, 14,920 kW of 180 °C steam and 56,325 kW of cooling water. This is taken as the base case.

Table 3. Existing HEN of the EG process.

Stream	C1	C2	C3	C4	C5	C6	C7	C8	C9	C10	C11	C12	Cooling Water
H1	79,275	12,742											
H2		64,257											36,253
H3				2829									615
H4			22,118										
H5			10,607										
H6			4534										
H7													4873
H8					2092								810
H9						282							26
H10													9292
H11													448
H12													1493
H13													2127
H14													389
180 °C steam			8748					6080				93	
213 °C steam							21,866		16,707	933	1061		

Heat loads of the matches are given in kW.

To retrofit the existing HEN for improved heat recovery and reduced steam requirements, all possible matches are considered. Table 4 shows the minimum permissible temperature differences for matches between process and utility streams.

Table 4. Minimum temperature differences (°C) for different matches.

Stream	C1	C2	C3	C4	C5	C6	C7	C8	C9	C10	C11	C12	Cold Utility
H1	10	10	10	10	10	10	10	10	10	10	10	10	5
H2	10	3.39	10	10	10	10	10	10	10	10	10	10	2
H3	10	10	10	10	10	10	10	10	10	10	10	10	3
H4	10	10	7.2	10	10	10	10	10	10	10	10	10	5
H5	10	10	5	10	10	10	10	10	10	10	10	10	5
H6	10	10	10	10	10	10	10	10	10	10	10	10	5
H7	10	10	10	10	10	10	10	10	10	10	10	10	4
H8	10	10	10	10	10	10	10	10	10	10	10	10	15
H9	10	10	10	10	10	5	10	10	10	10	10	10	6
H10	10	10	10	10	10	10	10	10	10	10	10	10	10
H11	10	10	10	10	10	10	10	10	10	10	10	10	10
H12	10	10	10	10	10	10	10	10	10	10	10	10	27
H13	10	10	10	10	10	10	10	10	10	10	10	10	89
H14	10	10	10	10	10	10	10	10	10	10	10	10	106
Hot utility	10	10	33.7	10	10	10	21	42	33	28	17	15	-

For process–process matches, a minimum temperature difference of 10 °C is assumed. However, a smaller value is used if the existing match has a smaller temperature difference. For such matches (H2–C2, H4–C3, H5–C3 and H9–C6), the minimum temperature difference is set to the smaller of the temperature differences at the cold and hot ends of the exchanger. For process–utility matches, the minimum temperature differences are set according to the current operating conditions or based on practical considerations [26]. To avoid a complex HEN retrofit, the following assumptions are made:

- No stream splitting is allowed;
- Any pair of hot and cold process streams has at most one match throughout the superstructure;
- The condenser, evaporator and reboiler streams each have at most one match with other process streams.

These constraints are given in Equations (31)–(35).

$$\sum_{j\in \mathbf{J}} z_{ijk} \leq 1 \quad \forall i \in \mathbf{I}, k \in \mathbf{ST} \tag{31}$$

$$\sum_{i\in \mathbf{I}} z_{ijk} \leq 1 \quad \forall j \in \mathbf{J}, k \in \mathbf{ST} \tag{32}$$

$$\sum_{k\in \mathbf{ST}} z_{ijk} \leq 1 \quad \forall i \in \mathbf{I}, j \in \mathbf{J} \tag{33}$$

$$\sum_{j\in \mathbf{J}} \sum_{k\in \mathbf{ST}} z_{ijk} \leq 1 \quad \forall i \in \{\text{H10–H14}\} \tag{34}$$

$$\sum_{i\in \mathbf{I}} \sum_{k\in \mathbf{ST}} z_{ijk} \leq 1 \quad \forall j \in \{\text{C7–C12}\} \tag{35}$$

Solving the HEN model considering all possible matches in seven stages yields the results in Table 5. More details of this solution are given in Table S2. The minimum utility requirements are determined to be 13,710 kW of 213 °C steam, 26,302 kW of 180 °C steam and 40,849 kW of cooling water. This can be achieved in the minimum number of units of 33 (20 exchangers, five heaters and eight coolers). The main feature of this optimized HEN is that the outlet recycle gas (H1) is used to heat the evaporator (C7) and reboiler streams C8 and C9. In addition, condenser streams H11-H14 are used to heat the inlet recycle gas (C1) and recycled water (C2). Compared with the base case, the total steam demand decreases by 28% (despite a 76% increase in the load of 180 °C steam), while the cooling water demand decreases by 27%.

Table 5. Optimized HEN of the EG process.

Stream	C1	C2	C3	C4	C5	C6	C7	C8	C9	C10	C11	C12	Cooling Water
H1	46,223		12,374	482			10,151	6080	16,707				
H2		75,135											25,374
H3				2346	98								999
H4			22,118										
H5			10,607										
H6	4508	27											
H7													4873
H8			908		1994								
H9						232							26
H10													9292
H11		344											103
H12		1493											
H13	1956												171
H14	379												10
180 °C steam	26,209											93	
213 °C steam							11,715			933	1061		

Heat loads of the matches are given in kW.

Figure 2 shows the variation of minimum utility requirements with the number of stages in the superstructure. Increasing the number of stages does not result in significant reductions in steam and cooling water demands. The optimized HENs have the common feature that H1 is used to heat C7–C9. Therefore, the design with the fewest (33) heat exchange units (i.e., the one in Table 5) was selected for further evaluation. To achieve the theoretical reduction in total steam demand of 15,476 kW, the HEN retrofit entails 14 new matches and two expanded matches. Making all these changes to the current HEN might be prohibitively expensive. Moreover, it was found that several of the new matches (H1–C7, H1–C8, H1–C9, H6–C1, H13–C1 and H14–C1) could cause EO penetration, which will adversely affect the MEG product. Such matches are deemed risky and should be forbidden, thus rendering the solution infeasible. It was then suggested that the condensers, evaporator and reboilers should be excluded from the heat integration, despite the large potential steam saving.

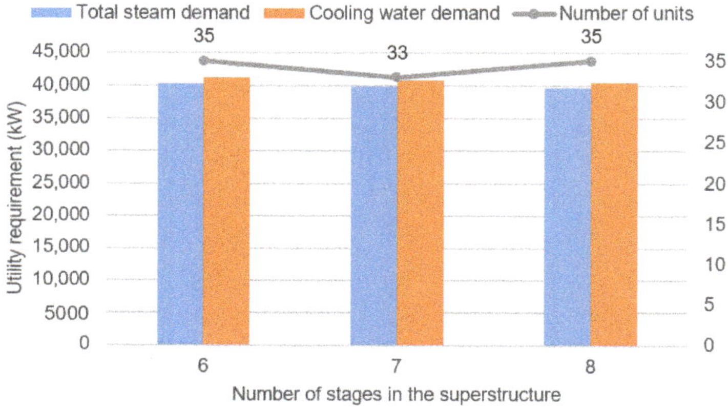

Figure 2. Minimum utility requirements versus the number of stages.

4.2. Analysis without Condensers, Evaporator and Reboilers

Having excluded H10–H14 and C7–C12 from the analysis, the process stream data are updated as in Table 6. It was found that the heat capacity of the recycle gas varies significantly for large temperature changes and cannot be taken as a constant, as shown in Figure 3. The variation in heat capacity of recycle gas streams ($Cp_{\dagger}$) with temperature (T) can be correlated as:

$$Cp_{\dagger} = a_{0,\dagger} + a_{1,\dagger}T \quad \dagger \in \{\mathrm{H1}, \mathrm{C1}\} \tag{36}$$

where $a_{0,\dagger}$ and $a_{1,\dagger}$ are constants determined by correlating the heat capacity data: $a_{0,\mathrm{H1}}$ = 0.396767 and $a_{1,\mathrm{H1}}$ = 0.000578 for 60 °C $\leq T \leq$ 222 °C (R^2 = 0.999); $a_{0,\mathrm{C1}}$ = 0.399877 and $a_{1,\mathrm{C1}}$ = 0.000555 for 45 °C $\leq T \leq$ 193 °C (R^2 = 0.997). Thus,

$$\Delta H_{\dagger} = \int_{T_1}^{T_2} Cp_{\dagger} dT = \left[a_{0,\dagger}T + \frac{a_{1,\dagger}T^2}{2}\right]_{T_1}^{T_2} \quad \dagger \in \{\mathrm{H1}, \mathrm{C1}\} \tag{37}$$

where $\Delta H_{\dagger}$ is the enthalpy change from T_1 to T_2.

Equations (1)–(4), (13) and (14) are then modified for H1 and C1 by replacing "$Cp\Delta T$" with "ΔH," which involves nonlinear terms in Equations (3), (4), (13) and (14). Thus, the HEN model becomes an MINLP, which is solved using solver BARON in GAMS. Some of the process stream data have also been revised in response to the change in heat capacity of the recycle gas.

Figure 4 shows the revised base-case conditions. The details were determined by solving the HEN model considering only the existing matches. The resulting heat loads and stream temperatures agree with the updated operating conditions in the plant. Excluding the utility requirements of H10–H14 and C7–C12, the process requires 9277 kW of 180 °C steam and 46,089 kW of cooling water. For improved heat recovery and reduced steam demand, the existing HEN is optimized considering all feasible matches. Forbidden matches such as H6-C1, H8-C1 and H9-C1 can be included using Equation (38).

$$z_{ijk} \leq M_{ij} \quad \forall i \in \mathbf{I}, j \in \mathbf{J}, k \in \mathbf{ST} \tag{38}$$

where M_{ij} is a binary parameter denoting allowable (M_{ij} = 1) and forbidden matches (M_{ij} = 0). To avoid a complex HEN retrofit, the following assumptions hold:

- No stream splitting is allowed (Equations (31) and (32));
- Any pair of hot and cold process streams has at most one match (Equation (33)).

Table 6. Updated process stream data.

Stream	Description	Flowrate (t/h)	Inlet Temperature (°C)	Outlet Temperature (°C)	Heat Capacity (J/g/°C)	Latent Heat (J/g)	Phase Change Fraction
H1	Outlet recycle gas	981	222	60	$f_{H1}(T)$	-	0
H2	Recycled water	1150	107.4	31	4.393	-	0
H3	Recycled water	37.1	116.14	35	4.393	-	0
H4	Process condensate	373.7	106.3	52	4.226	-	0
H5	Process vapor	18.74	124	124	-	2194.508 (at 124 °C)	1
H6	Process vapor	8.4	171	139	1.925	2092 (at 171 °C)	1
H7	Absorbent	354.1	45	33.1	4.163	-	0
H8	MEG product	48.5	138.2	60	2.761	-	0
H9	DEG product	5.1	123	38	2.552	-	0
C1	Inlet recycle gas	981	45.5	192.7	$f_{C1}(T)$	-	0
C2	Recycled water	1189.2	45.8	102.38	4.155	-	0
C3	EO solution	400	44.8	146.3	4.393	-	0
C4	Recycled water	37.62	32	100	4.226	-	0
C5	Process condensate	30	30.7	90.1	4.226	-	0
C6	Process condensate	12	40	60	4.226	-	0

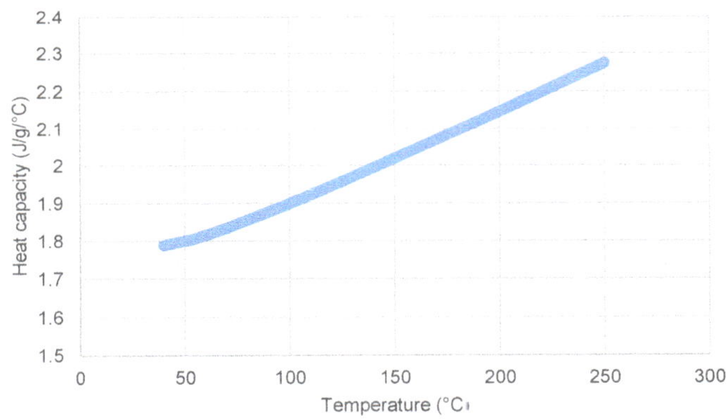

Figure 3. Heat capacity of the recycle gas.

In addition, to make use of the existing matches and keep down the number of network modifications, Equation (39) is used.

$$\sum_{k \in \mathbf{ST}} z_{ijk} = 1 \quad \forall (i,j) \in \{(\text{H1},\text{C1}),(\text{H1},\text{C2}),(\text{H2},\text{C2}),(\text{H3},\text{C4}),(\text{H4},\text{C3}),(\text{H5},\text{C3}),(\text{H6},\text{C3}),(\text{H8},\text{C5}),(\text{H9},\text{C6})\} \tag{39}$$

This constraint enables the existing matches to be preferred when determining the minimum number of units, if the same steam saving can be achieved.

Solving the HEN model considering all feasible matches in six stages yields the results in Figure 5. The minimum utility requirements are determined to be 0 kW of 180 °C steam and 36,812 kW of cooling water. This can be achieved in the minimum number of units of 15 (10 exchangers, no heater and five coolers). The main feature of this optimized HEN is that instead of using steam, the outlet recycle gas (H1) is used to heat the EO solution (C3). Compared with the base case in Figure 4, the steam demand decreases by 100%, while the cooling water demand decreases by 20%. The cost benefit associated with this steam saving is calculated to be 7.5 million USD/y, assuming steam costs 0.101 USD/kWh and the plant operates for 8000 h/y [26].

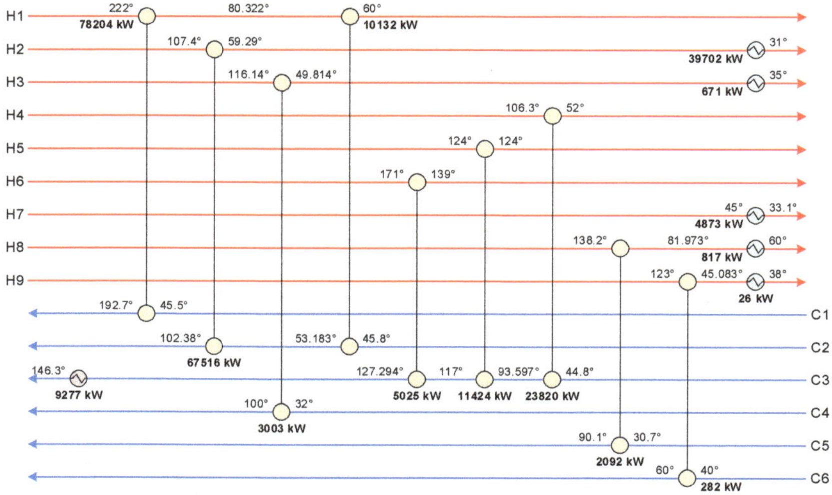

Figure 4. Existing HEN of the EG process (updated).

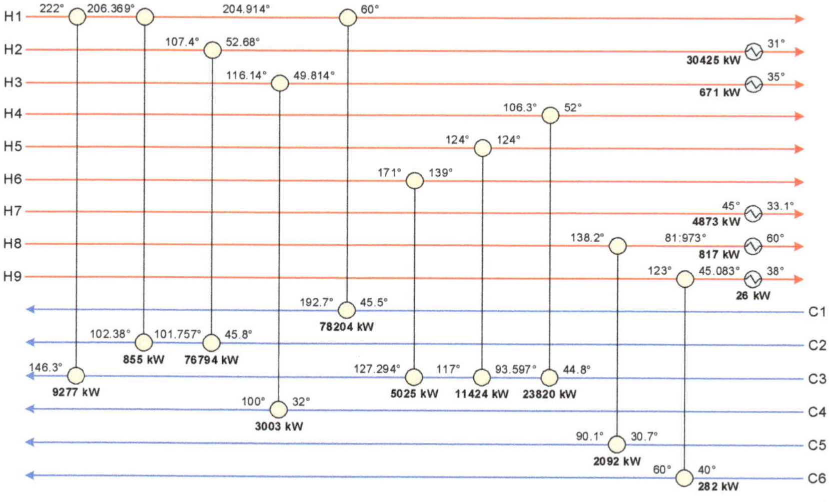

Figure 5. Optimized HEN of the EG process (without condensers, evaporator and reboilers).

This solution was further evaluated for its feasibility. To achieve the theoretical reduction in steam demand of 9277 kW, the HEN retrofit entails adding a new match (H1–C3) and modifying two existing ones for decreased temperature differences (H1–C1) and increased heat transfer (H2–C2). The total retrofit cost is estimated at 24.2 million USD by the process engineers. Dividing the total retrofit cost by the annual cost saving gives a payback period of 3.2 y. On the other hand, the decrease in the logarithmic mean temperature difference of match H1–C1 (from 32 to 13.3 °C) requires a three-times larger heat transfer area for the same heat duty. Although the design in Figure 5 does not pose a hazard to the EG process, the plant has no space to accommodate the required area increase. Therefore, such a retrofit project cannot be implemented. The process engineers then suggested seeking an alternative solution that does not require additional heat transfer area for match H1–C1.

To avoid additional heat transfer area required for match H1–C1, the minimum temperature difference between H1 and C1 is set to 29.3 °C (instead of 10 °C) according to the base-case conditions. Additionally, it is desirable to have no more than three hot process streams matched with C1, if its heating demand cannot be satisfied solely by H1. This constraint is given in Equation (40).

$$\sum_{i \in \mathbf{I}} \sum_{k \in \mathrm{ST}} z_{i,\mathrm{C1},k} \leq 3 \tag{40}$$

Solving the HEN model with the revised $\Delta T^{\min}_{\mathrm{H1,C1}}$ and the constraint on the number of matches of C1 gives the results in Figure 6. The minimum utility requirements are determined to be 829 kW of 180 °C steam and 37,641 kW of cooling water. This can be achieved in the minimum number of units of 18 (11 exchangers, one heater and six coolers). In this HEN design, H1 is used to heat C3 instead of using H5, which is used to heat C1 together with H4 and H1. Due to practical constraints, the steam saving decreases from 100 to 91%, while the cooling water demand increases by 2% (from 36,812 to 37,641 kW). It should be noted that match H5–C1 is a gas–gas match, for which the current minimum temperature difference of 10 °C may be too small and may lead to a large heat transfer area. Therefore, an additional case is considered where $\Delta T^{\min}_{\mathrm{H5,C1}}$ is revised upwards to 20 °C to avoid a too large area requirement. Solving the HEN model with the revised $\Delta T^{\min}_{\mathrm{H5,C1}}$ gives the results in Figure S1. With the same HEN structure as in Figure 6 but changes in some heat loads and temperatures, the steam saving decreases from 91 to 50%, while the cooling water demand increases from 37,641 to 41,441 kW. The results comparison indicates the trade-off between energy and capital. So far, both solutions in Figure 6 and Figure S1 have been taken as alternatives being evaluated by the process engineers for technical and economic feasibility. Although the complete economic assessments are not available yet, the payback period is likely to be shorter than 3.2 y because modifications for match H1-C1 are no longer needed. The capital cost of modifying match H1–C1 (including the equipment and auxiliaries) is 20.1 million USD, accounting for 83% of the total retrofit cost for the design in Figure 5.

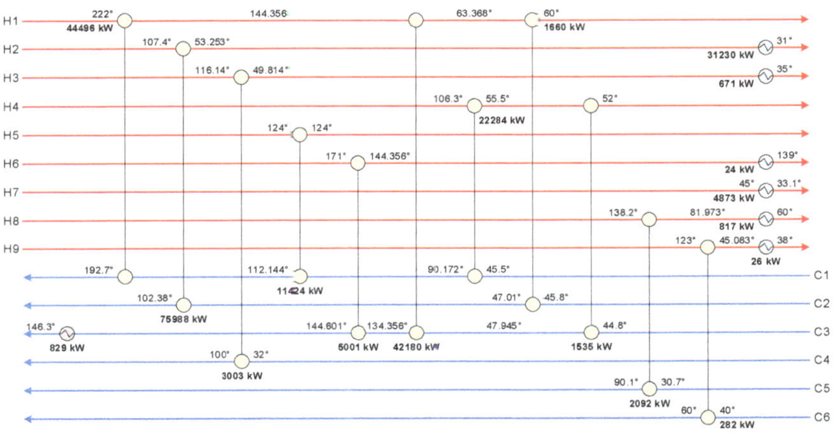

Figure 6. Alternative HEN of the EG process.

5. Conclusions

A superstructure-based mathematical optimization model for the synthesis of HENs has been presented in this paper. This HEN model considers three types of process streams involving temperature and phase changes, and has been applied to an industrial case study of optimizing the heat recovery system of an EG process. The results obtained show a compromise between steam savings and process feasibility, and how the proposed model can be modified to reflect practical considerations such as stream matching constraints and variable heat capacities as a function of temperature. Although in the preliminary analysis a substantial steam saving of 15,476 kW (28%) can be achieved, which is equivalent to a cost saving of 12.5 million USD/y, the solution involves forbidden matches that pose a hazard to the MEG product and is not going to be implemented. In the analysis without the condensers, evaporator and reboilers, despite the space limitation in the plant, the achievable steam saving of up to 8448 kW (91%) is still considerable and the solutions are currently being evaluated in more detail. The corresponding cost saving could double if carbon taxes are considered. Future work will extend the optimization approach to other EG plants on the same site for improved heat recovery. The focus will also be on the optimization of utility systems, particularly those with time-dependent supply and demand, for which multiperiod planning is indicated.

Supplementary Materials: The following supporting information can be downloaded at: https://www.mdpi.com/article/10.3390/pr11020321/s1, Figure S1: alternative HEN of the EG process ($\Delta T^{min}_{H5,C1} = 20$ °C); Table S1: stream matches and temperatures of the existing HEN; Table S2: stream matches and temperatures of the optimized HEN (preliminary analysis).

Author Contributions: Conceptualization, J.-Y.L.; methodology, J.-Y.L.; software, J.-Y.L. and P.-Y.C.; validation, J.-Y.L. and P.-Y.C.; formal analysis, J.-Y.L.; investigation, J.-Y.L.; resources, J.-Y.L.; data curation, J.-Y.L.; writing—original draft preparation, J.-Y.L.; writing—review and editing, J.-Y.L. and P.-Y.C.; visualization, J.-Y.L. and P.-Y.C.; supervision, J.-Y.L.; project administration, J.-Y.L.; funding acquisition, J.-Y.L. All authors have read and agreed to the published version of the manuscript.

Funding: This research was funded by the National Science and Technology Council (NSTC) of R.O.C., grant number 111-2221-E-027-003.

Institutional Review Board Statement: Not applicable.

Informed Consent Statement: Not applicable.

Data Availability Statement: The data presented in this study are available on request from the corresponding author.

Acknowledgments: The authors thank Nan Ya Plastics Corporation for financial support from an industry–academia collaboration project (grant number 211A058) with National Taipei University of Technology.

Conflicts of Interest: The authors declare no conflict of interest. The funders had no role in the design of the study; in the collection, analyses, or interpretation of data; in the writing of the manuscript; or in the decision to publish the results.

Appendix A. Nomenclature

Notation used in the HEN model is given below.

Appendix A.1. Indices and Sets

$i \in \mathbf{I}$	process hot streams
$i \in \mathbf{I}^1$	process hot streams involving only sensible heat
$i \in \mathbf{I}^2$	process hot streams involving only latent heat
$i \in \mathbf{I}^3$	process hot streams involving both sensible and latent heat
$j \in \mathbf{J}$	process cold streams
$j \in \mathbf{J}^1$	process cold streams involving only sensible heat
$j \in \mathbf{J}^2$	process cold streams involving only latent heat
$j \in \mathbf{J}^3$	process cold streams involving both sensible and latent heat
$k \in \mathbf{K}$	temperature locations
$k \in \mathbf{ST}$	stages

Appendix A.2. Parameters

Cp_i	heat capacity of hot process stream i (J/g/°C)
Cp_j	heat capacity of cold process stream j (J/g/°C)
F_i	flowrate of hot process stream i (t/h)
F_j	flowrate of cold process stream j (t/h)
Q_i^U	upper limit to the heat transfer from hot process stream i (kW)
Q_{ij}^U	upper limit to the heat transfer from hot process stream i to cold process stream
Q_j^U	upper limit to the heat transfer to cold process stream j (kW)
T_{cu}^{out}	outlet temperature of the cold utility (°C)
T_{hu}^{out}	outlet temperature of the hot utility (°C)
T_i^{in}	inlet temperature of hot process stream i (°C)
T_i^{out}	outlet temperature of hot process stream i (°C)
T_j^{in}	inlet temperature of cold process stream j (°C)
T_j^{out}	outlet temperature of cold process stream j (°C)
α_i	phase change fraction of hot process stream i
α_j	phase change fraction of cold process stream j
$\Delta T_{hu,j}^{min}$	minimum temperature difference for hot utility-process matches (°C)
$\Delta T_{i,cu}^{min}$	minimum temperature difference for process-cold utility matches (°C)
ΔT_{ij}^{min}	minimum temperature difference for process-process matches (°C)
ε	a sufficiently small value (°C)
λ_i	latent heat of hot process stream i (J/g)
λ_j	latent heat of cold process stream j (J/g)
Γ	a sufficiently large value (°C)

Appendix A.3. Variables

q_i^{cu}	cold utility requirement for hot process stream i (kW)
$q_i^{\Lambda,cu}$	condensation heat load of hot process stream i in the cooler (kW)
q_{ijk}	heat transfer from hot process stream i to cold process stream j in stage k (kW)
q_{ik}^{Λ}	condensation heat load of hot process stream i in stage k (kW)
$q_j^{\Lambda,hu}$	evaporation heat load of cold process stream j in the heater (kW)
q_j^{hu}	hot utility requirement for cold process stream j (kW)
q_{jk}^{Λ}	evaporation heat load of cold process stream j in stage k (kW)
t_{jk}	temperature of cold process stream j at temperature location k (°C)
y_{ik}	binary indicating if the inlet temperature of hot process stream i at stage k is greater than or equal to the condensation temperature
y_{jk}	binary indicating if the outlet temperature of cold process stream j at stage k is greater than or equal to the evaporation temperature
z_i^{cu}	binary indicating if hot process stream i uses the cold utility
z_{ijk}	binary indicating if hot process stream i transfers heat to cold process stream j in stage k
z_j^{hu}	binary indicating if cold process stream j uses the hot utility

References

1. International Energy Agency (IEA). *Net Zero by 2050: A Roadmap for the Global Energy Sector*; IEA: Paris, France, 2021.
2. Klemeš, J.J.; Kravanja, Z. Forty years of Heat Integration: Pinch Analysis (PA) and Mathematical Programming (MP). *Curr. Opin. Chem. Eng.* **2013**, *2*, 461–474. [CrossRef]
3. Kemp, I.C. *Pinch Analysis and Process Integration: A User Guide on Process Integration for the Efficient Use of Energy*, 2nd ed.; Butterworth-Heinemann: Oxford, UK, 2007.
4. Yee, T.F.; Grossmann, I.E. Simultaneous optimization models for heat integration—II. Heat exchanger network synthesis. *Comput. Chem. Eng.* **1990**, *14*, 1165–1184. [CrossRef]
5. Huang, K.F.; Karimi, I.A. Simultaneous synthesis approaches for cost-effective heat exchanger networks. *Chem. Eng. Sci.* **2013**, *98*, 231–245. [CrossRef]
6. Beck, A.; Hofmann, R. A novel approach for linearization of a MINLP stage-wise superstructure formulation. *Comput. Chem. Eng.* **2018**, *112*, 17–26. [CrossRef]
7. Ziyatdinov, N.N.; Emel'yanov, I.I.; Chen, Q.; Grossmann, I.E. Optimal heat exchanger network synthesis by sequential splitting of process streams. *Comput. Chem. Eng.* **2020**, *142*, 107042. [CrossRef]
8. Liu, Z.; Yang, L.; Yang, S.; Qian, Y. An extended stage-wise superstructure for heat exchanger network synthesis with intermediate placement of multiple utilities. *Energy* **2022**, *248*, 123372. [CrossRef]
9. Čuček, L.; Boldyryev, S.; Klemeš, J.J.; Kravanja, Z.; Krajačić, G.; Varbanov, P.S.; Duić, N. Approaches for retrofitting heat exchanger networks within processes and Total Sites. *J. Clean. Prod.* **2019**, *211*, 884–894. [CrossRef]
10. Pavão, L.V.; Costa, C.B.B.; Ravagnani, M.A.S.S. Heat exchanger networks retrofit with an extended superstructure model and a meta-heuristic solution approach. *Comput. Chem. Eng.* **2019**, *125*, 380–399. [CrossRef]
11. Cheng, S.-L.; Chang, C.-T.; Jiang, D. A game-theory based optimization strategy to configure inter-plant heat integration schemes. *Chem. Eng. Sci.* **2014**, *118*, 60–73. [CrossRef]

12. Nair, S.K.; Guo, Y.; Mukherjee, U.; Karimi, I.A.; Elkamel, A. Shared and practical approach to conserve utilities in eco-industrial parks. *Comput. Chem. Eng.* **2016**, *93*, 221–233. [CrossRef]
13. Kachacha, C.; Farhat, A.; Zoughaib, A.; Tran, C.T. Site wide heat integration in eco-industrial parks considering variable operating conditions. *Comput. Chem. Eng.* **2019**, *126*, 304–320. [CrossRef]
14. Pavão, L.V.; Miranda, C.B.; Costa, C.B.B.; Ravagnani, M.A.S.S. Efficient multiperiod heat exchanger network synthesis using a meta-heuristic approach. *Energy* **2018**, *142*, 356–372. [CrossRef]
15. Hofmann, R.; Panuschka, S.; Beck, A. A simultaneous optimization approach for efficiency measures regarding design and operation of industrial energy systems. *Comput. Chem. Eng.* **2019**, *128*, 246–260. [CrossRef]
16. Ibrić, N.; Ahmetović, E.; Kravanja, Z. Simultaneous optimization of water and energy within integrated water networks. *Appl. Therm. Eng.* **2014**, *70*, 1097–1122. [CrossRef]
17. Ibrić, N.; Ahmetović, E.; Kravanja, Z.; Maréchal, F.; Kermani, M. Simultaneous synthesis of non-isothermal water networks integrated with process streams. *Energy* **2017**, *141*, 2587–2612. [CrossRef]
18. Huang, K.; Karimi, I.A. Work-heat exchanger network synthesis (WHENS). *Energy* **2016**, *113*, 1006–1017. [CrossRef]
19. Li, J.; Demirel, S.E.; Hasan, M.M.F. Building Block-Based Synthesis and Intensification of Work-Heat Exchanger Networks (WHENS). *Processes* **2019**, *7*, 23. [CrossRef]
20. Vikse, M.; Watson, H.A.J.; Barton, P.I.; Gundersen, T. Nonsmooth formulation for handling unclassified process streams in the optimization of work and heat exchange networks. *Ind. Eng. Chem. Res.* **2019**, *58*, 9526–9539. [CrossRef]
21. Yang, R.; Zhuang, Y.; Zhang, L.; Du, J.; Shen, S. A thermo-economic multi-objective optimization model for simultaneous synthesis of heat exchanger networks including compressors. *Chem. Eng. Res. Des.* **2020**, *153*, 120–135. [CrossRef]
22. Hamedi, H.; Karimi, I.A.; Gundersen, T. Simulation-based approach for integrating work within heat exchange networks for sub-ambient processes. *Energy Convers. Manag.* **2020**, *203*, 112276. [CrossRef]
23. Li, M.; Zhuang, Y.; Li, W.; Dong, Y.; Zhang, L.; Du, J.; Shen, S. A surrogate-based optimization framework for simultaneous synthesis of chemical process and heat exchanger network. *Chem. Eng. Res. Des.* **2021**, *170*, 180–188. [CrossRef]
24. Ponce-Ortega, J.M.; Jiménez-Gutiérrez, A.; Grossmann, I.E. Optimal synthesis of heat exchanger networks involving isothermal process streams. *Comput. Chem. Eng.* **2008**, *32*, 1918–1942. [CrossRef]
25. Watanapanich, S.; Li, S.-T.; Lee, J.-Y. Optimal integration of organic Rankine cycles into process heat exchanger networks: A simultaneous approach. *Energy Convers. Manag.* **2022**, *260*, 115604. [CrossRef]
26. Syu, J.-R.; R & D Center, Nan Ya Plastics Corporation, New Taipei City, Taiwan; Lee, Y.-C.; R & D Center, Nan Ya Plastics Corporation, New Taipei City, Taiwan. Personal communication, 2022.

Disclaimer/Publisher's Note: The statements, opinions and data contained in all publications are solely those of the individual author(s) and contributor(s) and not of MDPI and/or the editor(s). MDPI and/or the editor(s) disclaim responsibility for any injury to people or property resulting from any ideas, methods, instructions or products referred to in the content.

processes

MDPI

Article

Data-Driven Conjugate Heat Transfer Analysis of a Gas Turbine Vane

Hao Cui, Lang Wang, Xueying Li * and Jing Ren

Department of Energy and Power Engineering, Tsinghua University, Beijing 100084, China
* Correspondence: li_xy@tsinghua.edu.cn

Abstract: Cooling structures of gas turbine blades have become more complex to achieve a better cooling effect. Therefore, heat transfer analysis tools with higher accuracy and efficiency are required to verify the effectiveness of cooling designs and continuously improve the design. In this work, a data-driven method is combined with a decoupled conjugate heat transfer analysis. The analysis object is a typical air-cooled gas turbine first-stage vane with film cooling, impingement cooling, and pin-fin cooling. In addition, a conventional 3-D conjugate heat transfer simulation of the vane was executed for contrast. Results show that this method shortens the time of the heat transfer analysis process significantly and ensures accuracy. It proves that the data-driven method is effective for the evaluation of a modern gas turbine cooling design and is an improvement compared to the traditional three-dimensional heat transfer analysis method.

Keywords: gas turbine; conjugate heat transfer; film cooling; data-driven; CFD

Citation: Cui, H.; Wang, L.; Li, X.; Ren, J. Data-Driven Conjugate Heat Transfer Analysis of a Gas Turbine Vane. *Processes* **2022**, *10*, 2335. https://doi.org/10.3390/pr10112335

Academic Editors: Pei Liu, Ming Liu and Xiao Wu

Received: 21 October 2022
Accepted: 7 November 2022
Published: 9 November 2022

Publisher's Note: MDPI stays neutral with regard to jurisdictional claims in published maps and institutional affiliations.

Copyright: © 2022 by the authors. Licensee MDPI, Basel, Switzerland. This article is an open access article distributed under the terms and conditions of the Creative Commons Attribution (CC BY) license (https://creativecommons.org/licenses/by/4.0/).

1. Introduction

For higher thermal efficiency and larger specific work, gas turbines' inlet temperature rises rapidly and exceeds the allowable temperature of even the best superalloys. The turbine blades of the first several stages must be provided with effective thermal protection to avoid failures. For example, first-stage vanes of a modern gas turbine usually apply the compound scheme with internal impingement cooling, pin-fin cooling, and external film cooling. A complex geometry results in a complex flow, which increases the difficulty of predicting the temperature distribution of gas turbine blades [1,2].

The temperature distribution of the blades is an important criterion for a cooling scheme's validity during the designing period of a turbine. The peak temperature should not exceed the metal temperature tolerance limit, and the temperature gradient should not be too large to avoid excessive thermal stress on the surface. The solution of the temperature distribution can be abstracted as a heat conduction problem of an infinite one-dimensional plate under the third kind of boundary conditions, convection heat transfer with mainstream hot gas on the outer surface and coolant on the inner surface. The final time-average temperature distribution relies on both conductivity and thermal boundary conditions [3,4]. This is due to the temperature of the blades influencing the flow of the surrounding fluids and may cause the thermal boundary condition to change. This bidirectional influence also makes the prediction of the temperature field more difficult.

A conventional method to obtain the detailed temperature distribution is the conjugate heat transfer calculation. The Navier-Stokes equation is solved in the fluid region, and the Fourier equation is solved in the solid regions. The coupling of two regions is achieved by a common wall temperature and the same heat flux through the intersurface. Bohn et al. [5] developed a conjugate calculation code and used it to calculate the temperature distribution of the convection-cooled nozzle guide vane Mark II. The heat transfer boundary conditions on the air-cooled surface are prescribed with the experimental data; the temperature on the outer surface is fully determined by the conjugate calculation.

Finally, the maximum difference in temperature on the outer surface is less than 2%. The conjugate calculation of a film-cooled vane with the same code was also carried out [6]. The results from different film flow angles provided a valuable reference for a film-cooled vanes' design. Takahashi et al. simulated a real turbine vane with a typical 3-D structure using a commercial conjugate code. The results proved that the conjugate method is useful for the prediction of blade temperature depending on various conditions of operation. Determining the flow and heat transfer in one code is quite convenient. However, the more complicated cooling structure makes the 3-D conjugate heat transfer analysis time-consuming, which is not quite suitable for the optimization design that requires a quick evaluation for a large number of design schemes. It is necessary to develop a relatively accurate cooling-scheme evaluation method that saves time and resources.

In order to make the calculation flexible enough for different accuracy and efficiency demands, a decoupled method was developed, which allows applying different analysis methods to solve different domains. Bonini and Andreini [7,8] developed a cooling system simulation tool combining external CFD, internal flow network code, and a finite element conductive model and applied it to the first rotor blade of the MS5002E gas turbine. Temperature profiles from the simulation tool and experiment matched well. Chowdhury et al. [9] developed a model to calculate the distribution of the coolant mass flow rate and the metal temperatures of a turbine blade, using the mass and energy balance equations at given external and internal boundary conditions. With the application of predicting the temperature distribution of the GE E3 stage 1 blade, this model proved, with reasonable accuracy, if the correct boundary conditions were applied. Ngetich et al. [10] also used the decoupled method to analyze a double-walled turbine blade. The blade was divided into basic units connecting with each other whose internal and external boundary conditions were assigned by correlations. The results from the decoupled method, which include film, metal effectiveness, and coolant mass flow consumption, have been found to closely match results available from the open literature. These decoupled analyses validated the results of each region to timely obtain convinced final results, although the outer iteration between regions is complex [11,12].

In the decoupled analysis method, the models applied in the domains are the key to the final accuracy. The development of the experimental measurement technique [13–15] offers a great opportunity for data-driven models based on two-dimensional experimental data, which enables reliable and efficient predictions. This work developed a data-driven method to predict the temperature of gas turbine vanes based on the decoupled heat transfer analysis. The decoupled method is mainly employed for its ability to customize intersurface conditions. The structure of the code allows the flow and heat transfer to interact with each other in a timely manner. The validation of the method was carried out on a typical air-cooled first-stage guide vane. The accuracy and efficiency were assessed by comparing the data-driven method with the fully conjugate calculation andthe experimental data.

2. Data-Driven CHT Method

This section will introduce the data-driven method in detail. Section 2.1. contains the core strategy of the data-driven method concerning the division of regions and the interactiveness between the regions. Section 2.2 introduces the experimental data source and the processing operations. Finally, Section 2.3. describes the integration of the database with the CHT algorithm.

2.1. General Simulation Strategy

The strategy of the data-driven method mainly includes the division method of the regions and the iterative structure of the code.

The outer interface of fluid and solid is the key point in heat transfer analysis. Two sets of domains separated by the outer interface of the vane were chosen for the following analysis. Domain 1 is the passage of hot gas, which is used for the mainstream flow calculation. Domain 2 contains the solid region of the vane and the fluid region inside the vane where coolant flows. The calculation of domain 1 is able to offer mainstream information, which is necessary for heat transfer with film cooling. The other information is obtained from the calculation of Domain 2 The position relationship of the two domains is shown in Figure 1.

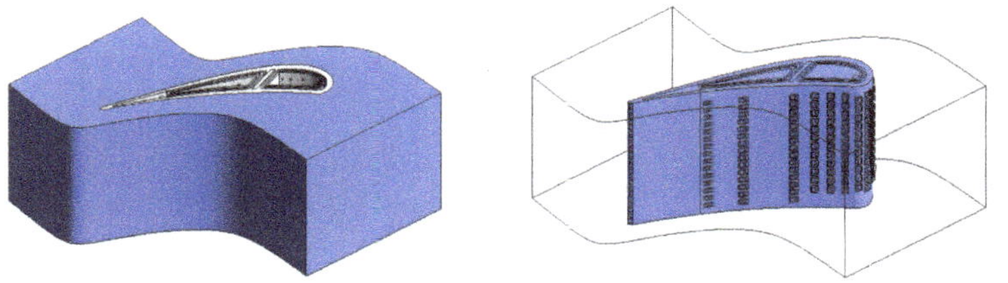

Figure 1. Domain division: (**a**) Domain 1; (**b**) Domain 2.

The interaction between the domains follows the rules in Figure 2. Three parts are for different regions, roughly. In part 1, a simulation of mainstream passage without film cooling on the vane's surface is carried out. Key parameters of the flow and heat transfer near the vane's surface were exported for the next parts' calculations. For the vane and the coolant, a simulation of Domain 2 is carried out to obtain the initial internal flow parameters of the vane. A hypothetical external thermal boundary condition is used in the first simulation of Domain 2 because the actual condition is unknown yet. In part 2, the external thermal boundary condition that consists of the heat transfer coefficient (HTC) and the heat transfer driving temperature (T_{aw}) were generated from the results of the first part. HTC h is a product of HTC without film cooling (h_0) and the HTC enhancement factor (h/h_0). Equations (1) and (2) were applied for h_0's calculation at the leading edge [16,17],

$$\frac{Nu_0}{Re^{0.5}} = 1.01 + 2.624\frac{Tu(Re)^{0.5}}{100} - 3.07\left(\frac{Tu(Re)^{0.5}}{100}\right)^2, \tag{1}$$

$$Nu(\theta) = Nu_0\left(1 - \left(\frac{\theta}{90}\right)^3\right), -80^\circ < \theta < 80^\circ, \tag{2}$$

and the flat board correlation, Equation (3), is for h_0 on the pressure surface and suction surface [17].

$$Nu = 0.0288Re^{4/5}Pr^{1/2} \tag{3}$$

h/h_0 is determined by Baldauf's method [18]. T_{aw} is calculated by Equation (4) and is assigned to be the heat transfer driving temperature,

$$T_{aw} = T_\infty - \eta(T_\infty - T_c). \tag{4}$$

Adiabatic film effectiveness η's distribution on the vane's surface is generated by the data-driven method using results from part 1, which will be introduced later in Section 2.2.

In part 3, with boundary conditions from part 2, another CFD simulation of Domain 2 is used to obtain new internal flow parameters These parameters usually differ from those in part 1 due to the different boundary assignments. Therefore, the previous parameters in part 1 were replaced by newly obtained ones, and then part 2 was executed again with new parameters. Such iterations go on until the average temperature variation is less than 0.01 K.

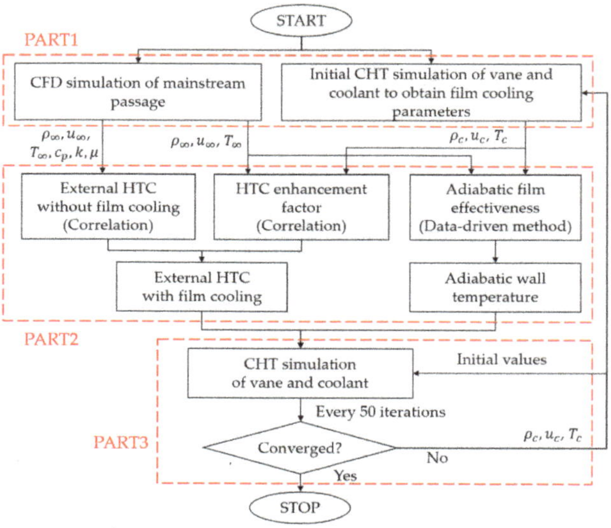

Figure 2. Iterative strategy of the decoupled CHT method.

The three parts construct a data-driven method for the heat transfer of the vane's outer surface. The parameter transfer and iterative calculations ensure the rationality of the final temperature distribution.

2.2. Experimental Database of Film Cooling Effectiveness

The data-driven method in this work considers the film effectiveness distribution on the vane's surface, a result of the joint influence of multiple units. The basic unit is a single film jet. Each film jet influences its downstream area of a certain width at different degrees, which is determined by the geometry and flow parameters [19]. To describe these influences accurately, a series of data under different sensitive experimental parameters was referenced.

The first step in the data-driven method is to establish a database. Experimental data of a single hole were measured from Qin's experiment with pressure-sensitive paint (PSP) [20,21]. PSP is an effective surface measurement method that provides a detailed two-dimensional distribution of adiabatic film effectiveness. Its measurement error range is acceptable and known. In this database, data per experimental condition were reshaped to a 61 × 171 matrix corresponding to an area of flow direction coordinates (*x/D*) ranging from −1 to 16 and lateral coordinates (*y/D*) ranging from −3 to 3. The total number of data matrices is 440. The experimental conditions were recorded, and their ranges are listed in Table 1. The orthogonal experimental samples provide excellent support for the prediction of film effectiveness based on multiple sensitive parameters.

Table 1. Experimental conditions in the database.

Experimental Conditions	Value
Hole shape (-)	Cylindrical, Fan-shaped
Compound angle (°)	0, 45
Blowing ratio (-)	0.5, 0.75, 1.0, 1.2, 1.5
Density ratio (-)	1.0, 1.5
Non-dimensional curvature of the external surface (-)	−1/30, 0, 1/30
Non-dimensional pressure gradient along the mainstream flow direction ($\times 10^{-7}$)	−3.0, 0, 4.8, 7.3, 10

A sample of adiabatic film effectiveness data matrices is plotted in Figure 3. Since the area affected by the film is a long and narrow strip, the measured data only contains its main area of influence near the hole. The solution to avoid this disadvantage will be introduced in Section 2.3.

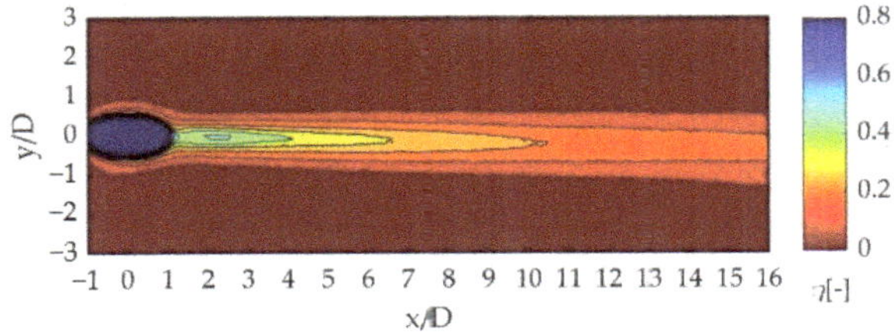

Figure 3. A sample of experimental data matrix.

2.3. Integration of Database with CHT Algorithm

After simulations in part 1, which were introduced in Section 2.1, the key parameters in Table 1 should be calculated as an input to query the film effectiveness. The multidimensional linear interpolation is implemented according to input parameters. In addition, there is a problem to consider in that the actual coverage of a single hole's influence is larger than the area's existing data value. To extract the 2D distribution law and expand it to a larger area, the non-linear least squares regressions were applied with queried data. The detailed method of data extension and the integration with the CHT algorithm is introduced in Section 2.3.

Equation (5) describes the 2-D distribution of cylindrical holes' film effectiveness in the form of Goldstein's correlation [22],

$$\eta(x,z)_{cylindrical} = a(x)\exp\left(-\left(\frac{z-b}{c}\right)^2\right). \tag{5}$$

When x is assigned, lateral film effectiveness obeys the Gaussian distribution. Strictly, c is also a function of x, but when x is large. The change in c with x is small. Thus, it is considered a constant in the far downstream area. Equation (6) can be deduced from the definition of the lateral averaged film effectiveness ($\overline{\eta}(x)$) and the integral of $\eta(x,z)_{cylindrical}$,

$$\overline{\eta}(x) = \frac{a(x)c\sqrt{\pi}}{P/D}. \tag{6}$$

Combining Equations (5) and (6) to eliminate $a(x)$, $\eta(x,z)_{cylindrical}$ is written as Equation (7),

$$\eta(x,z)_{cylindrical} = \frac{\overline{\eta}(x)(P/D)}{c\sqrt{\pi}}\exp\left(-\left(\frac{z-b}{c}\right)^2\right). \tag{7}$$

$\overline{\eta}(x)$ follows the correlation form in Equation (8). Therefore, the final form of $\eta(x,z)_{cylindrical}$ is Equation (9):

$$\overline{\eta}(x) = \frac{1}{d+e(x/D)^f} \tag{8}$$

$$\eta(x,z)_{cylindrical} = \frac{P/D}{c\sqrt{\pi}(d+e(x/D)^f)}\exp\left(-\left(\frac{z-b}{c}\right)^2\right) \tag{9}$$

If b, c, d, e, and f in Equation (9) are known, the distribution is uniquely defined. The parameters d, e, and f are obtained by regression on $\overline{\eta}(x)$ data of x/D ranging from 4 to 13, while b and c are decided by regression on $\eta(x,z)_{cylindrical}$ data of x/D ranging from 12 to 13.

Then, Equation (9) is used to calculate a film effectiveness distribution where *x/D* is larger than 12.

To make the combination result smooth, the final film effectiveness, where *x/D* is between 12 and 13, adopts an interpolation from the data queried and the distribution calculated.

The film effectiveness of a fan-shaped hole distributes a little differently from a cylindrical hole. Due to the expansion at the exit, there are usually two peak points in the lateral distribution of film effectiveness. Therefore, the form of $\eta(x,z)_{shaped}$ is described by the superposition of two Gaussian distributions as Equation (10). Equations (6) and (8) are also suitable for fan-shaped holes. The parameters are determined in the same way as cylindrical holes.

$$\eta(x,z)_{shaped} = \frac{1}{2}a(x)\left(\exp\left(-\left(\frac{z-b}{c}\right)^2\right) + \exp\left(-\left(\frac{z+b}{c}\right)^2\right)\right). \tag{10}$$

Finally, Sellers's Equation [23], as below, defined the superposition rules when there are overlaps between film holes' influence areas:

$$\eta = 1 - \prod_{i=1}^{n}(1 - \eta_i(x)). \tag{11}$$

After the superposition, a complete film effectiveness covering the outer surface of the vane is formed, as shown in Figure 4a,b. The generated film effectiveness distribution is basically consistent with the expected distribution. On the pressure surface, film effectiveness increases along the flow direction because multi-row holes bring coolant accumulation. On the suction surface, film effectiveness keeps declining along the flow direction without continuous coolant supplement.

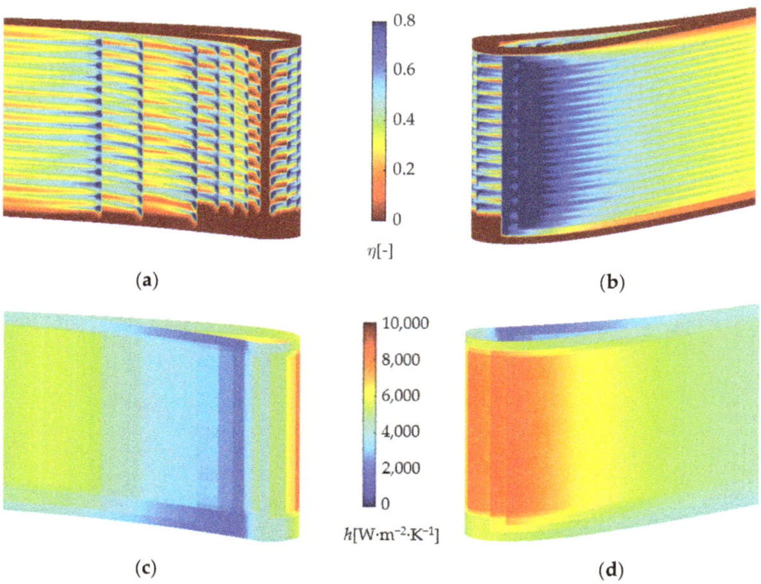

Figure 4. Boundary conditions generated by the data-driven method: (**a**) Film effectiveness on pressure surface (-); (**b**) Film effectiveness on suction surface (-); (**c**) HTC on pressure surface (W·m^{-2}·K^{-1}); (**d**) HTC on suction surface (W·m^{-2}·K^{-1}).

The heat transfer enhancement factor (h/h_0) is calculated according to single holes' flow parameters. Superposition rules similar to film effectiveness are applied to form a global distribution in Figure 4c,d.

These surface distributions are written to a (.csv) file to be loaded in preprocessing as a boundary condition. The distribution is recalculated in each new iteration of the data-driven CHT analysis as the key parameters changes.

3. Calculation Analysis

In this section, numerical calculation methods and techniques are introduced, including governing equations, geometric models, mesh selection, calculation settings, etc.

3.1. Governing Equations and Computational Technique

Fluid simulation in this work uses the steady-state RANS method. Three-dimensional steady-state, compressible Reynolds averaged Navier-Stokes equations are solved in the fluid domain. The Solver version is CFX-Solver 19.1. The continuity equation and the momentum equations can be written as

$$\frac{\partial \rho}{\partial t} + \nabla \cdot (\rho \boldsymbol{U}) = 0 \tag{12}$$

and

$$\frac{\partial (\rho \boldsymbol{U})}{\partial t} + \nabla \cdot (\rho \boldsymbol{U} \otimes \boldsymbol{U}) = -\nabla p + \nabla \cdot \tau + S_M, \tag{13}$$

respectively. In Equation (13), τ is the stress tensor, and S_M is the momentum source term. This work deals with heat transfer, so the total energy equation is also used:

$$\frac{\partial (\rho h_{tot})}{\partial t} - \frac{\partial p}{\partial t} + \nabla \cdot (\rho \boldsymbol{U} h_{tot}) = \nabla \cdot (\lambda \nabla \mathrm{T}) + \nabla \cdot (\boldsymbol{U} \cdot \tau) + \boldsymbol{U} \cdot S_M + S_E. \tag{14}$$

h_{tot} is the total enthalpy, and S_E is the energy source term. In this work, both momentum and energy source terms can be ignored.

In the solid domain, the following thermal equation is solved instead of the total energy equation:

$$\frac{\partial (\rho h)}{\partial t} + \nabla \cdot (\rho \boldsymbol{U}_S h) = \nabla \cdot (\lambda \nabla \mathrm{T}) + S_E. \tag{15}$$

The velocity of the solid $\boldsymbol{U}_s$ is zero in this work.

The SST k-ω turbulence model is used to close the N-S equations after Reynolds averaging. The γ-θ transitional model is set for the fluid domain to obtain a reasonable judgment of transition. The convergence criterion of the traditional coupled heat transfer calculation is that the residual is less than 10^{-5}, and the convergence criterion of the data-driven decoupled calculation method is that the temperature error of the outer iteration is less than 0.01 K.

Two materials are defined for this simulation. The solid's thermal conductivity (λ_s) follows a linear expression as Equation (12).

$$\lambda s = 0.0115T + 9.9105 \tag{16}$$

The fluid is an ideal gas. Its thermal conductivity (λ_f) and dynamic viscosity (μ_f) are assigned by Sutherland's formula.

3.2. Vane's Geometry

One typical guide vane is selected for the validation of the data-driven method. It is modified from the Energy Efficient Engine's first stage guide vane [24] and has most of the advanced cooling structures. Figure 5 shows its cross-section geometry and cooling structures. The profiles of the sections remain the same along the blade height direction. The vane is fully covered by film cooling holes. Seven rows of cylindrical film holes with

25 degrees of radial inclination are set at the leading edge. Two rows of fan-shaped holes are arranged near the leading edge of the suction surface. Three rows of cylindrical film holes with axial inclination along the axis are on the pressure surface, of which the two rows near the leading edge also have a compound angle. Meanwhile, the vane has a complex internal structure. There are two coolant cavities at different pressures inside the vane with impingement cooling. The cooling structure at the trailing edge includes several rows of pin-fins, which are more usually applied in modern vane designs.

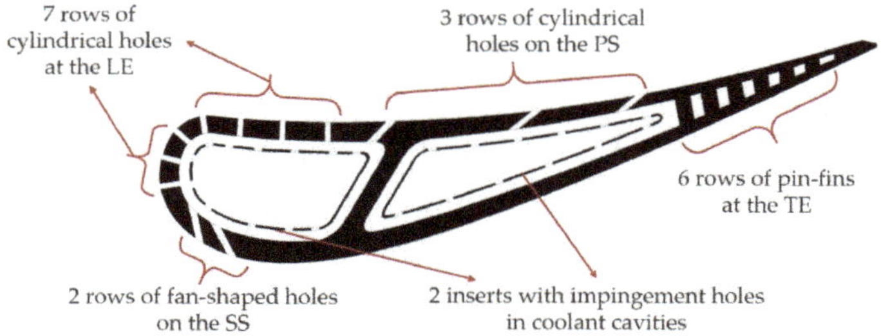

Figure 5. The geometry and cooling structure of a GE E3 1st Stage Vane.

3.3. Mesh Applied on Calculation

The computing mesh is generated according to the needs of iterative calculations. Three different sets of mesh in Figure 6 are output for certain purposes. The first mesh only contains mainstream passage and the second one includes coolant passage and vane solid. These two are prepared for the data-driven method. The third mesh contains whole computation domains with the mainstream passage, coolant passage, and vane solid. It is prepared for the conventional coupled CHT of the vane together with the mainstream passage.

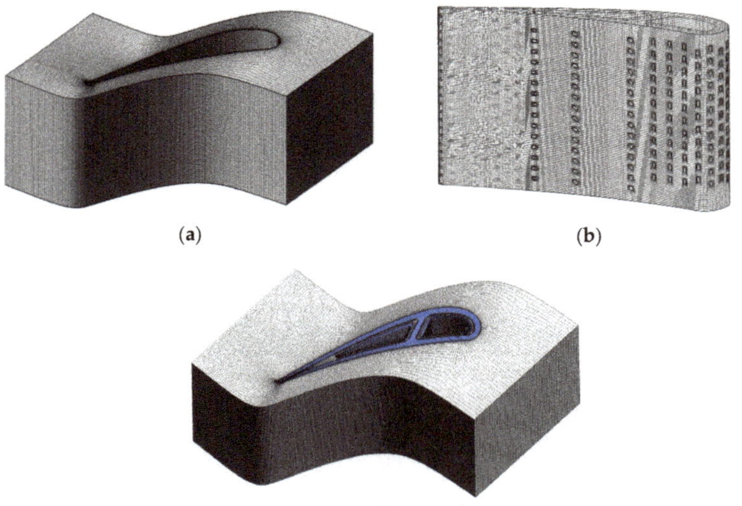

(a) (b) (c)

Figure 6. Mesh for data-driven heat transfer analysis and validation: (**a**) Mesh 1; (**b**) Mesh 2; (**c**) Mesh 3.

For the fluid–solid interfaces and other no-slip walls, the fluid grids nearby are refined to adapt to the turbulence model applied.

The grid independence test is performed against the computational domain, which corresponds to Mesh 3. Three sets of grids with different degrees of density are respectively used for the calculation, and the cell numbers and calculation results are listed in Table 2.

Table 2. Grid independence test.

Grid Density	Cell Number	Mean Temperature of Vane's Outer Surface
Low	11,050,194	1208.26 K
Medium	11,834,831	1208.68 K
High	12,830,966	1209.52 K

The change in mesh density has little influence on the calculation results. Thus, the medium grid density is employed to determine the final grid amount applied. Mesh 1 has 2,099,916 hexahedral cells in total. Most cells of Mesh 2 are also hexahedral except for some solid areas near the film hole where the geometry changes dramatically. The total number of cells in Mesh 2 is 7,422,585. Mesh 3 is not simply a joint of Mesh 1 and Mesh 2. The mesh of the mainstream passage near the exit of the film holes is refined, and the total number of cells exceeds 11,834,831.

3.4. Boundary Set-Ups

The flow parameters are set up according to the design report of the Energy Efficient Engine turbine. For the mainstream inlet, the total pressure is 2.526 MPa, and the total temperature is 2012 K. The Mach number of the mainstream outlet is 0.78. The outlet static pressure recalculated with the isentropic flow hypothesis is 1.51 MPa. The front cavity's inlet has a higher total pressure of 2.588 MPa. The rear cavity's inlet has a total pressure of 2.576 MPa. The total temperature at both coolant inlets is 883 K. All boundary conditions about flow are included in Table 3. The hub and shroud are isothermal no-slip walls since the heat transfer of these surfaces is not a concern in this work. Two side surfaces in tangential directions are rotating periodic interfaces. The interface between the fluid and solid is a no-slip wall with heat transfer.

Table 3. Boundary conditions setup.

Boundary Position	Parameter	Value
Mainstream inlet	Total pressure	2.526 MPa
Mainstream inlet	Total temperature	2012 K
Mainstream inlet	Turbulent intensity	0.1
Mainstream outlet	Static pressure	1.51 MPa
Front coolant cavity's inlet	Total pressure	2.588 MPa
Front coolant cavity's inlet	Total temperature	883 K
Rear coolant cavity's inlet	Total pressure	2.576 MPa
Rear coolant cavity's inlet	Total temperature	883 K

4. Results

This section qualitatively analyzes the effect of the data-driven method through the contour map of the calculation results. It then quantitatively evaluates the improvement of the data-driven method compared with the traditional method through the indicators of accuracy and efficiency.

4.1. Temperature Distribution Results

Figure 7 shows the temperature distribution of the vane's external surface with the conventional CHT method. Referring to the empirical flow pattern of the cascade and film jet, the temperature distribution reasonably reflects the effect of the cooling structure. Film cooling protects most areas of the external surface, except for the top and bottom edges of the suction surface. These areas are controlled by vortices from the endwall, while no film cooling is set on the endwall in this simulation. Therefore, the part close to the middle section of the vane reflects the effect of the blade cooling design. By comparing the temperature distribution on the pressure surface and suction surface, the influence of the film hole type on the cooling effect can be significantly observed. In the case of similar hole spacing, fan-shaped expansion can make the cooling downstream of the hole more uniform. On the pressure surface, the film holes with the compound angle made the coolant accumulate on top of the surface, so the surface temperature decreased along the radial direction. The distribution trend of the traditional CHT is believable due to the features above and the integrity of its analysis domain, but the accuracy of its numerical value is still yet to be discussed.

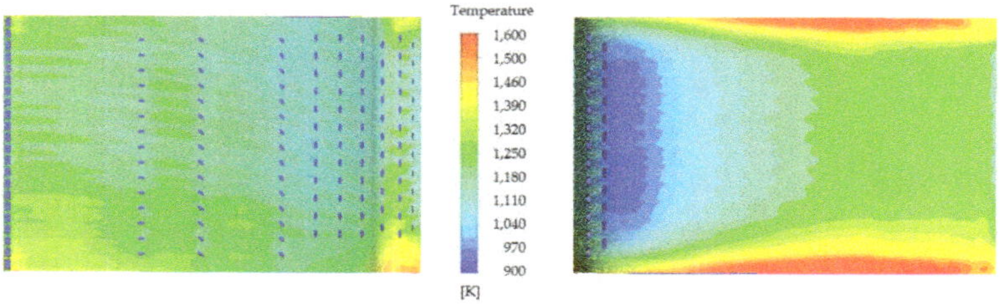

Figure 7. Temperature distribution of vane's external surface in conventional method.

Temperature distribution on the external surface of the vane from the data-driven method is shown in Figure 8 with the same legend as Figure 7. Due to the detailed thermal boundary prediction from the data-driven method, the final temperature distribution presents a detailed result similar to that from the conventional CHT. It can obviously be found on the pressure surface that the existence of film holes makes the downstream temperature appear as a band, which means the downstream temperature of the hole is significantly lower than the temperature of the area without air film on both sides. Downstream of the fan-shaped film holes on the suction surface, the uniformity of the temperature distribution of the blade spanwise is consistent with the traditional conjugate calculation, which well captures the cooling characteristics of fan-shaped holes.

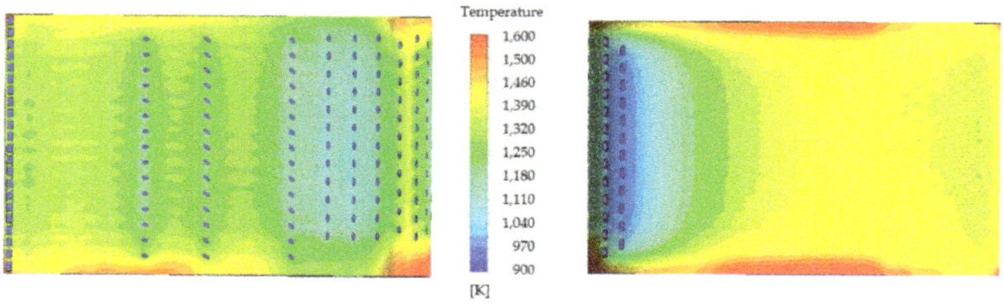

Figure 8. Temperature distribution of the vane's external surface in data-driven method.

The overall temperature in Figure 8 is slightly higher than that in Figure 7. Considering that the difference between the two cases lies in the calculation of film cooling, the main reason for the above phenomenon is that the film effectiveness predicted by the data-driven method is relatively lower. When the cooling effect of film on the surface is weakened, the reduction in temperature by internal cooling is more obvious in the cloud image. The low-temperature area on the pressure surface and suction surface near the trailing edge is caused by the enhanced heat transfer of the pin-fins inside the trailing edge. The low-temperature area near the hole outlet is caused by the heat transfer of the hole's inner wall.

4.2. Performance Evaluation of the Data-Driven Method

This section will analyze in detail the improvement of data-driven methods compared to traditional methods in terms of accuracy and efficiency.

4.2.1. Accuracy

The temperature distribution of the middle section is picked out to form a line diagram in Figure 9. The abscissa is the flow direction arc length calculated from the leading edge, and the ordinate is the temperature of the blade's metal outer surface. The heat transfer analysis result in the E3 design report is also presented in this figure as a reference. The mean temperature and general trend from the data-driven method are closer to those from the design reports compared to the conventional method.

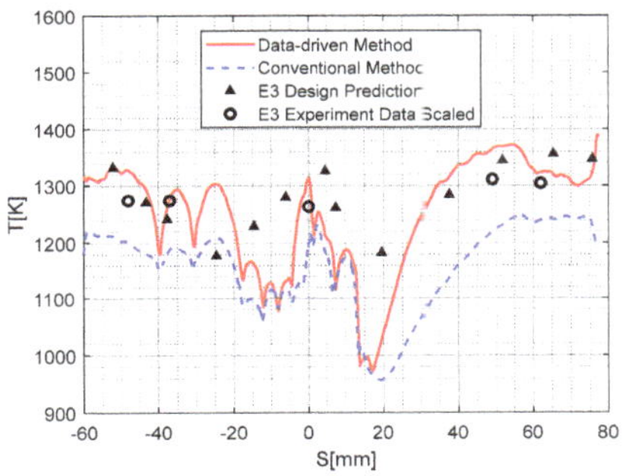

Figure 9. External surface temperature distribution of the middle section [24,25].

At the arc length position of the experimental data, the temperature calculation errors of the two methods are calculated. Results are shown in Table 4. The maximum error of the data-driven method is about 53.3 K, corresponding to a relative error of 4.1%. The maximum error of the conventional method is about −92.7 K, corresponding to a relative error of −7.1%. The minus sign means that the calculation result is lower than the test result. For the conventional method, the low prediction temperature may be mainly due to the high prediction of the downstream film effectiveness by CFD. By citing the experimental data of film effectiveness, the data-driven method avoids the inherent problem of low wall temperature caused by CFD and effectively improves the accuracy of temperature prediction.

Table 4. Calculation error of the middle section temperature.

Parameter	Data-Driven	Conventional
Maximum Error (K)	53.3	−92.7
Maximum Relative Error (%)	4.1	−7.1
Mean Error (K)	34.5	−75.7
Mean Relative Error (%)	2.7	−5.9

4.2.2. Computational Cost

The computations are both made with 36 processes on the same server. The data-driven method takes 5 h and 21 min, while the conventional CHT method takes 17 h and 49 min, which is three times larger than the data-driven one. This is more obvious in Table 5 with the same units. The application of the data-driven method can save 69.97% of heat transfer analysis time. High efficiency is a significant advantage of the data-driven method. The computational cost will be less if this method is spread to predict the internal boundary conditions.

Table 5. Time cost.

Parameter	Value
Data-driven method's time cost (min)	321
Data-driven method's time cost (min)	1069
Percentage decreased by simplification (-)	69.97%

5. Conclusions

At present, the heat transfer prediction of gas turbine blades with a complex cooling structure is difficult and time-consuming. Under this background, a data-driven conjugate analysis method is proposed in this paper. Different from the traditional coupling CHT method, this method uses a decoupled method, which computes different domains with different methods. The data transfer between the domains is realized through the iteration of the outer cycle. In the process of the outer iteration, the data-driven prediction model of film cooling is applied at the conjugate boundary. The data-driven prediction model of film cooling is based on the film cooling unit experimental database under the influence of multiple parameters, and the data is modified and extended to predict the film cooling effect applied to the cascade surface.

In this paper, a typical first-stage vane based on the E^3 turbine is selected for validation. The temperature distribution calculated by the data-driven method is similar to that of the conventional method, which can correctly reflect the influence of the typical cooling structure on the solid wall temperature. Compared with the conventional method, the temperature distribution of the blade mid-section calculated by the data-driven method is in better agreement with the experimental data. This means that the data-driven method effectively avoids the overestimation of the film cooling effectiveness of the conventional CHT, and its prediction accuracy can support the effectiveness evaluation of the cooling scheme. In addition, compared with the traditional coupled CHT, the analysis efficiency of the data-driven method is also significantly improved. This reduction in heat transfer analysis time enables more iterations of cooling designs in the same amount of time, and further efficiency improvements in this direction will hopefully support the optimization design based on higher prediction accuracy.

Author Contributions: Conceptualization, supervision, X.L. and J.R.; investigation, methodology, validation, H.C. and L.W.; data curation, visualization, H.C.; writing—original draft preparation, H.C.; writing—review and editing, X.L. and J.R. All authors have read and agreed to the published version of the manuscript.

Funding: This research was funded by National Science and Technology Major Project, grant number J2019-III-0007-0050.

Data Availability Statement: Not applicable.

Conflicts of Interest: The authors declare no conflict of interest.

References

1. Bunker, R.S. Evolution of Turbine Cooling. In Proceedings of the ASME Turbo Expo 2017: Turbomachinery Technical Conference and Exposition, Volume 1: Aircraft Engine; Fans and Blowers; Marine; Honors and Awards, Charlotte, NC, USA, 26–30 June 2017. [CrossRef]
2. Han, J. Turbine Blade Cooling Studies at Texas A&M University: 1980–2004. *J. Thermophys. Heat Transf.* **2006**, *20*, 161–187. [CrossRef]
3. Heidmann, J.D.; Kassab, A.J.; Divo, E.A.; Rodriguez, F.; Steinthorsson, E. Conjugate Heat Transfer Effects on a Realistic Film-Cooled Turbine Vane. In Proceedings of the ASME Turbo Expo 2003, Collocated with the 2003 International Joint Power Generation Conference, Volume 5: Turbo Expo 2003, Parts A and B, Atlanta, GA, USA, 16–19 June 2003. [CrossRef]
4. Zecchi, S.; Arcangeli, L.; Facchini, B.; Coutandin, D. Features of a Cooling System Simulation Tool Used in Industrial Preliminary Design Stage. In Proceedings of the ASME Turbo Expo 2004: Power for Land, Sea, and Air, Volume 3: Turbo Expo 2004, Vienna, Austria, 14–17 June 2004. [CrossRef]
5. Bohn, D.; Bonhoff, B.; Schrinenbom, H. Combined Aerodynamic and Thermal Analysis of a Turbine Nozzle Guide Vane. In Proceedings of the 1995 Yokohama International Gas Turbine Congress, Yokohama, Japan, 22–27 October 1995.

6. Bohn, D.E.; Becker, V.J.; Kusterer, K.A. 3-D Conjugate Flow and Heat Transfer Calculations of a Film-Cooled Turbine Guide Vane at Different Operation Conditions. In Proceedings of the ASME 1997 International Gas Turbine and Aeroengine Congress and Exhibition, Volume 3: Heat Transfer; Electric Power; Industrial and Cogeneration, Orlando, FL, USA, 2–5 June 1997. [CrossRef]
7. Bonini, A.; Andreini, A.; Carcasci, C.; Facchini, B.; Ciani, A.; Innocenti, L. Conjugate Heat Transfer Calculations on GT Rotor Blade for Industrial Applications: Part I—Equivalent Internal Fluid Network Setup and Procedure Description. In Proceedings of the ASME Turbo Expo 2012: Turbine Technical Conference and Exposition, Volume 4: Heat Transfer, Parts A and B, Copenhagen, Denmark, 11–15 June 2012. [CrossRef]
8. Andreini, A.; Bonini, A.; Da Soghe, R.; Facchini, B.; Ciani, A.; Innocenti, L. Conjugate Heat Transfer Calculations on GT Rotor Blade for Industrial Applications: Part II—Improvement of External Flow Modeling. In Proceedings of the ASME Turbo Expo 2012: Turbine Technical Conference and Exposition, Volume 4: Heat Transfer, Parts A and B, Copenhagen, Denmark, 11–15 June 2012. [CrossRef]
9. Chowdhury, N.H.K.; Zirakzadeh, H.; Han, J. A Predictive Model for Preliminary Gas Turbine Blade Cooling Analysis. *J. Turbomach.* **2017**, *139*, 091010. [CrossRef]
10. Ngetich, G.C.; Murray, A.V.; Ireland, P.T.; Romero, E. A Three-Dimensional Conjugate Approach for Analyzing a Double-Walled Effusion-Cooled Turbine Blade. *J. Turbomach.* **2019**, *141*, 011002. [CrossRef]
11. Amaral, S.; Verstraete, T.; Braembussche, R.V.; Arts, T. Design and Optimization of the Internal Cooling Channels of a High Pressure Turbine Blade—Part I: Methodology. *J. Turbomach.* **2010**, *132*, 021013. [CrossRef]
12. Verstraete, T.; Amaral, S.; Braembussche, R.V.; Arts, T. Design and Optimization of the Internal Cooling Channels of a High Pressure Turbine Blade—Part II: Optimization. *J. Turbomach.* **2010**, *132*, 021014. [CrossRef]
13. Zhang, L.J.; Jaiswal, R.S. Turbine nozzle endwall film cooling study using pressure-sensitive paint. *J. Turbomach.* **2001**, *123*, 730–738. [CrossRef]
14. Gao, Z.; Wright, L.M.; Han, J. Assessment of Steady State PSP and Transient IR Measurement Techniques for Leading Edge Film Cooling. In Proceedings of the ASME 2005 International Mechanical Engineering Congress and Exposition, Heat Transfer, Part A, Orlando, FL, USA, 5–11 November 2005; pp. 467–475. [CrossRef]
15. Gao, Z.; Han, J. Influence of Film-Hole Shape and Angle on Showerhead Film Cooling Using PSP Technique. *J. Heat Transf.* **2009**, *131*, 061701. [CrossRef]
16. Lowery, G.W.; Vachon, R.I. The Effect of Turbulence on Heat Transfer from Heated Cylinders. *Int. J. Heat Mass Transf.* **1975**, *18*, 1229–1242. [CrossRef]
17. Kreith, F.; Manglik, R.M. *Principles of Heat Transfer*, 8th ed.; Cengage Learning: Boston, MA, USA, 2018; pp. 342+362.
18. Baldauf, S.; Scheurlen, M.; Schulz, A.; Wittig, S. Heat flux reduction from film cooling and correlation of heat transfer coefficients from thermographic measurements at enginelike conditions. *J. Turbomach.* **2002**, *124*, 699–709. [CrossRef]
19. Bogard, D.G.; Thole, K.A. Gas Turbine Film Cooling. *J. Propul. Power* **2006**, *22*, 249–270. [CrossRef]
20. Qin, Y.; Ren, J.; Jiang, H. Effects of Streamwise Pressure Gradient and Convex Curvature on Film Cooling Effectiveness. In Proceedings of the ASME Turbo Expo 2014: Turbine Technical Conference and Exposition, Volume 5B: Heat Transfer, Düsseldorf, Germany, 16–20 June 2014. [CrossRef]
21. Qin, Y.; Li, X.; Ren, J.; Jiang, H. Effects of compound angle on film cooling effectiveness with different streamwise pressure gradient and convex curvature. *Int. J. Heat Mass Transf.* **2015**, *86*, 482–491. [CrossRef]
22. Goldstein, R.J. Film Cooling. *Adv. Heat Transf.* **1971**, *7*, 321–379. [CrossRef]
23. John, P.; Sellers, J.R. Gaseous Film Cooling with Multiple Injection Stations. *AIAA J.* **1963**, *1*, 321–379. [CrossRef]
24. Halila, E.E.; Lenahan, D.T.; Thomas, T.T. *Energy Efficient Engine High Pressure Turbine Test Hardware Detailed Design Report*; Report No.: NASA CR-167955; NASA: Washington, DC, USA, 1982; pp. 23–34.
25. Stearns, E.M. *Energy Efficient Engine Core Design and Performance Report*; Report No.: NASA CR-168069; NASA: Washington, DC, USA, 1982; pp. 357–361.

processes

MDPI

Article

Research on Regional Short-Term Power Load Forecasting Model and Case Analysis

Kang Qian [1], **Xinyi Wang** [2] **and Yue Yuan** [1,*]

[1] College of Energy and Electrical Engineering, Hohai University, Nanjing 211100, China; qiankang@jspdi.com.cn

[2] China Energy Engineering Group Jiangsu Power Design Institute Co., Ltd., Nanjing 211102, China; wangxinyi@jspdi.com.cn

* Correspondence: yyuan@hhu.edu.cn; Tel.: +86-139-5168-3948

Abstract: Integrated energy services will have multiple values and far-reaching significance in promoting energy transformation and serving "carbon peak and carbon neutralization". In order to balance the supply and demand of power system in integrated energy, it is necessary to establish a scientific model for power load forecasting. Different algorithms for short-term electric load forecasting considering meteorological factors are presented in this paper. The correlation between electric load and meteorological factors is first analyzed. After the principal component analysis (PCA) of meteorological factors and autocorrelation analysis of the electric load, the daily load forecasting model is established by optimal support vector machine (OPT-SVM), Elman neural network (ENN), as well as their combinations through linear weighted average, geometric weighted average, and harmonic weighted average method, respectively. Based on the actual data of an industrial park of Nantong in China, the prediction performance in the four seasons with the different models is evaluated. The main contribution of this paper is to compare the effectiveness of different models for short-term electric load forecasting and to give a guideline to build the proper methods for load forecasting.

Keywords: short-term electric load forecasting; meteorological factors; optimized support vector machine; Elman neural network; combined model

Citation: Qian, K.; Wang, X.; Yuan, Y. Research on Regional Short-Term Power Load Forecasting Model and Case Analysis. *Processes* **2021**, *9*, 1617. https://doi.org/10.3390/pr9091617

Academic Editors: Pei Liu, Ming Liu and Xiao Wu

Received: 1 July 2021
Accepted: 31 August 2021
Published: 8 September 2021

Publisher's Note: MDPI stays neutral with regard to jurisdictional claims in published maps and institutional affiliations.

Copyright: © 2021 by the authors. Licensee MDPI, Basel, Switzerland. This article is an open access article distributed under the terms and conditions of the Creative Commons Attribution (CC BY) license (https://creativecommons.org/licenses/by/4.0/).

1. Introduction

Integrated energy services will have multiple values and far-reaching significance in promoting energy transformation and serving "carbon peak and carbon neutralization". Before the implementation of integrated energy services, in order to balance the supply and demand of the power system, it is necessary to establish a scientific model for power load forecasting. The balance of supply and demand in the power system plays an important role in regional economic and social development. To improve the quality of power supply, it is important to analyze various characteristics of regional electric load and establish a proper model for the short-term electrical load forecasting [1].

Many factors affect electric load forecasting, including social-economic factors, such as population and industrial structure, and meteorological conditions [2,3]. Therein, complex and changeable meteorological factors are essential factors in load forecasting and have the greatest impact on forecast results [2]. Forecasting that considers meteorological factors can improve forecasting accuracy.

According to relevant statistics, in large cities with large population, residential power consumption in summer can reach 50% or even higher of the total power load. The prediction of residential power load plays an important role in regional energy allocation, energy-saving control, and power grid reliability. At the same time, residential power load shows the characteristics of complexity and uncertainty, which is closely related to temperature, humidity, wind speed, week type, building function, building area, industrial structure, and other factors.

In short-term load forecasting, the processing methods of meteorological factors mainly include the following ideas: (1) Do not consider the influence of meteorological factors; (2) Only the influence of single meteorological factor shall be considered; (3) Considering the influence of multiple meteorological factors without coupling effect; (4) Considering the influence of meteorological index formed by coupling multiple meteorological factors; (5) Consider the influence of real-time meteorological factors. In the actual prediction process, the appropriate treatment scheme is selected according to local conditions in combination with the actual local weather and load conditions.

Load forecasting can be divided into ultra-short-term, short-term, medium-term, and long-term according to different purposes. Short-term load forecasting refers to daily load forecasting and weekly load forecasting. It is necessary to fully study the law of power grid load changes and analyze the relevant factors of load changes for arrangement of daily scheduling plan or weekly scheduling plan.

Extensive research has been done on load forecasting, and the algorithms can be divided into two groups. Traditional statistical algorithms including the time series model [4–6] and the linear regression model [7] are the early and widely developed methods for load forecasting. It is modeled only based on load time series, which is computationally fast. However, short-term load forecasting is affected by many factors as mentioned above. When the weather changes greatly or the holidays occur, these models fail to forecast accurately [8]. Therefore, the second load forecasting algorithm that considers meteorological factors came into being. It takes into account weather and calendar factors to make the forecast results closer to the actual situation. These modern intelligent algorithms include the artificial neural network (ANN) [9–11], the expert system [12], support vector machine(SVM) [13–16], etc. Though these individual forecasting models obtain satisfactory performance, each model has unavoidable drawbacks, and their performance varies under different conditions, which results in risks and limitations in the practical applications [17]

Thus, it is possible to consider combining different individual forecasting models. In 1969, Bates and Granger made a systematic study on the combination forecasting method for the first time and then aroused great interest in combination forecasting models [18] According to the pros and cons of individual algorithms, the combination of these algorithms can reduce the risks of the individual method and achieve better forecasting performance. Various combinations are proposed, such as the BP neural network model combined with the gray theory model based on variable weights [19], different types of neural network combinations based on non-positive constraint theory [20], SVM combined with chaos theory [21], etc. These researches of combination models mainly focus on the determination of the optimal weights to improve the forecast accuracy, and, meanwhile avoid falling into local optimal solution, lower convergent rate, and increasing modeling complexity [22–24].

In addition, some recent papers have also proposed some short-term load forecasting methods: a novel hybrid forecasting algorithm is proposed in [25]. The proposed hybrid forecasting method is based on locally weighted support vector regression (LWSVR) and the modified grasshopper optimization algorithm (MGOA). Obtaining the appropriate values of LWSVR parameters is vital to achieving satisfactory forecasting accuracy. Therefore, the MGOA is proposed in this paper to optimally select the LWSVR's parameters. The proposed MGOA can be derived by presenting two modifications on the conventional GOA in which the chaotic initialization and the sigmoid decreasing criterion are employed to treat the drawbacks of the conventional GOA. There are also other related papers, e.g., the work in [26] proposes an ultra-short-term power load forecasting method based on two-layer xgboost (extreme gradient boosting) algorithm. The work in [27] proposes a short-term power load forecasting model based on similar day selection and multi-integration combination. In [28], particle swarm optimization (PSO) is proposed to optimize the neural network model of quantum weighted gated circulating unit to predict short-term power load. The work in [29] introduces K-means clustering analysis technology based on cosine similarity to predict short-term power load. The work in [30] proposes a load forecasting method based on improved deep belief network algorithm.

It can be seen that there are many short-term load forecasting methods, such as time series model, multiple linear regression, support vector machine, and so on. Compared with other traditional algorithms, support vector machine has better prediction effect for small samples and has the advantages of strong learning ability and high precision in processing high-dimensional data. However, the prediction effect of support vector machine is closely related to the quality of parameter estimation. At present, the commonly used grid parameter optimization methods have a large amount of calculation and limited parameter estimation accuracy. Using an intelligent optimization algorithm to find the global optimal parameter combination has gradually become an effective way to improve the application effect of support vector machine. In light of this, by coupling a simulated annealing particle swarm optimization algorithm and support vector machine, an optimized support vector machine model is constructed to study the short-term load forecasting effect under different meteorological variable input conditions and analyze the influence of meteorological factors on load forecasting.

In the field of power load forecasting, static neural network, mainly represented by BP neural network, is most widely used. This kind of model actually transforms dynamic time modeling into static space modeling, resulting in its insufficient ability to capture the highly complex time-varying characteristics of load. Therefore, Elman neural network with dynamic adaptive time-varying characteristics is applied after adding a receiving layer on BP neural network.

For the short-term load forecasting of regional power system, the problems existing in the current research include the following.

1. In the practical application of power load forecasting models, most models and methods cannot predict the load according to different regions and different utilization types, and there is a lack of intelligent forecasting methods that can effectively adapt to a variety of conditions and improve the accuracy.
2. The current research lacks in-depth analysis of regional basic situation, economy, population and industrial structure, analysis of regional power load characteristics, understanding of environmental and meteorological factors, and detailed analysis of the impact of environmental and meteorological factors on load forecasting.
3. In the actual dispatching of power system, seasonal power shortage occurs from time to time, especially in summer and winter. Therefore, it is necessary to strengthen the research on seasonal power load forecasting in order to meet the power supply demand more reasonably.

This paper presents some combined models for short-term load forecasting. It integrates simulated annealing OPT-SVM and ENN through time-varying weights selection. Compared with other traditional algorithms, support vector machines have better prediction effects for small samples. The advantages of strong learning ability and high accuracy when processing high-dimensional data are also very prominent. The Elman neural network has better generalization and prediction ability for nonlinear and high-frequency time series data. The proposed models can take into account regional power load characteristics and their correlation with meteorological factors such as temperature, humidity, wind speed, and sunshine. On the one hand, they can improve the adaptability of the power dispatching department to regional power load changes; on the other hand, they can improve the accuracy of load forecasting to meet the demand of regional power load. The accuracy of short-term electrical load forecasting based on single OPT-SVM and ENN varies in different periods. The weighted combination of the results of different forecasting models in each period could result in higher accuracy of load forecasting. To this end, the time history of the contribution of each single model to the combined forecasting is studied. The effectiveness of the combination forecasting model is analyzed from the statistical accuracy so as to provide a useful reference for deepening the application of the time-varying weight combination model in short-term load forecasting.

The innovation of our paper is to find out a time-varying weight combination model which takes the results of two single models of OPT-SVM and Elman NN as input, studies the time history changes of each single model's contribution to the combined forecasting, and uses statistics methods to analyze the effectiveness of the combined model, providing a useful reference for deepening the application of time-varying weight combination models in short-term load forecasting.

The rest of this paper is organized as follows. The experimental data are described in Section 2. In Section 3, the mathematical principles and processes of the proposed models are described in detail. Furthermore, Section 4 presents and discusses the forecasting results. Finally, the proposed models are summarized in Section 5.

2. Data Preparation

Under the influence of meteorological and other factors, regional power load often presents complex time-varying characteristics of continuity, nonlinearity, and multifrequency nesting. Therefore, in order to establish a scientific and accurate power load forecasting model, it is necessary to understand the time-varying characteristics of regional power load, analyze the correlation between power load and meteorological factors, and provide a priori knowledge for selecting the key influencing factors of power load forecasting.

The basic data of this paper come from the actual load demand data of an industrial park in Nantong, Jiangsu, China from 1 March 2016 to 28 February 2017. These data contain the following.

1. Load demand data with a time interval of 15 min for industries, residents, and businesses; then, the hourly load demand and daily load demand are derived by accumulation, and the daily load curve is given in Figure 1;
2. Meteorological data (e.g., temperature, relative humidity, wind speed, evaporation, and surface temperature) of every hour from Nantong Weather Station.

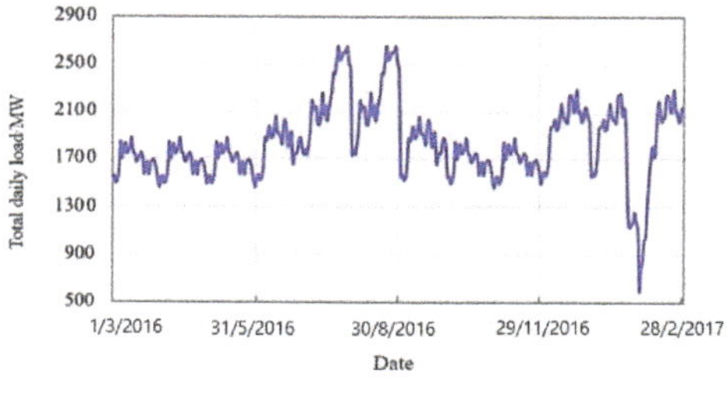

Figure 1. Daily load curve of the studied park.

The studied industrial park area is about 170 km^2, with relatively concentrated distribution, simple terrain, and less-fluctuated climate, so one weather station can effectively cover the area.

The first three-quarters of the remaining load sequence length after excluding the pre-ordered autocorrelation load is used as the training sample, and the last one-quarter as the prediction verification sample; the modeling period and forecasting period are determined according to the four seasons and the load autocorrelation time lag as follows:

1. Spring: March 5 to May 11 as modeling period, totaling 68 days, and May 11 to June 3 as forecasting period, totaling 23 days;
2. Summer: June 11 to August 12 as modeling period, totaling 63 days, and August 13 to September 3 as forecasting period, totaling 22 days;
3. Autumn: September 12 to November 11 as modeling period, totaling 63 days, and November 12 to December 3 as forecasting period, totaling 21 days;

4. Winter: December 6 to February 6 (2017) as modeling period, totaling 63 days, and February 7 to February 28 as forecasting period, totaling 21 days.

During the preprocessing of the raw data, horizontal and vertical processing methods are used here to deal with abnormal data [31], and the Least squares method is applied for curve fitting to obtain the missing data. The power load forecasting model established should consider the influence of meteorological factors, and the meteorological factors, such as temperature, humidity, precipitation, sunshine days, wind speed, and electricity load, have different dimensions and units and cannot be used directly. Therefore, it is necessary to standardize all relevant factors so that all data are in the same order of magnitude.

Furthermore, the different types of meteorological data are normalized as follows:

$$Y = \frac{I_i - I_{min}}{I_{max} - I_{min}} \tag{1}$$

where Y is the normalized data, and I_i, I_{min}, and I_{max} are the current sample value, minimum, and maximum of the training data, respectively.

3. Optimal Combined Model Considering Meteorological Factors

This paper improves the results of short-term load forecasting by the steps as shown in Figure 2.

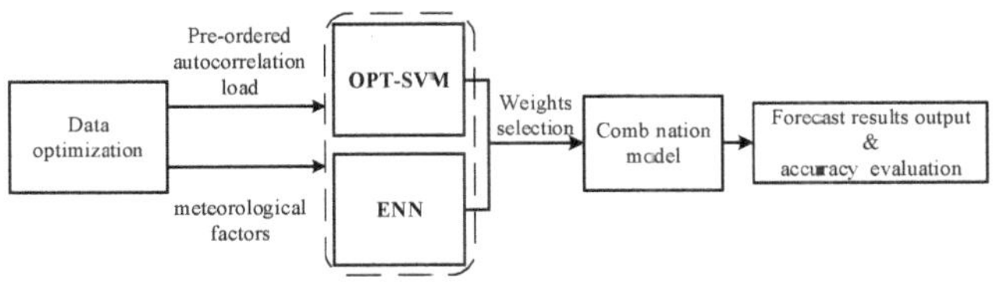

Figure 2. Framework of the proposed method.

3.1. Data Optimization

The pre-ordered autocorrelation load and comprehensive meteorological factors are taken as the input of the forecasting model. Therefore, the processing of the data is introduced as follows.

3.1.1. Correlation Analysis

First of all, the correlation analysis between electric load and meteorological factors is necessary. Pearson correlation coefficient is a measure of linear correlation between two sets of data, which has a value between −1 and 1. It is defined as

$$\rho_{X,Y} = \frac{\text{cov}(X,Y)}{\sigma_X \sigma_Y} = \frac{E[(X-\mu_X)(Y-\mu_Y)]}{\sigma_X \sigma_Y} \tag{2}$$

where $\rho_{X,Y}$ represents the overall correlation coefficient, X and Y are two variables, $\text{cov}(X,Y)$ is the covariance of the variables X and Y, μ_X and μ_Y are the expectations of the variables X and Y, and σ_X and σ_Y are the standard deviations of the variables X, Y.

3.1.2. Principal Component Analysis (PCA)

There could exist correlations between different meteorological factors so that multiple meteorological variables may contain redundant information. PCA can be used to extract main feature factors from multiple meteorological factors for dimensionality reduction. It replaces the original variables with fewer new variables and enables these new variables to retain the original full variable information as much as possible. The detailed steps of PCA can be found in [32].

3.1.3. Autocorrelation Analysis of Electrical Load

The pre-ordered autocorrelation load is taken as the input of the proposed load forecasting model in this paper, and the autocorrelation time lag needs to be determined. The pre-ordered autocorrelation load time lag is determined according to the autocorrelation function (ACF) in [33].

3.2. Model Description

The proposed combined model integrates two individual models (i.e., OPT-SVM and ENN) based on the time-varying optimal weight selection. OPT-SVM and ENN both have good forecasting performance.

3.2.1. Optimal Supporting Vector Machine (OPT-SVM)

The basic idea of SVM algorithm is to construct an optimal hyperplane from the sample space or feature space to make the margin between different types of data points as far as possible. Given the sample sets $\{x_i, y_i\}, i = 1, 2, \ldots, n$, x_i is the input sample, y_i is the output sample, and n is the total number of samples. The nonlinear load forecasting SVM model is expressed as

$$f(x) = \omega^T \varphi(x) + b \tag{3}$$

where $\varphi(x)$ denotes the nonlinear mapping from the input space to high-dimensional feature space. ω is the weight vector and b is constant. They can be obtained by the principle of structural minimization, and the optimization target can be expressed as follows:

$$\frac{1}{2}\omega^2 + c\sum_{i=1}^{n} |y_i - (\omega, \varphi(x_i)) - b|_\varepsilon \tag{4}$$

with c as penalty factor.

Introducing slack variables ξ_i and ξ_i^* into (4), it turns into

$$\frac{1}{2}\|\omega\|^2 + c\sum_{i=1}^{n} (\xi_i + \xi_i{}^*), \text{ s.t.} \begin{cases} y_i - (\omega^T(x_i) + b) \leq \varepsilon + \xi_i{}^* \\ (\omega^T \varphi(x_i + \mathrm{b})) - y_i \leq \varepsilon + \xi_i \\ \xi_i, \xi_i{}^* \geq 0, i = 1, \ldots, n \end{cases} \tag{5}$$

By introducing Lagrange multiplier α, we transform Equation (5) into

$$f(x, a) = \sum_{i=1}^{n} (\alpha_i{}^* - \alpha_i) K(x_i, x) + b \tag{6}$$

where $K(x, x_i)$ is the kernel function. Here, we choose Gauss radial basis function (GRBF) as kernel function for its good analytical performance. GRBF is expressed as

$$K(x_i, x_j) = e^{-\gamma \|x_i - x_j\|^2} \tag{7}$$

where γ is a tunable parameter.

Error penalty parameter c and kernel function parameter γ are the two key parameters to the SVM. Compared with other traditional algorithms, support vector machines have better prediction effects for small samples. The advantages of strong learning ability and high accuracy when processing high-dimensional data are also very prominent. However, the prediction effect of support vector machines is closely related to the quality of parameter estimation. Currently, the commonly used gridding parameter optimization method has a large amount of calculation and limited parameter estimation accuracy. The use of intelligent optimization algorithms to find the global optimal parameter combination has gradually become an improved support vector and the most effective way to attain the machine application effect. Particle Swarm Optimization (PSO), as an excellent representative algorithm with clear principles and simple calculations, has been widely used in solving optimization problems. PSO is a swarm intelligence evolutionary computation technology based on iterative optimization. Furthermore, the simulated annealing algorithm is introduced to the PSO algorithm to increase its global optimization ability. The specific steps of the algorithm can refer to the work in [34]. Here, we utilize PSO to solve the optimal SVM parameters to avoid lengthy calculations. Because of this, this chapter builds an optimized support vector machine model by coupling the simulated annealing particle swarm optimization algorithm and support vector machine to study the effect of short-term load forecasting under different meteorological variable input conditions and analyze the influence of meteorological factors on load forecasting.

According to the description of simulated annealing PSO and SVM above, the algorithm of OPT-SVM for electric load forecasting proceeds as follows:

1. Take the first three-quarters of the load sequence as the training sample, and the last one-quarter as the prediction verification sample. Determine the modeling period and forecast period in terms of the division of seasons and autocorrelation load time lag.
2. Initialize the parameters of simulated annealing PSO algorithm, including number of particles L, particle dimension D, learning factors c_1, c_2, initial temperature T_0, cooling rate α_c, and maximum number of iterations M.
3. Taking the maximum certainty coefficient of SVM as the objective function, use the simulated annealing PSO algorithm to optimize the penalty factor c and the kernel function parameter γ to obtain the best parameter combination.
4. Assign the best parameter combination (c, γ) to SVM to predict the load in the first period of the training period; after the prediction completed, take the measured load in the first period as a known value, and continue to predict the load in the second period with pre-ordered autocorrelation load and meteorological factors as the input of the second period, and repeat forward until the end of the forecast period.
5. De-normalize the simulated value to obtain the predicted load value.

3.2.2. Elman Neural Network (ENN)

ENN is based on the basic structure of the BP neural network and adds a "feedback". Such an internal feedback network increases its ability to deal with dynamic information of the network itself, which means the system can adapt to the time-varying characteristics and that it is more suitable to establish the forecast model of time series.

ENN includes an input layer, hidden layer, context layer, and output layer. The connection of the input layer, the hidden layer, and the output layer is similar to the BP neural network. The input layer performs signal transmission and the output layer plays a weighting role. The context layer is used to memorize the output value of the hidden layer unit at the previous moment, and is usually considered as a delay operator network. The structure diagram is shown in Figure 3.

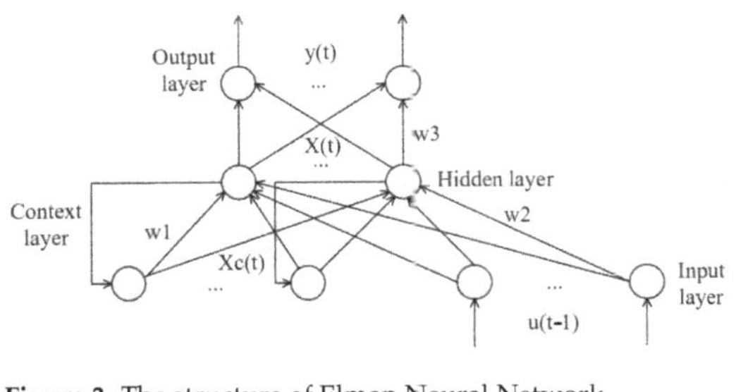

Figure 3. The structure of Elman Neural Network.

The mathematical model of ENN is

$$x(k) = f(\omega_1 x_c(k) + \omega_2 i(k-1)) \tag{8}$$

$$x_c(k) = a \cdot x_c(k-1) + x(k-1) \tag{9}$$

$$y(k) = g(\omega_3 x(k)) \tag{10}$$

where y is the m-dimensional output vector, x is the n-dimensional hidden layer vector, i is the r-dimensional input vector, x_c is the n-dimensional feedback state vector, a is the self-connected feedback gain factor, ω_1 is connection weight between the context layer and the hidden layer, ω_2 is the connection weight of the input layer and the hidden layer, ω_3 is the connection weight of the hidden layer and the output layer, $g(.)$ is the transfer function of the output neuron, and $f(.)$ is the transfer function of the hidden layer neuron.

Like the BP neural network, the number of hidden neurons is the core parameter to determine the model structure. A trial and error process is needed to decide the optimal number of hidden neurons. Once the model structure is settled, the connection weights between different layers are determined by self-learning using the gradient descent method. The training and prediction process by ENN is shown in Figure 4.

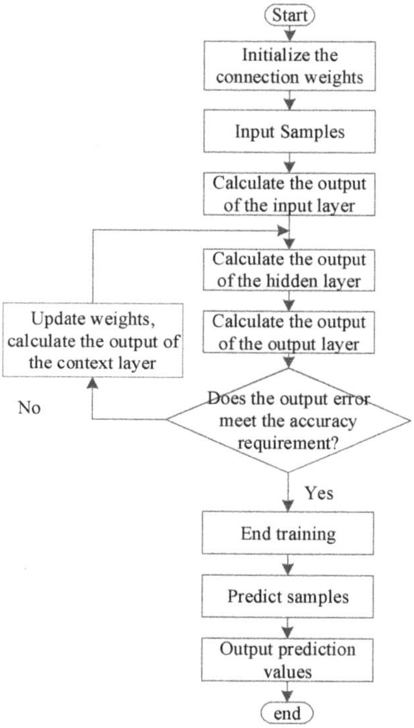

Figure 4. The flowchart of electrical load forecast by ENN.

3.2.3. Combined Forecasting Model

For the same prediction problem, the optimized combination of multiple different prediction models can effectively improve the prediction accuracy of the model under certain conditions. The combination model development includes three key steps: (1) selection of proper individual models, (2) construction of the mathematical expression of the combination from multiple individual models, and (3) solving the weight of each single model through optimization algorithms.

As the OPT-SVM and ENN are selected here as the candidates for the combined model, the mathematical expressions of the combination model are then supposed to be determined. We focuses on three forms of expressions: linear weighted average, geometric weighted average, and harmonic weighted average; they are expressed as follows:

$$f_{t,c} = \alpha_1(t) f_{1,t} + \alpha_2(t) f_{2,t} + \cdots + \alpha_m(t) f_{m,t} \tag{11}$$

$$f_{c,t} = f_{1,t}^{\alpha_1(t)} f_{2,t}^{\alpha_2(t)} \cdots f_{m,t}^{\alpha_m(t)} \tag{12}$$

$$f_{c,t} = \frac{1}{\frac{\alpha_1(t)}{f_{1,t}} + \frac{\alpha_2(t)}{f_{2,t}} + \cdots + \frac{\alpha_m(t)}{f_{m,t}}} \tag{13}$$

where t represents the t-th forecast period; f_c is the combined prediction result; $f_1, \cdots, f_m$ are forecast values of m individual prediction models; $\alpha_1, \cdots, \alpha_m$ are the weights of individual models, respectively; and $\alpha_1 + \alpha_2 + \cdots + \alpha_m = 1, \alpha_i \geq 0, i = 1, 2, \ldots, m$.

Solving the time-varying weights Can be regarded as an optimization problem as follows:

$$\min Q = Q(\alpha_1, \alpha_2, \cdots, \alpha_m), \ s.t. \begin{cases} \alpha_1 + \alpha_2 + \cdots + \alpha_m = 1 \\ \alpha_i \geq 0, i = 1, 2, \ldots, m \end{cases} \tag{14}$$

where Q is the objective function that indicates the combined forecast accuracy corresponding to a certain set of weights. In this paper, the Mean absolute percentage error (MAPE) is used as the optimization objective function.

In the t-th forecast period, assign all the data before the t-th period, including simulated values of the modeling period, the predicted values of the forecast period, and the actual measured values of the corresponding period, to Equation (14). Equation (14) is then solved by the sequential quadratic programming method. In the $(t+1)$-th period, the predicted value and measured value of $(t+1)$-th period are incorporated into the known knowledge of the weight solution problem by real-time learning of the latest prediction experience, thereby updating the weights of individual models.

4. Results and Discussion

4.1. Correlation Analysis Results of Meteorological Factors and Daily Load in Four Seasons

Figure 5 shows the correlation analysis results between electric load and meteorological variables. The dependent variable is the electric load, and the independent variables are the meteorological factors. The color bars in the right of each figure represent the range of the correlation coefficient, and the numbers in the figure represent the correlation coefficient. When the correlation coefficient is positive, it shows a trend toward the upper right side; when the correlation coefficient is 1, it is a straight line; when the correlation coefficient is negative, it shows a trend toward the lower left side. The smaller the correlation coefficient, the closer the graph is to a circle. The first row in the figure shows the correlation coefficients between electric load and various meteorological factors. The other rows in the figure are the correlations between various meteorological factors.

It can be seen from the correlation analysis results that the correlation between summer load and meteorological factors the most obvious, followed by autumn and winter loads, and the correlation between spring load and meteorological factors is slightly weaker. In general, air temperature (surface temperature) has the most significant impact on the seasonal load. Besides, among meteorological factors, the correlation between air temperature and the surface temperature reaches 0.9, and the correlation between relative humidity and evaporation in spring, summer, and autumn is also relatively high. It would cause multi-collinearity, which should be carefully treated.

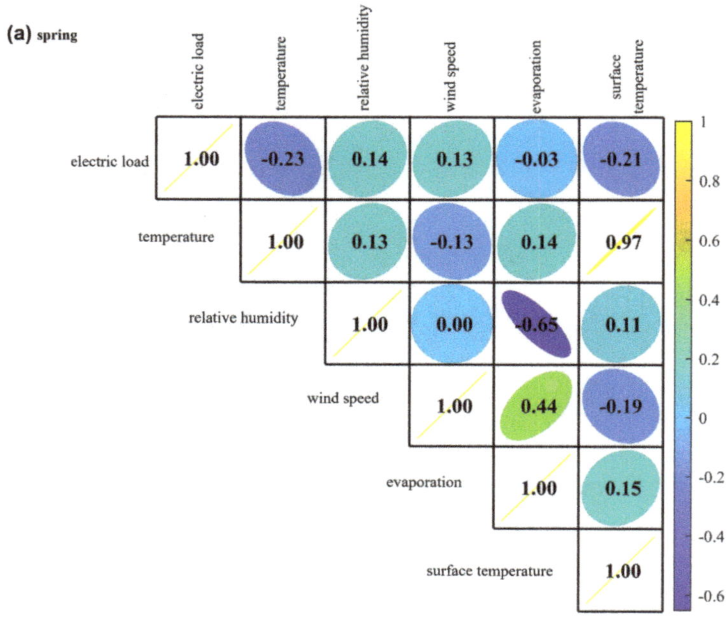
(a) spring
electric load
temperature
relative humidity
wind speed
evaporation
surface temperature
1.00 -0.23 0.14 0.13 -0.03 -0.21
1.00 0.13 -0.13 0.14 0.97
1.00 0.00 -0.65 0.11
1.00 0.44 -0.19
1.00 0.15
1.00
1 0.8 0.6 0.4 0.2 0 -0.2 -0.4 -0.6

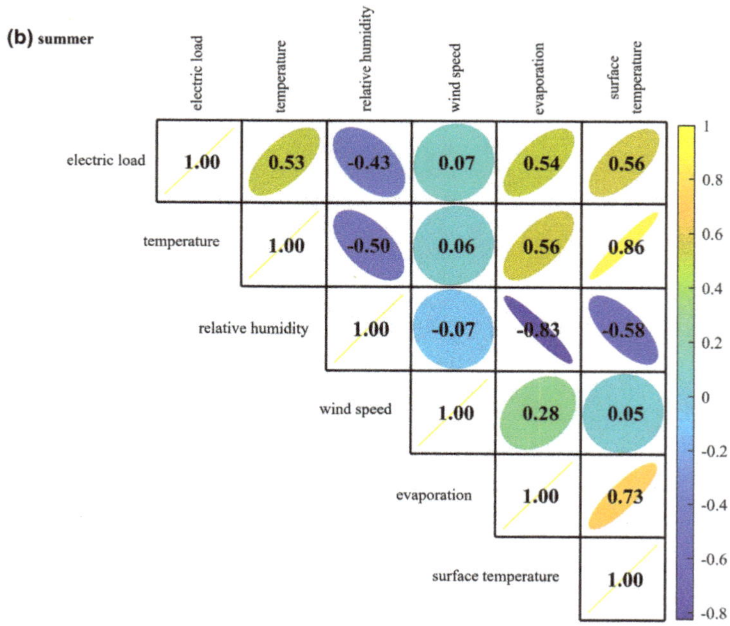
(b) summer
electric load
temperature
relative humidity
wind speed
evaporation
surface temperature
1.00 0.53 -0.43 0.07 0.54 0.56
1.00 -0.50 0.06 0.56 0.86
1.00 -0.07 -0.83 -0.58
1.00 0.28 0.05
1.00 0.73
1.00
1 0.8 0.6 0.4 0.2 0 -0.2 -0.4 -0.6 -0.8

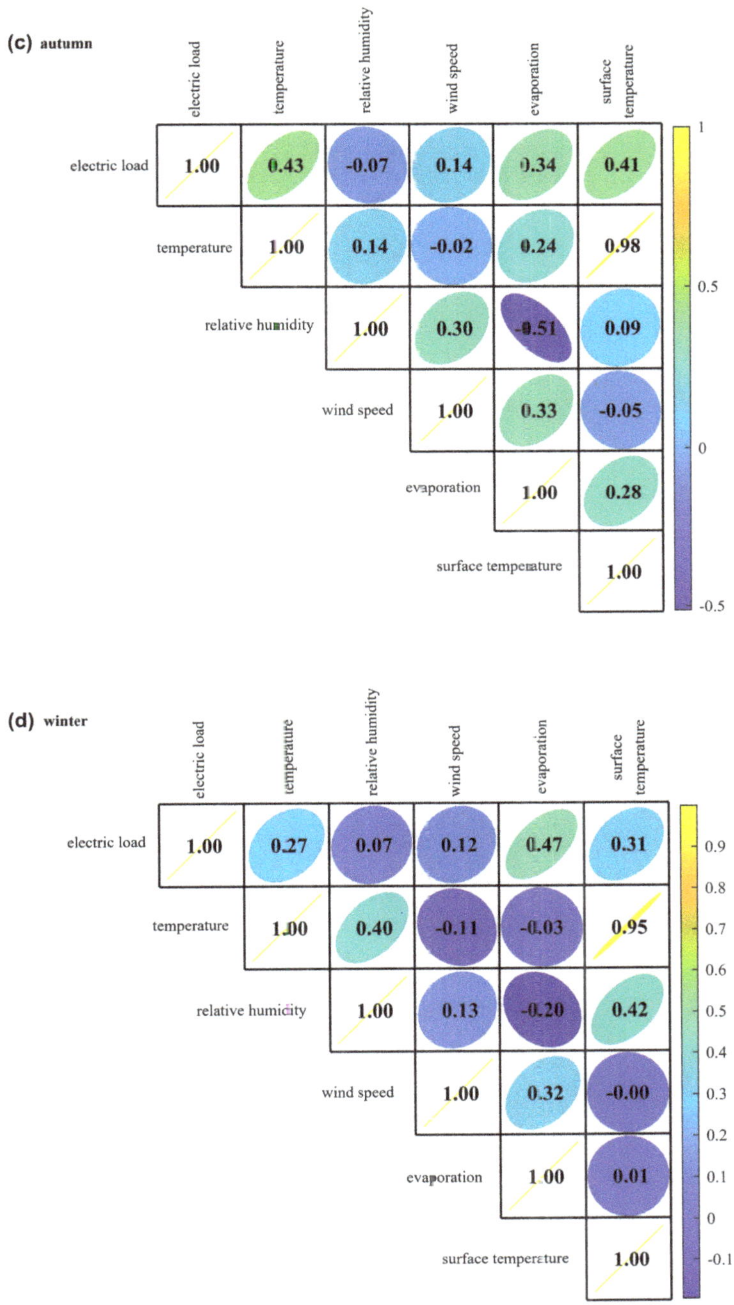

Figure 5. Correlation analysis of daily load and meteorological factors in different seasons. (**a**) Spring (March 1 to June 3); (**b**) Summer (June 4 to September 3); (**c**) Autumn (September 4 to December 3); (**d**) Winter (December 4 to February 28 of the following year).

4.2. Analysis of Time-Varying Characteristics of the Resident Load

The total regional power load is composed of three parts: commercial area, industrial area, and residential area, among which the hourly power load of commercial area and industrial area has obvious periodicity, and the influence of environmental factors is relatively small. Resident electric load is difficult to predict due to the influence of people's lifestyle and weather factors. This paper mainly studies the daily load forecasting, and tries to forecast the hourly residential load with an additional example.

The hourly load of typical daily residents in the four seasons in the region is shown in Figure 6 (select April 15 in spring, August 21 in summer, October 22 in autumn, and February 14 in winter as typical days). It can be seen from the figure that the hourly load characteristics and trends of residents on typical days in different seasons are relatively similar, and the peak load is mainly concentrated at 20 o'clock.

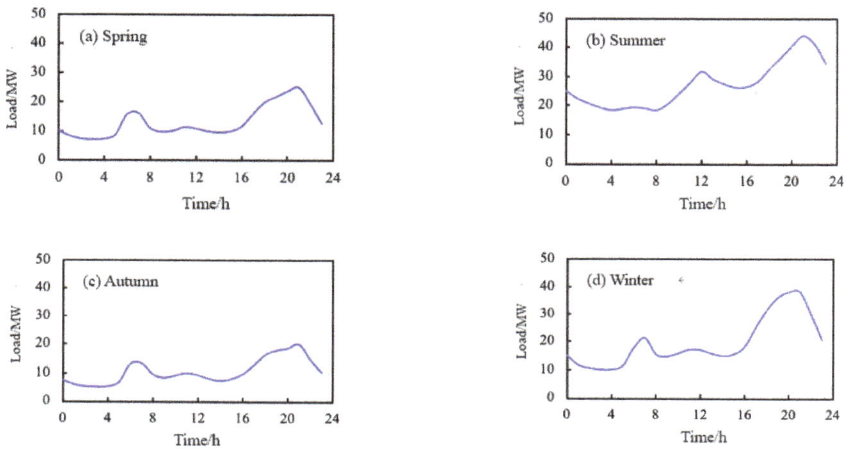

Figure 6. Hourly load changes of typical daily residents in an industrial park in different seasons.

The hourly load trend of residents on a typical day is similar. The peak electricity consumption is mainly concentrated from 17:00 to 22:00. As this period is the time when residents are off work, the load in the residential area reaches the peak of the day.

Due to weather changes, residential areas have increased air conditioning in summer, air conditioning and heating in winter, and the hourly load peak and valley values of residents in summer and winter are higher than those in spring and autumn. The peak difference between summer and spring is about 15 MW. The peak difference in autumn is nearly 20 MW.

4.3. PCA Results of the Meteorological Factors

PCA is conducted on five meteorological variables of temperature, wind speed, relative humidity, evaporation, and surface temperature on a daily scale in the four seasons The results are presented in Table 1.

Taking winter as an example, the comprehensive meteorological factor (the first principal component) contains 41.27% of information. The normalized variables temperature relative humidity, and surface temperature have similar coefficient values greater than 0.5 indicating that these three variables are closely related to the meteorological conditions. At the same time, the new factor almost contains the main information of each representative variable, which can be used as a new meteorological feature to represent the meteorological factor. Only the first principal component is used as the comprehensive meteorological factor inputting the proposed model.

Table 1. The first principal component coefficients and contribution of meteorological factors on daily scales.

Variables	Spring	Summer	Autumn	Winter
Air temperature	0.69	0.43	0.49	0.55
Relative humidity	0.15	−0.44	0.73	0.62
wind speed	−0.09	0.09	0.15	−0.02
Evaporation	0.04	0.54	−0.15	−0.16
Surface temperature	0.70	0.56	0.42	0.54
Contribution/%	49.75	64.06	39.64	41.27

4.4. Pre-Ordered Autocorrelation Load Time Lag Results

The autocorrelation function values of daily load in the four seasons are shown in Figure 7. The blue line in the figure represents the critical value of the 95% significance level of the autocorrelation function (the critical value is inversely proportional to the length of the time series). According to the length of the entire load sequence of the four seasons, we checked the critical value table of the correlation coefficient to determine the critical values to be 0.205, 0.209, 0.210, and 0.211. When the autocorrelation function value is greater than the critical value, it indicates that the time lag load is significantly correlated with the target period load; otherwise, the autocorrelation is not significant. The maximum time lag where the positive autocorrelation function value is greater than the critical value is taken as the pre-ordered autocorrelation load time lag. As a result, the pre-ordered autocorrelation load time lag is four days in spring, seven days in summer and autumn, and three days in winter.

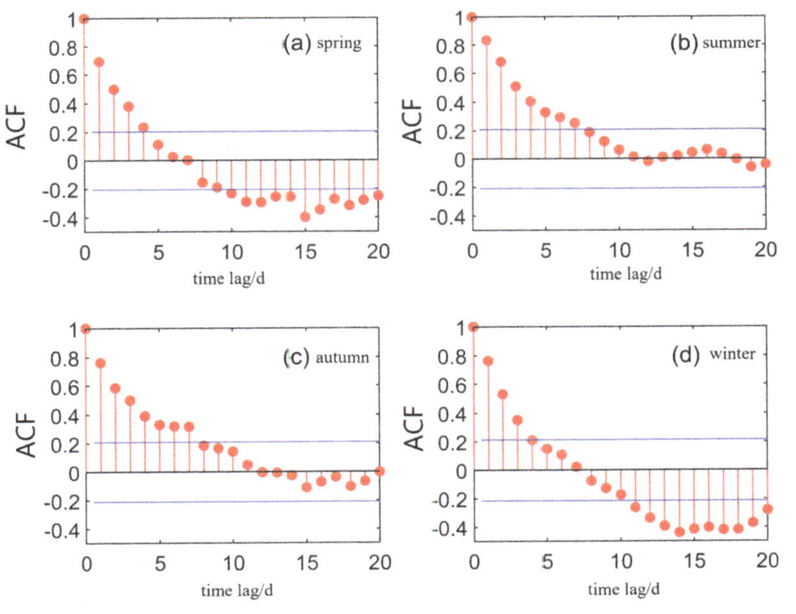

Figure 7. ACF of daily load in four seasons.

4.5. Evaluation Criteria

In this paper, MAPE and maximum absolute percentage error (max APE) are selected to evaluate the predictive performance of the model. The good stability of MAPE can be used as a benchmark for evaluation criteria [35]. They are defined as

$$MAPE = \frac{1}{n}\sum_{i=1}^{n} \frac{|\hat{y}_i - y_i|}{y_i} \times 100\,\% \tag{15}$$

$$\max\_APE = \max\left(\frac{|\hat{y}_i - y_i|}{y_i} \times 100\,\%;\ i = 1,2,\ldots,n\right) \tag{16}$$

where n is the number of samples, $\hat{y}_i$ is the predicted value, and y_i is the actual value.

4.6. Load Forecasting Results

First, the forecasting results obtained by individual models are summarized and compared in Table 2 and Figure 8. Figure 8 shows the statistical error results of the OPT-SVM and Elman model for the four-season daily load in simulation period and forecast period. In the figure, the cross x represents the average position, the horizontal line in the box represents the median, the upper and lower boundaries of the box are the 75% and 25% quantiles, and the upper and lower ends of the whiskers line represent the maximum and minimum values except for outliers. It can be seen that for the daily load in spring, summer, and autumn, the average error, 75%, and 25% quantile of ENN are significantly smaller than OPT-SVM; and in winter, the average error and 75% quantile of ENN are larger than that of OPT-SVM. Therefore, ENN has a better forecasting performance for the daily load in spring, summer, and autumn, while OPT-SVM is more suitable for the daily load forecasting in winter.

Table 2. Forecasting accuracy comparison of OPT-SVM and ENN.

Season	Model	Modeling Stage			Prediction Stage		
		Sequence Length (d)	MAPE(%)	max_APE(%)	Sequence Length (d)	MAPE(%)	max_APE (%)
Spring	OPT-SVM	68	1.98	2.90	23	1.96	3.10
	ENN	68	1.08	7.90	23	1.54	8.02
Summer	OPT-SVM	63	3.87	6.29	22	5.69	24.36
	ENN	63	2.69	12.23	22	3.90	26.41
Autumn	OPT-SVM-PCA	63	2.54	4.12	21	1.81	4.15
	ENN	63	1.58	5.85	21	2.04	8.10
Winter	OPT-SVM-PCA	63	3.50	9.70	21	2.80	4.80
	ENN	63	4.28	72.49	21	3.21	9.74

For the hourly load forecast of residents in summer and winter shown in Figures 9 and 10, the addition of meteorological factors (OPT-SVM-T and OPT-SVM-PCA) to the pre-autocorrelation load helps to improve the prediction accuracy. Here, OPT-SVM-T refers to a load forecasting model that considers the preorder autocorrelation load and adds air temperature as input. Support vector machine model: (1) Load forecasting model (OPT-SVM) that does not consider meteorological variables and takes pre-order autocorrelation load as input; (2) Load forecast model that takes pre-order autocorrelation load and temperature as input (OPT-SVM-T); (3) The load forecasting model (OPT-SVM-PCA) that considers the pre-order load and adds the comprehensive meteorological variable factor (this article uses the first principal component after the principal component analysis of multiple meteorological variables as the comprehensive meteorological factor) as the input. Among them, the use of pre-autocorrelation load and comprehensive weather and OPT-SVM-PCA with the factor as the input has a better forecasting effect on the summer resident load, and the OPT-SVM-T taking into account the temperature factor is more effective for improving the accuracy of the winter resident load forecasting.

1. For the hourly load of residents in summer, the Elman-PCA, which adds comprehensive weather factors, has higher accuracy in load forecasting than the other two models. For the hourly load of residents in winter, the Elman-T model has a better application effect.
2. For the hourly load of residents in summer and winter, both Elman-PCA and Elman-T are better than the optimized support vector machine model under the same input conditions. Here, Elman-T refers to Elman neural network that considers the pre-order autocorrelation load and adds the temperature as input to the load forecasting model.
3. From the time-by-period error analysis, it is found that the forecast accuracy of the load forecasting model changes with time, and models with lower average forecasting accuracy in a few periods also contain useful information that can help improve the forecasting effect.

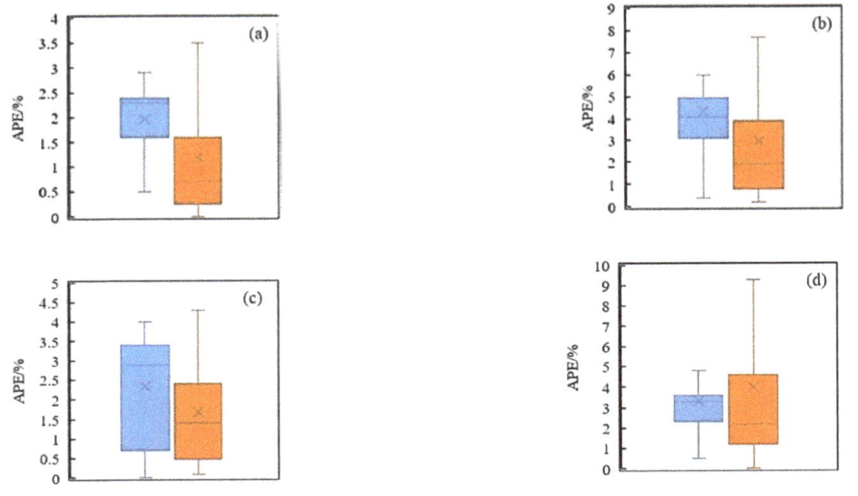

Figure 8. Statistical chart of total load error in four seasons. (**a**) Spring (March 1 to June 3); (**b**) Summer (June 4 to September 3); (**c**) Autumn (September 4 to December 3); (**d**) Winter (February 4 to February 28 of the following year).

The forecast accuracy of the individual models changes over time. Even the models with lower average forecasting accuracy in a few periods also contain useful information that can help improve the forecasting accuracy. As the prediction accuracy of individual models have a good positive correlation with their respective weights, and the weights can also reflect the contribution of the predicted value of each individual model to the combined forecasting model. Based on the results above, we then analyze the time-varying weights of every single model.

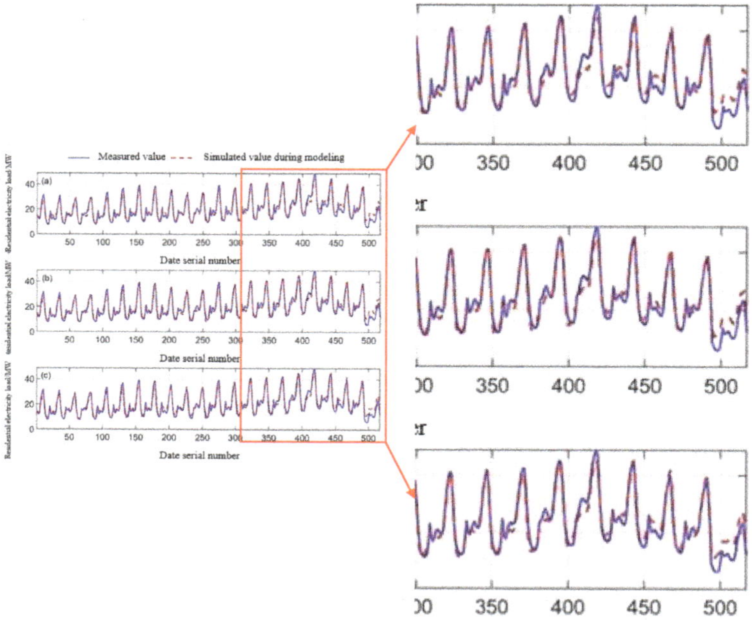

Figure 9. Optimized support vector machine model to predict the hourly load of residents in winter: (**a**) OPT-SVM (**b**) OPT-SVM-T (**c**) OPT-SVM-PCA.

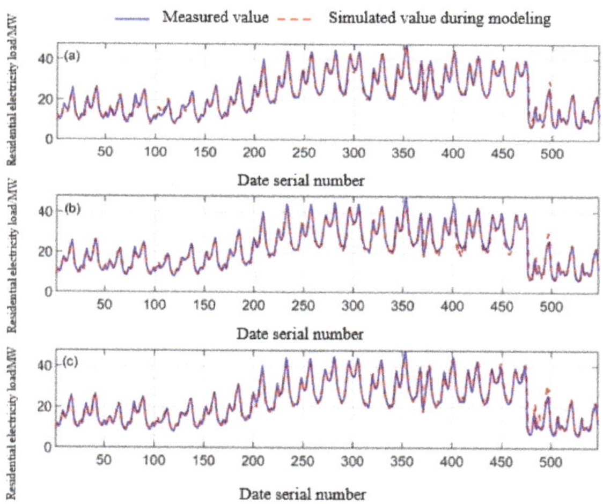

Figure 10. Hourly power load forecasting in residential area in summer: (**a**) OPT-SVM (**b**) OPT-SVM-T (**c**) OPT-SVM-PCA.

Figure 11 shows the time-varying weights of individual models through linear weighted average daily load forecasting in the four seasons. It can be seen from Figure 11 that ENN weights are all higher than OPT-SVM in each forecast period in spring, summer, and autumn, indicating that ENN forecast accuracy is generally higher than OPT-SVM in these three seasons; in winter, the performance of the two prediction model reverses: OPT-SVM has better prediction effect and correspondingly greater weights than ENN. Besides, it can be seen that the time-varying characteristics of the weights of OPT-SVM and ENN in different seasons vary: (1) Before the 13th day in spring, the weights of a single model remain basically stable. After that, the weights of ENN show a small drop while the weights of OPT-SVM increase. (2) During the summer period, the weights of the two models in the combined model have a small change, especially in the late forecast period, the ENN weights stay stable and close to 1. (3) In the autumn forecast period, with the accumulation of forecast experience, the ENN weights slowly decrease while OPT-SVM weights increase accordingly. (4) The ENN weights and OPT-SVM weights in the winter forecast period are between 0.4–0.5 and 0.5–0.6, respectively. According to the time-varying characteristics, ENN weights and OPT-SVM weights can be divided into three segments in winter, and both weights in the middle segment are getting closer.

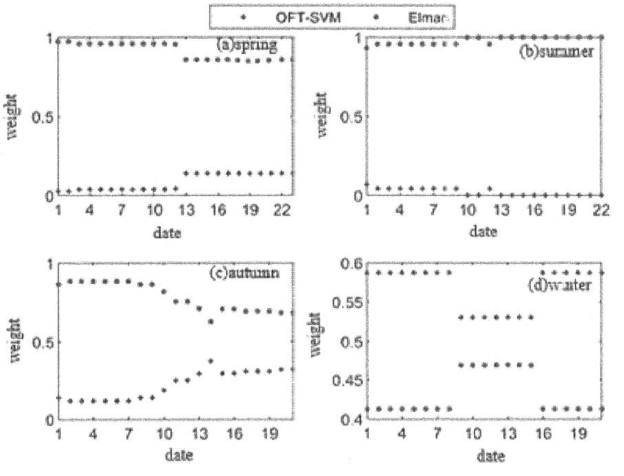

Figure 11. Time-varying weights of individual models through linear weighted average daily load forecasting.

Figure 12 shows the time-varying weights of individual models through harmonic weighted average daily load forecasting in the four seasons. Comparing Figure 11, we can see that although the mathematical expressions of the combined prediction are different, the time-varying characteristics of the weights of ENN and OPT-SVM have not changed significantly. Due to space limitations, the time-varying weights of the geometric weighted average method are no longer given.

Finally, the prediction accuracy of the individual models and combined models is compared in Table 3. The following observations have been noted. (1) For load forecasting in spring, the MAPEs and maxAPEs of the three combined forecasting models are generally close to each other; the MAPEs of the three combined models are lower than that of OPT-SVM and ENN, and the maxAPEs of combined models are between two individual models. (2) For the daily load forecasting in summer, the MAPEs and maxAPEs of the three combined models are close to or slightly lower than that of OPT-SVM and ENN. (3) For the daily load forecasting in autumn, the MAPEs and maxAPEs of the combined models for daily load forecast are both greater than the individual models. (4) In winter, the combined forecasting models significantly decrease the MAPEs while the maxAPEs of the combined models are between that of OPT-SVM and ENN.

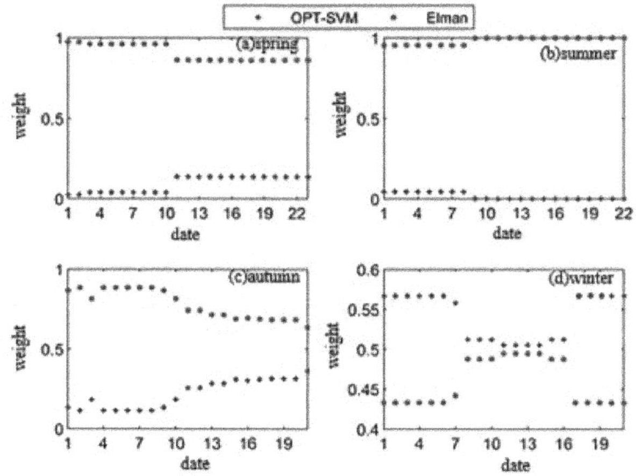

Figure 12. Time-varying weights of individual models through harmonic weighted average daily load forecasting.

Table 3. Prediction result comparison of the individual models and combined models.

Season	Model	Prediction Sequence Length	MAPE/%	max_APE/%
Spring	OPT-SVM	23	1.96	2.88
	ENN	23	1.54	7.53
	Linear weighted average	23	1.49	7.34
	Geometric weighted average	23	1.49	7.34
	Harmonic weighted average	23	1.48	7.34
Summer	OPT-SVM	22	5.68	22.59
	ENN	22	3.80	25.01
	Linear weighted average	22	3.82	25.01
	Geometric weighted average	22	3.82	25.00
	Harmonic weighted average	22	3.81	25.01
Autumn	OPT-SVM	21	1.80	3.96
	ENN	21	2.03	7.57
	Linear weighted average	21	1.85	5.93
	Geometric weighted average	21	1.86	5.94
	Harmonic weighted average	21	1.87	5.95
Winter	OPT-SVM	21	2.80	4.25
	ENN	21	3.20	9.31
	Linear weighted average	21	2.54	6.10
	Geometric weighted average	21	2.55	6.23
	Harmonic weighted average	21	2.56	6.35

We further rank the prediction errors of five models in each prediction period. Figure 13 shows the ranking results. The smaller the blue circle in the figure, the smaller the absolute percentage error. As shown in Figure 13, the number of periods with the smallest forecast error of the harmonic weighted average method is significantly higher than the other two combined forecasting models. Based on the above analysis, it can be concluded that the harmonic weighted average is the best model among the three combined models for daily load forecasting of the four seasons. For the daily load forecasting in spring and winter, the combined model can be used to archive better performance than a single model; for the daily load forecasting in summer, the accuracy of the combined models is basically the same as that of the individual models; for the daily load forecasting in autumn, the prediction accuracy of the combined models is between OPT-SVM and ENN.

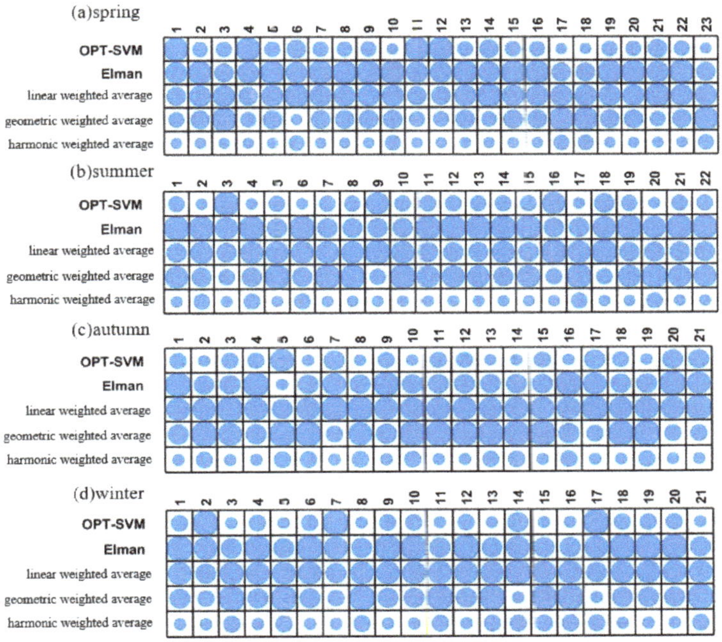

Figure 13. Prediction error ranking results for each period with different models. (**a**) Spring (March 1 to June 3); (**b**) Summer (June 4 to September 3); (**c**) Autumn (September 4 to December 3); (**d**) Winter (December 4 to February 28 of the following year).

Further research on the forecasting accuracy improvement of the combined model in each time period is of great significance for deepening the understanding of the effectiveness of combined forecasting. Figure 14 shows the forecast accuracy of the combined models for the total daily load of the four seasons compared with the single models. As can be seen from the Figure 14 compared with OPT-SVM, the harmonic weighted average has a positive error reduction value for multiple periods when predicting the total daily load in spring, and finally shows a greater improvement in the average prediction accuracy of OPT-SVM. Compared with Elman, only a few time period errors have been slightly reduced. At the same time, some time periods are affected by OPT-SVM, and the error is slightly increased. The combined effect causes the harmonic weighted average to be smaller than the OPT-SVM average error. For the total load of autumn, although the forecast error in some periods has also been effectively improved, the errors in the two periods are significantly increased, which means the average accuracy of the harmonic weighted average forecast is worse than OPT-SVM. Based on the above analysis, it can be seen that the improvement of the average forecast accuracy of the combined forecasting model can roughly reflect the overall situation of the forecast error reduction in each period. When the average forecasting accuracy is improved, it can be trusted that the combined forecasting model can get closer to the actual measurement in more time periods. When the average prediction accuracy is worse than the optimal single model, it can be considered how to make full use of the partial error to obtain the prediction result of the compressed period, so as to provide useful guidance for further improving the prediction accuracy.

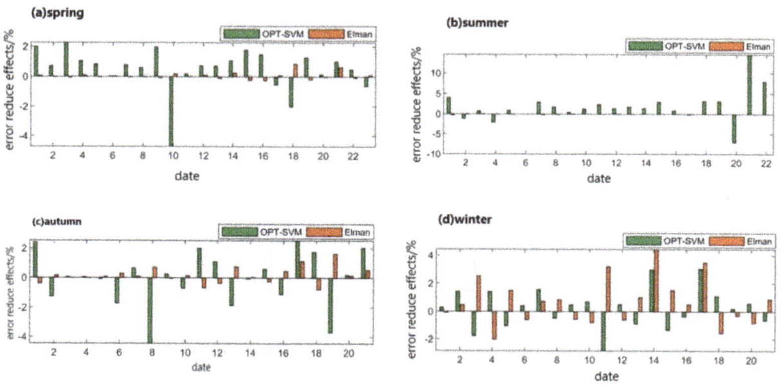

Figure 14. The forecast error reduction effects of the combined model by harmonic weighted average compared with the single models. (**a**) Spring (March 1 to June 3); (**b**) Summer (June 4 to September 3); (**c**) Autumn (September 4 to December 3); (**d**) Winter (December 4 to February 28 of the following year).

5. Conclusions

Electric load forecasting plays an important role in the power system. In order to balance the supply and demand of power system in integrated energy, it is necessary to establish a scientific model for power load forecasting. This paper takes an industrial park in Nantong as an example and establishes forecasting models that consider meteorological factors.

This paper establishes a prediction model considering meteorological factors, including an optimization vector machine model, Elman neural network model, and time-varying weight combination prediction model, that studies the impact of meteorological factors on power load forecasting and evaluates the applicability of the model. Based on the analysis of the variation characteristics of regional power load in different seasons and typical days, the correlation between meteorological factors such as temperature, relative humidity, wind speed, evaporation and surface temperature, and load is analyzed, so as to provide basis for selecting key meteorological factors. From the analysis results, the temperature has the greatest impact on the load in four seasons.

When the average forecasting accuracy is improved, the combined forecasting models can obtain prediction results closer to the actual measured values in more periods. The results show that the harmonic weighted average method is recommended for the optimal combined forecasting model. The first mock exam is better than the first mock exam model, especially for the total load in spring and winter. The average absolute percentage error (MAPE) of the forecast is lower than that of the single model.

Improving the accuracy of short-term regional power load forecasting is a long-term work. Its forecasting models are diverse and the influencing factors are very complex. This paper only analyzes the impact of some meteorological factors on short-term load forecasting. In the field of power load forecasting, whether from the exploration of load forecasting model or the comprehensive consideration of influencing factors, there are still many research contents and methods worthy of further test, which can be studied and analyzed from the following perspectives:

1. In this paper, considering the influence of meteorological factors, in the follow-up research, we can take into account the plot type, week type, and cultural activities, and establish a joint model with multiple influencing factors to investigate the impact of comprehensive influencing factors on regional short-term power load.
2. The follow-up research can consider prediction models and methods to make intelligent prediction for more different regions, different utilization types, and different environmental factors.

Author Contributions: K.Q. developed the concept, conceived the experiments, designed the study, and wrote the original manuscript. X.W., Y.Y. and K.Q. reviewed and edited the manuscript. All authors have read and agreed to the published version of the manuscript.

Funding: This work is supported by the research project "Research on typical mode and key technology of multi station integration based on rescurce intensive sharing" (CEEC2020-KJ07) of China energy Engineering group.

Acknowledgments: This work is supported by the research project "Research on typical mode and key technology of multi station integration based on resource intensive sharing" (CEEC2020-KJ07) of China energy construction group. At the same time, this project is also a part of the research project "Research on integrated energy system planning and business model" (GSKJ2-X03-2021) of China energy Engineering group planning and Design Co., Ltd.

Conflicts of Interest: The authors declare no conflict of interest.

References

1. Raza, M.Q.; Khosravi, A. A review on artificial intelligence based load demand forecasting techniques for smart grid and buildings. *Renew. Sustain. Energy Rev.* **2015**, *50*, 1352–1372. [CrossRef]
2. Fahad, M.U.; Arbab, N. Factor affecting short term load forecasting. *J. Clean Energy Technol.* **2014**, *2*, 305–309. [CrossRef]
3. Khatoon, S.; Singh, A.K. Effects of various factors on electric load forecasting: An overview. In Proceedings of the 2014 6th IEEE Power India International Conference (PIICON), Delhi, India, 5–7 December 2014; pp. 1–5.
4. Amjady, N. Short-term hourly load forecasting using time-series modeling with peak load estimation capability. *IEEE Trans. Power Syst.* **2001**, *16*, 498–505. [CrossRef]
5. Božić, M.; Stojanović, M.; Stajić, Z.; Floranović, N. Mutual information-based inputs selection for electric load time series forecasting. *Entropy* **2013**, *15*, 926–942. [CrossRef]
6. Pappas, S.S.; Ekonomou, L.; Karampelas, P.; Karamousantas, D.C.; Katsikas, S.K.; Chatzarakis, G.E.; Skafidas, P.D. Electricity demand load forecasting of the hellenic power system using an arma model. *Electr. Power Syst. Res.* **2010**, *80*, 256–264. [CrossRef]
7. Goia, A.; May, C.; Fusai, G. Functional clustering and linear regression for peak load forecasting. *Int. J. Forecast.* **2010**, *26*, 700–711. [CrossRef]
8. Liao, N.; Hu, Z.; Ma, Y.; Lu, W. Review of the short-term load forecasting methods of electric power system. *Power Syst. Prot. Control* **2011**, *39*, 147–152.
9. Park, D.C.; El-Sharkawi, M.A.; Marks, R.J.; Atlas, L.E.; Damborg, M.J. Electric load forecasting using an artificial neural network. *IEEE Trans. Power Syst.* **1991**, *6*, 442–449. [CrossRef]
10. Nose-Filho, K.; Lotufo, A.D.P.; Minussi, C.R. Short-term multinodal load forecasting using a modified general regression neural network. *IEEE Trans. Power Deliv.* **2011**, *26*, 2862–2869. [CrossRef]
11. Viegas, J.L.; Vieira, S.M.; Melício, R.; Mendes, V.M.F.; Sousa, J.M.C. Ga-ann short-term electricity load forecasting. In *Doctoral Conference on Computing, Electrical and Industrial Systems*; Springer: Berlin/Heidelberg, Germany, 2016; pp. 485–493.
12. Ho, K.-L.; Hsu, Y.-Y.; Chen, C.-F.; Lee, T.-E.; Liang, C.-C. Tsau-Shin Lai, and Kung-Keng Chen. Short term load forecasting of taiwan power system using a knowledge-based expert system. *IEEE Trans. Power Syst.* **1990**, *5*, 1214–1221.
13. Mohandes, M. Support vector machines for short-term electrical load forecasting. *Int. J. Energy Res.* **2002**, *26*, 335–345. [CrossRef]
14. Wang, J.; Zhou, Y.; Chen, X. Electricity load forecasting based on support vector machines and simulated annealing particle swarm optimization algorithm. In Proceedings of the 2007 IEEE International Conference on Automation and Logistics, Jinan, China, 18–21 August 2007; pp. 2836–2841.
15. Pai, P.-F.; Hong, W.-C. Forecasting regional electricity load based on recurrent support vector machines with genetic algorithms. *Electr. Power Syst. Res.* **2005**, *74*, 417–425. [CrossRef]
16. Niu, D.; Wang, Y.; Wu, D.D. Power load forecasting using support vector machine and ant colony optimization. *Expert Syst. Appl.* **2010**, *37*, 2531–2539. [CrossRef]
17. Li, W.-Q.; Chang, L. A combination model with variable weight optimization for short-term electrical load forecasting. *Energy* **2018**, *164*, 575–593. [CrossRef]
18. Bates, J.M.; Granger, C.W.J. The combination of forecasts. *J. Oper. Res. Soc.* **1969**, *20*, 451–468. [CrossRef]
19. Zhou, D.-Q.; Wu, B.-L. Optimization and power load forecasting of gray bp neural network model. *Power Syst. Prot. Control* **2011**, *39*, 65–69.
20. Xiao, L.; Shao, W.; Liang, T.; Wang, C. A combined model based on multiple seasonal patterns and modified firefly algorithm for electrical load forecasting. *Appl. Energy* **2016**, *167*, 135–153. [CrossRef]
21. Liu, Z. Short-Term Load Forecasting Based on Chaos Theory and Support Vector Machine. Master's Thesis, Yanshn University, Qinhuangdao City, China, 2017.
22. Zhang, J.; Wei, Y.-M.; Li, D.; Tan, Z.; Zhou, J. Short term electricity load forecasting using a hybrid model. *Energy* **2018**, *158*, 774–781. [CrossRef]

23. Singh, P.; Dwivedi, P. Integration of new evolutionary approach with artificial neural network for solving short term load forecast problem. *Appl. Energy* **2018**, *217*, 537–549. [CrossRef]
24. Tian, C.; Hao, Y. A novel nonlinear combined forecasting system for short-term load forecasting. *Energies* **2018**, *11*, 712. [CrossRef]
25. Elattar, E.E.; Sabiha, N.A.; Alsharef, M.; Metwaly, M.K.; Abd-Elhady, A.M.; Taha, I.B.M. Short term electric load forecasting using hybrid algorithm for smart cities. *Appl. Intell.* **2020**, *50*, 3379–3399. [CrossRef]
26. Sun, C.; Lyu, Q.; Zhu, S.; Zheng, W.; Cao, Y.; Wang, J. Ultra-short-term Power Load Forecasting Based on Two-layer XGBoost Algorithm Considering the Influence of Multiple Features. *High Volt. Eng.* **2021**, *47*, 2885–2898.
27. Jin, C.; Lu, X.; Xu, Y.; Liu, R.; Zhang, J. Short-term Load Forecasting Based on Similar Days and Multi-integration Combination. CSU-EPSA. Available online: https://kns.cnki.net/kcms/detail/detail.aspx?doi=10.19635/j.cnki.csu-epsa.000814 (accessed on 30 August 2021).
28. Wang, S.; Zhang, Z. Short-term Load Forecasting of Power System Based on Quantum Weighted GRU Neural Network. CSU-EPSA. Available online: https://kns.cnki.net/kcms/detail/detail.aspx?doi=10.19635/j.cnki.csu-epsa.000804 (accessed on 30 August 2021).
29. Xu, X.; Zhao, Y.; Liu, Z.; Li, L.; Lu, Y. Daily Load Characteristic Classification and Feature Set Reconstruction Strategy for Short-term Power Load Forecasting. *Power Syst. Technol.* Available online: https://kns.cnki.net/kcms/detail/11.2410.TM.20210705.0939.001.html (accessed on 30 August 2021).
30. Wang, J.; Zheng, J.; Wang, X.; Yu, J. Research on Power System Load Forecasting Based on Improved Deep Belief Network. CSU-EPSA. Available online: https://kns.cnki.net/kcms/detail/detail.aspx?doi=10.19635/j.cnki.csu-epsa.000781 (accessed on 30 August 2021).
31. Cao, Y.; Zhang, Z.J.; Zhou, C. Data processing strategies in short term electric load forecasting. In Proceedings of the 2012 International Conference on Computer Science and Service System, Nanjing, China, 11–13 August 2012; pp. 174–177.
32. Guo, X.-C.; Chen, Z.-Y.; Ge, H.-W.; Liang, Y.-C. Short-term load forecasting using neural network with principal component analysis. In Proceedings of the 2004 International Conference on Machine Learning and Cybernetics (IEEE Cat. No. 04EX826), Shanghai, China, 26–29 August 2004; Volume 6, pp. 3365–3369.
33. Sood, R.; Koprinska, I.; Agelidis, V.G. Electricity load forecasting based on autocorrelation analysis. In Proceedings of the 2010 International Joint Conference on Neural Networks (IJCNN), Barcelona, Spain, 18–23 July 2010; pp.1–8.
34. Sudibyo, S.; Murat, M.N.; Aziz, N. Simulated annealing-particle swarm optimization (sa-pso): Particle distribution study and application in neural wiener-based nmpc. In Proceedings of the 2015 10th Asian Control Conference (ASCC), Kota Kinabalu, Malaysia, 31 May–3 June 2015; pp.1–6.
35. Kong, X.; Li, C.; Zheng, F.; Wang, C. Improved deep belief network for short-term load forecasting considering demand-side management. *IEEE Trans. Power Syst.* **2019**, *35*, 1531–1538. [CrossRef]

processes

MDPI

Article

A Long-Term Decarbonisation Modelling and Optimisation Approach for Transport Sector Planning Considering Modal Shift and Infrastructure Construction: A Case Study of China

Chenxi Li, Pei Liu * and Zheng Li

State Key Lab of Power Systems, Department of Energy and Power Engineering, Tsinghua-BP Clean Energy Center, Tsinghua University, Beijing 100084, China; licx21@mails.tsinghua.edu.cn (C.L.); lz-dte@tsinghua.edu.cn (Z.L.)

* Correspondence: liu_pei@tsinghua.edu.cn

Abstract: Reducing direct carbon emissions in the transport sector is crucial for carbon neutrality. It is a considerable challenge to achieve substantial CO_2 emissions reductions while satisfying rapidly growing traffic demands. Previous studies cannot be applied directly in long-term planning for the transport sector with rapid demand growth. To bridge this gap, a multi-regional model is proposed in this paper to quantify the optimal decarbonisation path for the transport sector in order to save costs. Considering modal shift and infrastructure construction, this model regards the transport sector as a whole and China is taken as a case study. The results show that electricity and hydrogen will be the major fuels of the transport sector in the future, accounting for 45 percent and 25 percent of fuel demands in 2060. This means that the electricity used by the transport sector accounts for 10 percent of the electricity consumed by the whole of society. The results reflect that freight transport has reached a CO_2 emissions peak, while passenger transport will reach its own CO_2 emissions peak around 2041. Giving priority to decarbonisation in freight transport can save 5 percent of the transition cost. The results also suggest that modal shift can save at most 7 percent of the transition cost.

Keywords: transport decarbonisation; systematic analysis; modal shift; infrastructure; optimisation

Citation: Li, C.; Liu, P.; Li, Z. A Long-Term Decarbonisation Modelling and Optimisation Approach for Transport Sector Planning Considering Modal Shift and Infrastructure Construction: A Case Study of China. *Processes* **2022**, *10*, 1371. https://doi.org/10.3390/pr10071371

Academic Editor: Jiaqiang E

Received: 16 June 2022
Accepted: 12 July 2022
Published: 13 July 2022

Publisher's Note: MDPI stays neutral with regard to jurisdictional claims in published maps and institutional affiliations.

Copyright: © 2022 by the authors. Licensee MDPI, Basel, Switzerland. This article is an open access article distributed under the terms and conditions of the Creative Commons Attribution (CC BY) license (https://creativecommons.org/licenses/by/4.0/).

1. Introduction

The transport sector is one of the largest sources of direct CO_2 emissions, responsible for 24 percent of global CO_2 emissions (through fuel combustion) in 2019 [1]. Efficient decarbonisation measures are expected to be implemented in the transport sector to meet the 1.5 °C target and achieve carbon neutrality, or the risk of a costly and disorderly transition will increase [2]. Due to rapid populational and economic growth, traffic demands are expected to rise continuously, posing great challenges for the low-carbon transition of the transport sector. Irrational planning and reckless movements towards transition in the transport sector could lead to unacceptable expenses. The costs of developing a low-carbon transport sector can be reduced significantly by quantifying the optimal scale of different types of traffic volumes and infrastructure.

Current studies mostly focus on a single transport subsector. The decarbonisation of the road subsector has been broadly studied formerly, for the reason that CO_2 emissions from the road subsector account for 75% of CO_2 emissions from the transport sector [3]. Maduekwe et al. [4], Fan et al. [5] and Tsita et al. [6] applied LEAP to design sustainable road transport systems in Lagos, Nigeria, Beijing, China and Greece, respectively. In addition to planning for the road subsector, several studies have been conducted to focus on the influence of specific low-carbon policies or technologies in the road subsector. Virley examined the influence of high oil prices on road transport CO_2 emissions in the UK [7] and Brand et al. explored the CO_2 emission reduction potential of policy instruments that influence consumers' car preferences, such as the UK road tax [8], while Santos and Rembalski discussed UK subsidies for electric cars [9]. Dhar et al. explored the effect of electric cars on CO_2 emissions reduction in India using ANSWER-MARKAL [10]. Larsson et al. [11] and Li et al. [12] estimated the implications of fuel cell electric cars in Sweden and ASEAN countries separately, which have great potential to replace oil-fuelled cars in the future. Verger et al. overviewed the application of biomass-based fuels for heavy-duty transport [13] and Altun et al. examined the effect of using n-butanol in conventional diesel fuel–biodiesel blends on reducing engine exhaust emissions [14]. As emissions from aviation and shipping continue to grow rapidly, these hard-to-abate transport subsectors are also receiving increased attention [1]. Several studies verified the positive impact of carbon pricing on deepening emissions cuts in aviation [15,16]. Rehmatulla et al. analysed the obstacle of wind technologies on ships qualitatively and possible solutions [17], while Stolz et al. concentrated on sustainable fuels for shipping in Europe [18]. However, due to tight carbon budgets and premature utilisation of renewable technologies in each subsector, the discussion of decarbonisation in individual subsectors will inevitably result in high overall costs for the whole transport sector. It has been proved that the total costs may be significantly overestimated when decoupling a system into several subsectors [19]. Hence, to save transformation costs as much as possible, the transport sector should be viewed as a whole instead of being divided into separate subsectors. Routine analysis of the holistic transport sector is carried out with developed energy models, such as TIMES [20], LEAP [21] and EnergyPLAN. Zhang et al. compared the decarbonisation perspective of the transport sector in China and the USA using China-TIMES and US-TIMES [22], while Korkmaz et al. applied the TIMES PanEU model to establish a decarbonisation path for the transport sector in Europe [23]. LEAP and EnergyPLAN were used to evaluate the efficiency of low-carbon policies for the transport sector announced by South Korea [24] and to formulate emission-reduction policies for the transport sector in Iran [25].

Although the above research is not limited to only one subsector, modal shifts between different subsectors and the construction of infrastructures are not evaluated quantitatively. Modal shift refers to the use of vehicles from one subsector to replace the vehicles from another subsector and is also known as behaviour change. For instance, a passenger who is used to travelling by air may decide to travel by train. It has been proposed that modal shift is a means of achieving emissions reductions with extraordinary potential [26,27] and which it is necessary to consider for areas with high emission-reduction costs in the transport sector. Salvucci et al. applied elasticities of substitution to describe passenger transport modal shift and developed TIMES-DKEMS to assess the impact of policies on modal shift [28], while Strauss et al. evaluated the influence on decarbonisation in China of modal shift from air to high-speed trains [29]. Zhang et al. and Kurtuluş et al. analysed the impact and potential of modal shift on freight transport in Shenzhen, China, and in Turkey, respectively [30,31]. Similar to most of the previous studies considering modal shift, these studies did not pay enough attention to the construction of infrastructure, which is the main driver and constraint on modal selection [32]. Hickman et al. considered the transition path to low-carbon transport in London and Delhi [33], while Daly et al. reformed TIMES to optimise the costs of decarbonisation of the transport sector in Ireland and California [34]. Both sets of authors considered the construction of infrastructure, including rail tracks, and regarded the studied area as a point. However, since the length of rail tracks cannot be depicted as an integral multiple of the distance between two regions, the studies may have underestimated the cost of railway construction. Due to the fact that traffic demand in previous studies increased relatively slightly in the planning period [35–37], even if the description of infrastructure construction is simplified, there is little impact on the results. Nevertheless, the rapid or substantial growth of traffic demands cannot be avoided in the context of long-term planning in the transport sector. When considering modal shift, it is inevitable to describe the construction of infrastructure, while the majority of studies that have considered modal shift and infrastructure have greatly simplified the construction of distance-dependent infrastructures, such as railways.

According to the literature review, previous models cannot be applied directly to find the appropriate long-term low-carbon transition pathways for the transport sector. In this paper, a multi-regional model considering modal shift and infrastructure that can calculate the transmission pathway of the transport sector is proposed; the model covers four modes, including road, rail, aviation and shipping, both in passenger and freight transport. China is taken as a case study; its transport demand is considered to have large room for growth, with the transport sector accounting for about 9.12% of the country's total CO_2 emissions and 19.42% of the country's total end-use CO_2 emissions in 2019 [38]. Through the case study of China, by using this model, the following issues are expected to be resolved:

- The role of various fuels in transport in the movement towards low-carbon transition and the means of reducing CO_2 emissions in the terminal sector of transport;
- The means of minimizing the costs of the transition of the transport sector when faced with rapidly growing traffic demands;
- The benefits of favourable CO_2 reduction policies in different sectors when considering the transport sector as a whole.

The novelty of this work can be summarised in three points. Firstly, viewing the transport sector as an entirety, a multi-regional optimisation model using the superstructure method is established. This can be applied to analyse the long-term transition path of transport sectors in regions with snowballing traffic demands and unformed infrastructures and assess the construction of inter-regional infrastructures. Secondly, through the case study, a likely transitional route for the transport sector in China is described and the impact of different travel habits considering modal shift and diverse policies are analysed, which analysis can be used to explain the gravity of the prioritisation of emissions reduction in freight transport. Thirdly, the relationship between the power sector and the transport sector is pointed out quantitatively.

This paper is organised as follows: Section 1 clarifies the background of this study. Section 2 presents the method in detail. Section 3 describes the basic details of the case study. Section 4 presents the results and sensitivity analysis of essential parameters. Section 5 proposes policy recommendations and Section 6 summarises the main conclusions.

2. Methodology

2.1. Model Structure and Assumptions

A schematic diagram of the relevant concepts is shown in Figure 1 and the model structure is shown in Figure 2. Several concepts need to be clarified here. A city is the smallest unit in the spatial dimension of the model, which can be seen as a point. A region is composed of several cities. People and goods can move in a city or between different cities, which is regarded as inner-city and inter-city transport. Only trucks in freight transport, private cars, taxis, subways and buses in passenger transport are allowed to be utilised in cities. Inter-city transport includes communication between different cities in the same region and different cities in diverse regions. Transport between different cities in diverse regions can also be called inter-regional transport. All transport modes except taxis, buses and subways are allowed to be applied in inter-city transport. In this study, the construction of inter-regional rail tracks is evaluated significantly, which requires the description of the distances between regions, while the construction of rail tracks within one region is not considered. The distance between two regions is considered as the distance between the largest traffic point cities in the two regions in our model. For historical, social and economic reasons, the completeness of transport facilities varies between different cities. A traffic point city is a city with the most complete transport infrastructure and high passenger or freight throughput in a region.

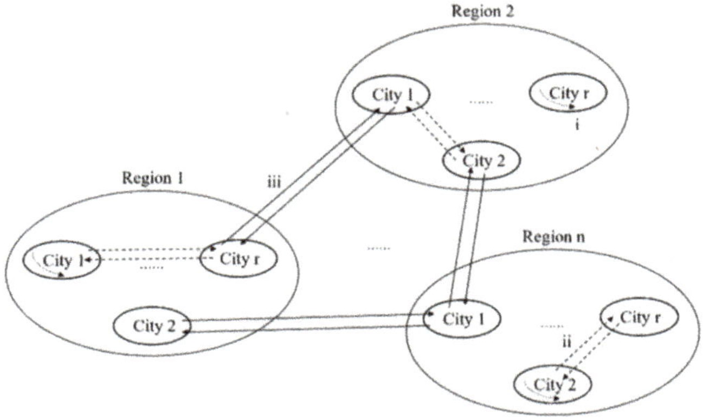

Figure 1. A schematic diagram of the relevant concepts. Line i represents inner-city transport, line ii and line iii represent inter-city transport, and line iii also represents inter-regional transport.

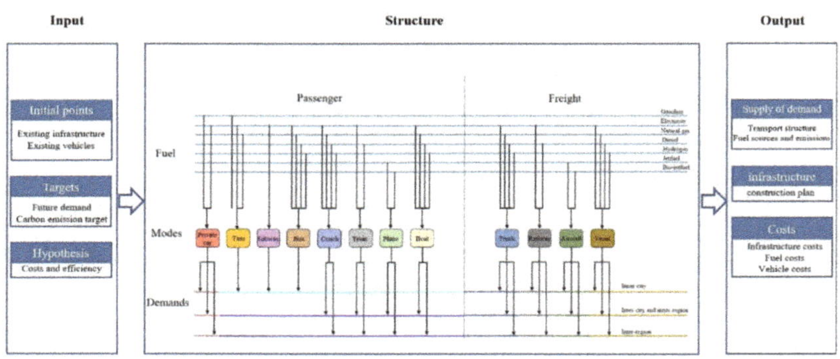

Figure 2. The framework of the model.

2.2. Mathematical Formulation

Mathematical formulas for this optimisation model are presented in this section. The sets t, ser, f and r denote time, traffic mode type (there being 12 modes, as presented in Figure 2), fuel type and region, respectively. What is more, r and rr are different names for the set region. Parameters are represented with uppercase letters and variables are represented with lowercase letters. The meanings of the different symbols are listed in Abbreviations.

2.2.1. Objective Function

The objective function of the transport optimisation model is to minimise the total cost of the transport sector from 2020 to 2060. The total cost is composed of the infrastructure cost, purchase and maintenance costs of both freight and passenger transport and fuel costs and is calculated using Equation (1). Four parts of the total cost are presented in detail in Equations (2)–(5). In this study, infrastructure, such as charging piles, subway tracks, rail tracks (including both ordinary rail and high-speed rail), airports and ports, are considered, as Equation (2) shows. The annual cost of each item of infrastructure includes two parts: new construction cost and maintenance cost. This is the same as the cost composition of vehicles, as shown in Equations (3) and (4).

$$cost = \sum_{t=2020}^{2060} \frac{infc_t + buyc_t + mainc_t + fuelc_t}{(1+I)^{t-2020}} \tag{1}$$

$$infc_t = chargec_t + metroc_t + railc_t + airportc_t + portc_t \quad (2)$$

$$buyc_t = \sum_{ser,f} BUY_{t,ser,f} * nnb_{t,ser,f} \quad (3)$$

$$mainc_t = \sum_{ser,f} MAIN_{t,ser,f} * \left(nb_{t,ser,f} - nnb_{t,ser,f}\right) \quad (4)$$

$$fuelc_t = \sum_{r,rr,ser,f} ts_{t,r,rr,ser,f} * TRFC_{t,ser,f} * FCOST_{t,r,f} \quad (5)$$

2.2.2. Supply and Demand Constraints

All demands in each transport mode are separated into two kinds: inescapable demands and changeable demands. It is considered that a certain proportion of demands are immutable in each traffic mode, which is the explanation of Equation (6). Only traffic modes with the same scope can convert mutually. In other words, in passenger transport, the modes of taxis, subways and buses can convert to each other, while the modes of coaches, trains, planes and boats can convert mutually. The four modes of freight transport can convert to each other. In this model, the participation of private cars in modal shift is not considered. Although there is a conversion between different modes, both freight and passenger transport should meet the requirement that total supply is not less than the total demand, as described in Equations (7)–(10). The constraint of travel time budget also needs to be considered in passenger transport. The total travel time budget can be computed using Equation (11), which has to meet the constraint of Equation (12), implying that the travel time budget for each person in a day is stationary every year, according to Tattini et al. [39]. Travel time budget makes the result of passenger transport mode conversion closer to reality.

$$\sum_f ts_{t,r,rr,ser,f} \geq TD_{t,r,rr,ser} * FIXPRO_{ser} \quad (6)$$

$$\sum_f \left(ts_{t,r,rr,taxi,f} + ts_{t,r,rr,subway,f} + ts_{t,r,rr,bus,f}\right) \geq TD_{t,r,rr,taxi} + TD_{t,r,rr,subway} + TD_{t,r,rr,bus} \quad (7)$$

$$\begin{gathered}\sum_f \left(ts_{t,r,rr,coach,f} + ts_{t,r,rr,train,f} + ts_{t,r,rr,plane,f} + ts_{t,r,rr,boat,f}\right) \geq \\ TD_{t,r,rr,coach} + TD_{t,r,rr,train} + TD_{t,r,rr,plane} + TD_{t,r,rr,boat}\end{gathered} \quad (8)$$

$$\sum_f nb_{t,private\ car,f} \geq NBD_{t,r} \quad (9)$$

$$\begin{gathered}\sum_f \left(ts_{t,r,rr,truck,f} + ts_{t,r,rr,railway,f} + ts_{t,r,rr,aircraft,f} + ts_{t,r,rr,vessel,f}\right) \geq \\ TD_{t,r,rr,truck} + TD_{t,r,rr,railway} + TD_{t,r,rr,aircraft} + TD_{t,r,rr,vessel}\end{gathered} \quad (10)$$

$$totalttb_t = \sum_{r,rr,pas,f} ts_{t,r,rr,pas,f} / SPEED_{pas,f} \quad (11)$$

$$totalttb_t = TTB * POP_t * 365 \quad (12)$$

2.2.3. Vehicle Constraints

For all traffic modes except private cars, the relationship between the number of vehicles and turnover described in Equation (13) should be satisfied. Private cars are generally not used commercially and it is difficult to determine the parameter $LOAD_{private\ car,f}$. In our model, it is assumed that the annual mileage of each private car is determined, avoiding the model becoming a nonlinear model. As a result, the traffic turnover of private cars in a year is shown in Equation (14). Parameter P means the average carrying capacity of a private car, which can be estimated by Equations (11) and (12) using the concept of TTB. The number of vehicles is presented in Equation (15). In this model, only oil-fuelled vehicles in the rail, aviation and shipping subsectors of passenger and freight transport can be decommissioned before their expiration date, implying that the number $retire_{t,ser,f}$ is fixed to 0 for other means of transport.

$$\sum_{r,rr} ts_{t,r,rr,ser,f} \leq LOAD_{ser,f} * nb_{t,ser,f} \tag{13}$$

$$\sum_{r,rr} ts_{t,r,rr,private\ car,f} = nb_{t,private\ car,f} * MILEAGE * P \tag{14}$$

$$nb_{t,ser,f} = nb_{t-1,ser,f} + nnb_{t,ser,f} - nnb_{t-tlt_{ser,f}-1,ser,f} - retire_{t,ser,f} \tag{15}$$

2.2.4. Infrastructure Constraints

Infrastructure needs to match the traffic demand, as calculated in Equations (16)–(23). Parameter PRO is the minimum ratio of the amount of infrastructure and traffic demand. Traffic demand can be the number of vehicles, the transport turnover or the capacity. The expiration date of infrastructure is not considered in order to simplify the model, for the reason that the lifespan of most transport facilities can be extended for a long time after regular maintenance. As a result, the number of specific transport facilities in year t equals the sum of newly added facilities in year t and maintaining facilities in year t − 1.

$$nbcharge_t \geq PRO_{charge} * \left(nb_{t,private\ car,elec} + nb_{t,taxi,elec} + nb_{t,bus,elec} + nb_{t,coach,elec} + nb_{t,truck,elec}\right) \tag{16}$$

$$nbmetro_{t,r} \geq PRO_{metro} * nb_{t,subway,r} \tag{17}$$

$$nbrail_{t,r,rr,f} * DISTANCE_{r,rr} \geq PRO_{rail,pas} * \left(ts_{t,r,rr,train,f} + ts_{t,rr,r,train,f}\right) / DISTANCE_{r,rr} \tag{18}$$

$$nbrail_{t,r,rr,f} * DISTANCE_{r,rr} \geq PRO_{rail,fre} * \left(ts_{t,r,rr,railway,f} + ts_{t,rr,r,railway,f}\right) / DISTANCE_{r,rr} \tag{19}$$

$$nbport_{t,r} \geq PRO_{port,pas} * \left(ts_{t,r,rr,boat,f} + ts_{t,rr,r,boat,f}\right) / DISTANCE_{r,rr} \tag{20}$$

$$nbport_{t,r} \geq PRO_{port,fre} * \left(ts_{t,r,rr,vessel,f} + ts_{t,rr,r,vessel,f}\right) / DISTANCE_{r,rr} \tag{21}$$

$$nbairport_{t,r} \geq PRO_{airport,pas} * \left(ts_{t,r,rr,plane,f} + ts_{t,rr,r,plane,f}\right) / DISTANCE_{r,rr} \tag{22}$$

$$nbairport_{t,r} \geq PRO_{airport,fre} * \left(ts_{t,r,rr,aircraft,f} + ts_{t,rr,r,aircraft,f}\right) / DISTANCE_{r,rr} \tag{23}$$

2.2.5. Maximum Speed of Modal Shift

In order to prevent a sudden increase in traffic supplementation, the maximum speed of modal shift is settled. This restriction has two practical implications. Firstly, the transformation of ideas takes time, meaning that the shift towards another mode cannot happen suddenly. Secondly, the number of vehicles that can be purchased is determined by the inventory, as it is impossible to satisfy an unexpected high demand immediately. The maximum speed of modal shift is set to be the increase in demand of one mode, which is depicted in Equation (24).

$$\Delta \sum_{r,rr} ts_{t,r,rr,ser,f} \leq \sum_{r,rr} \Delta TD_{t,r,rr,ser} \tag{24}$$

2.2.6. Policy Constraints

Total CO_2 emissions from the transport sector can be calculated using Equation (25). CO_2 emissions from the transport sector are subject to policy constraints and cannot exceed the specified upper limit, as shown in Equation (26).

$$ycarbon_t = \sum_{r,rr,ser,f} ts_{t,r,rr,ser,f} * TRFC_{t,ser,f} * FC_f \tag{25}$$

$$ycarbon_t \leq TRCO2_t \tag{26}$$

3. Case Study

China is used as a case study for the proposed model because of its huge economic growth potential allowing for explosive growth in traffic demand. The model of the case study is coded on the platform of the General Algebraic Modelling System (GAMS). Critical input data and fundamental assumptions are clarified in this section.

3.1. Basic Assumptions and Scenario Settings

The mainland of China, excluding Macao, Hong Kong and Hainan province, is considered in this case study and is divided into nine regions. The optimisation time interval is from 2020 to 2060; several variables for 2020 have been rectified according to published statistical data [40].

Several scenarios have been set in this case study, while crucial settings for the basic scenario discussed in Section 4 can be described as follows. Firstly, CO_2 emissions from the transport sector in China reach a peak in 2030, with no more than 1091 million tonnes and at most 100 million tonnes in 2060. At the same time, we set 2030–2035 as the platform period for decarbonisation. Secondly, modal shift in the freight sector is ignored. Although the validation of modal shift has been proven in China [31], conversion is difficult. Over the past decade, numerous policies of decarbonisation have been announced in the transport sector in China, but the shift towards low-carbon density in freight transport is not evident. The proportion of shipping has increased by only 0.7% [41]. Thirdly, electric vehicles in aviation and shipping subsectors can only be applied after 2045. The input traffic demands are shown in Tables 1 and 2.

Table 1. Demand for freight transport (billion tonne-kilometres).

	2030	**2040**	**2050**	**2060**
Truck	9169	12,286	14,979	16,898
Railway	4875	6523	7708	8237
Aircraft	39	59	78	89
Vessel	11,735	12,754	13,726	14,635

Table 2. Demand for passenger transport (billion passenger-kilometres).

	2030	**2040**	**2050**	**2060**
Taxi	247	293	317	316
Subway	309	419	514	582
Bus	1014	1298	1512	1634
Coach	607	803	967	1080
Train	1566	1938	2171	2273
Plane	1659	2704	3483	3714
Boat	10	11	12	12

3.2. Historical Quantity and Service Year of Different Means of Transportation

Historical quantities for various means of transport were obtained through statistical yearbooks for China [42–46] and public data. Remarkably, the number of vehicles increased significantly around 2010, which may result in an extensive retirement henceforth.

The service years of different types of vehicles are listed in Table A1 in Appendix A, with reference to the relevant national standards.

3.3. The Ceiling of Carrying Capacity of Vehicles and Infrastructure

Few studies have discussed the upper limit of the carrying capacity of vehicles and infrastructure in one year, making it difficult to determine the parameter $LOAD_{ser,f}$ and PRO in the case study. In this study, the determination of the above parameters refers to the current state of traffic development in China. Figures for the top ten crowded subway lines in China in 2019 were computed to settle the capacity of subways and figures for the top five busiest railway lines in China in 2019 were calculated to fix the capability of trains. The proportion of electric vehicles and charging piles is set at 3, according to the China Electric Vehicle Charging Infrastructure Promotion Alliance.

3.4. Investment Costs of Vehicles and Infrastructure

The current investment costs refer to public data, which are listed in Tables A2 and A3 in Appendix A. The costs of vehicles are expected to decline with improvements in technology. However, the future cost of infrastructure is much more complex. On the one hand, as with vehicles, the maturity of technology will lead to a decrease in costs; on the other hand, primary infrastructure is often constructed on flat terrain, while subsequent infrastructures will be constructed in relatively built-up areas, resulting in higher costs. After comprehensive consideration, we assume that infrastructure costs in different years remain constant except for the costs of charging piles.

3.5. Forecast of Costs of Fuels

The price of primary fuels is regarded as invariant in our case study. We do not assume an increase in the cost of fossil fuels or a decrease in the cost of electricity, as we consider carbon tax and subsidies to be matters of policy instead of prerequisite assumptions. In addition, the future output of fossil energy is uncertain, especially after the outbreak of coronavirus disease 2019 (COVID-19), which makes it more difficult to predict future prices [47,48]. Furthermore, the increase in renewable electricity and its supporting infrastructure will increase the cost of electricity, which will introduce uncertainty regarding the cost of electricity. However, considering the development of technologies, the cost of hydrogen and bio-jet-fuel is set to decline by 2% per year. The detailed costs are listed in Table A4.

3.6. Travel Time Budget

Schafer and Victor showed that, on average, a person spends about 1.1 h per day travelling across various societies and income classes [49]. However, this conclusion cannot be applied directly in our case study. Firstly, non-motorised travel modes, such as walking and biking, are not included in our case study due to a lack of statistics. Previous studies have shown that although non-motorised and motorised transport modes shift freely, the travel time budget for motorised transport modes is around 45 min [39], which can also be verified by statistics for China before 2020. Secondly, the influence of COVID-19 on the travel time budget of China cannot be ignored. During data calibration, it was found that the travel time budget in 2020 is far less than 45 min, owing to people being recommended to stay at home and restrict their movements. It is unknown how long the epidemic will affect China. However, the travel time budget in 2020 is the shortest in the next 40 years. Consequently, the travel time budget in our case study is constrained between the travel time budget in 2020 and 45 min. The average speed of different transport modes is listed in Table A5.

3.7. The Proportion of Specific Requirements in Passenger Transport

The review of scientific literature did not review any conclusions on the proportion of specific requirements in passenger transport. According to the characteristics of different transport modes, we set different values. The proportion of specific requirements for taxi, coach and plane is set as 0.6, while others are set as 0.4. Since this parameter is determined subjectively, we discussed its influence on this lately.

4. Results and Discussion

4.1. Optimal Structure of the Freight Sector

According to the results, the proportion of turnover powered by oil decreases in all kinds of modes of freight transport in China, as shown in Figure 3. This makes clear that although it is nearly a decade away from the peak of CO_2 emissions, the proportion of oil used in freight transport has no room to ascend. The role natural gas plays in road and shipping subsectors is divergent. Natural gas is a transitional fuel in trucks, while it plays a role as an alternative energy source until 2060 in vessels. Although the proportion of natural gas vehicles in vessels increases slowly, gas-fuelled turnover rises apparently with the enlargement of shipping demands. This shows the inevitability of the development of LNG powered ships. Gas-fuelled turnover in the shipping subsector can reach 4123 billion tonne-kilometres in 2060, accounting for 28.2%.

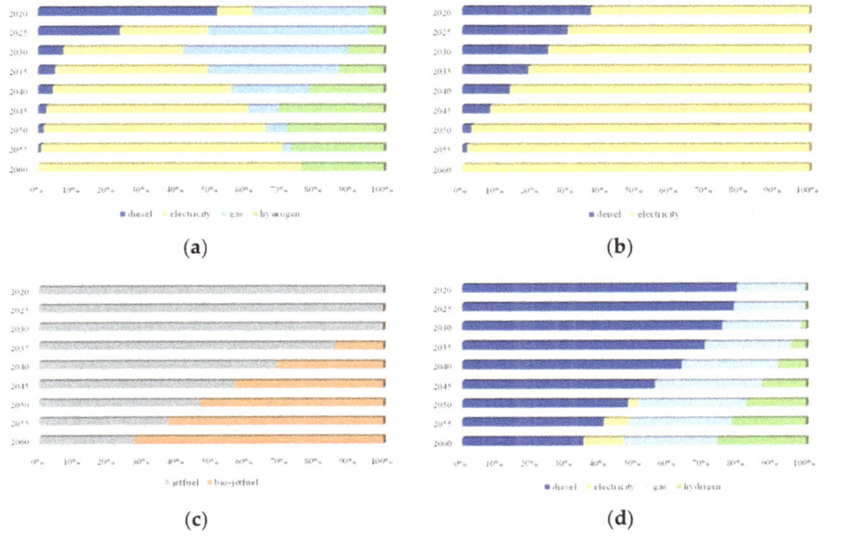

Figure 3. The structure of each freight transport subsector. (**a**) The structure of truck turnover. (**b**) The structure of railway turnover. (**c**) The structure of aircraft turnover. (**d**) The structure of vessel turnover.

As CO_2-free fuels, electricity and hydrogen are the future for freight transport in China. Hydrogen can only be applied in road and shipping subsectors. Hydrogen-powered trucks are expected to carry 15 times more turnover than in 2020 and electric trucks are expected to carry 21 times more turnover than in 2020. Hydrogen-powered vessels complement the reduction in the number of diesel-powered vessels partially due to carbon reduction, but hydrogen cannot replace diesel due to the high costs. Electricity will be one of the major energy sources for the future freight transport sector, providing 35%, 75% and 0% of energy needs in 2030 and 76%, 100% and 12% in 2060 for the road, railway and shipping subsectors.

4.2. Optimal Structure of the Passenger Sector

Figures 4 and 5 describe the inner-city passenger turnover and the structure of taxi and bus turnover. It can be observed, unexpectedly, that the turnover of buses rises incredibly due to modal shift. The reason for the slow growth of subway turnover is that the high cost of track construction limits the development of subways. However, bus turnover does not increase continuously. It reaches a peak in 2049 and then declines. Excessive bus travelling will add to travel time budgets, which is not in conformity with the law of constant travel time budgets. One unanticipated finding from Figure 5 is that oil use in both taxi and bus turnover increases first and then decreases, while the proportion of electricity shows an opposite trend. A possible explanation for this might be that passenger transport contributes less to CO_2 emissions and that the pressure to decarbonise is lighter. For the sake of saving costs, the freight transport should decarbonise first, while passenger transport should reduce its CO_2 emissions later. The decrease in the proportion of electricity-powered turnover implies that developing electric vehicles immediately is not a necessity, due to the high cost of electric vehicles and charging piles. This result is also reasonable. At present, many electric vehicles are promoted by demonstration projects or subsidies with which we are not concerned in our case study. The results suggest that electric vehicles are not economically competitive without subsidies at present. The trend for gas-fuelled taxi turnover changes three times. It was found that there was a boom in gas-fuelled taxies in the past decade and a rapid decline soon after, causing a downward trend from 2020 to 2035. After 2035, natural gas plays an essential role as a transitional fuel in taxi transport. Taxies are supposed to have 100% access to electricity in 2060.

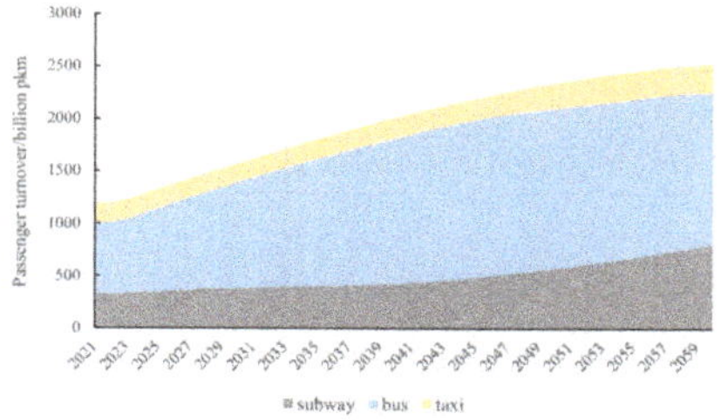

Figure 4. Inner-city passenger turnover.

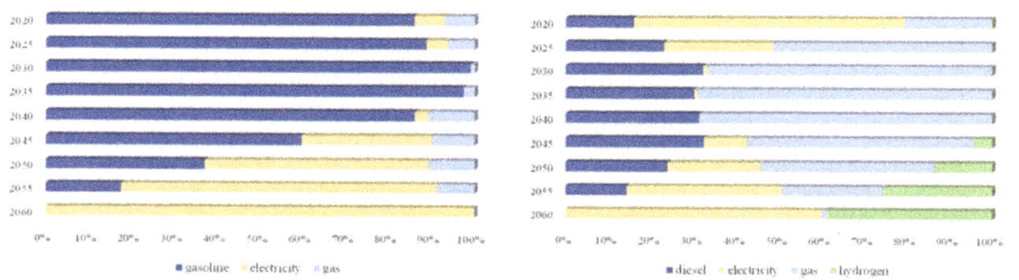

(a) (b)

Figure 5. The structure of each inner-city passenger transport mode. (**a**) The structure of taxi turnover. (**b**) The structure of bus turnover.

The results for inter-city passenger turnover are shown in Figure 6 and the structure of each inter-city passenger transport mode is presented in Figure 7. Despite a large expansion in the aviation subsector in Figure 6, the rail subsector has replaced part of the demand. Compared with our forecast, the optimisation results for train turnover are higher. The optimisation result is 634 billion pkm higher in 2030 and 1894 billion pkm higher in 2060 than the prediction. Meanwhile, no new rail track construction is required at the resolution of our case study. Among the infrastructures considered, only airports in the eastern region of China are supposed to be expanded by 38% by 2060. These results suggest that the inter-city transport infrastructure of China is relatively advanced. A huge sum of money for building new infrastructure can be saved if existing infrastructures are reasonably utilised.

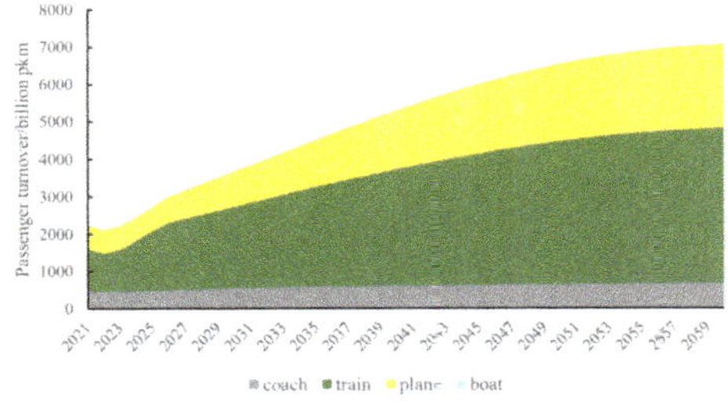

Figure 6. Inter-city passenger turnover.

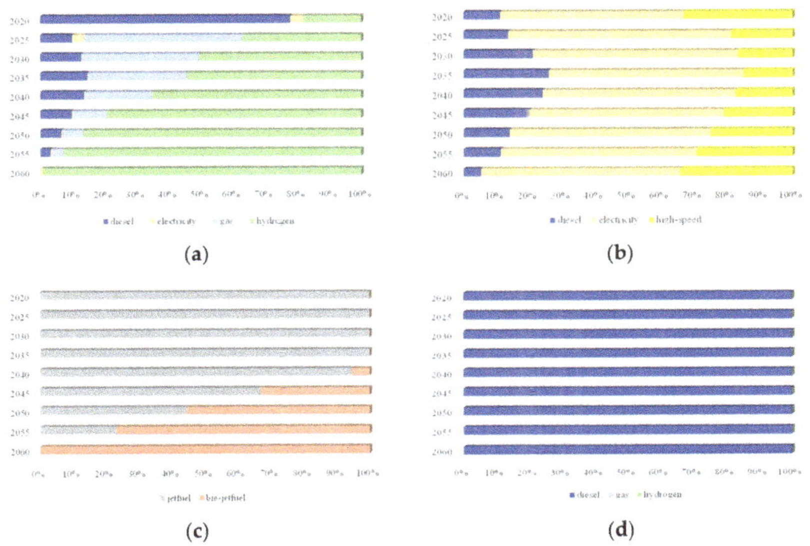

Figure 7. The structure of each inter-city passenger transport mode. (**a**) The structure of coach turnover. (**b**) The structure of train turnover. (**c**) The structure of plane turnover. (**d**) The structure of boat turnover.

Similar to inner-city passenger transport, the use of fossil energy will not decline immediately. Although the use of diesel in the road subsector decreases significantly between 2020 and 2025, the use of natural gas increases largely at the same time. Gas-fuelled road transport reaches a peak at 289 billion pkm in 2024. The reason for the decline of electricity-powered road passenger vehicles is similar to the reason for that of electric taxis and buses. However, hydrogen seems to be competitive in coaches. Too many electric vehicles will lead to a surge in charging piles, increasing the total cost significantly. Among all hydrogen-powered road vehicles from passenger and freight transport, the cost per unit turnover of coaches is the lowest, making hydrogen more competitive than electricity in coaches. Contrary to expectations, diesel train turnover increases until 2040. However, this has little impact on CO_2 emissions, owing to the low carbon density of railway passenger transport. CO_2 emissions contributed by diesel-fuelled trains reach a peak at 5.1 million tonnes in 2039, which is only 0.5% of the total emissions of the transport sector for that year. Compared to aviation freight transport, air passenger transport has to eliminate jet-fuel-powered planes in 2060. There is no need to transform the fuel used in shipping passenger transport, for the reason that shipping passenger turnover is too low to be of concern. It is calculated that the highest shipping passenger turnover will not exceed 5 million pkm.

4.3. Optimal Structure of Private Cars

Figure 8 provides the optimisation results for the number of private cars. The number of gasoline-fuelled private cars increases until 2047 and then falls steeply, meaning that electric private cars are not fully economic competitive presently. What is more, it can be found that there is no need to heavily promote hybrid electric vehicles (HEVs), which is consistent with reality. Compared with electric cars, HEVs are only 10% cheaper and generate nearly 0.6 tonnes more CO_2 emissions per year. HEVs are only technological transitional products from gasoline cars to electric cars. After 2047, the number of private electric vehicles improves speedily, with 103 million in 2050 and 542 million in 2060, while gasoline-fuelled private cars are not allowed to be purchased.

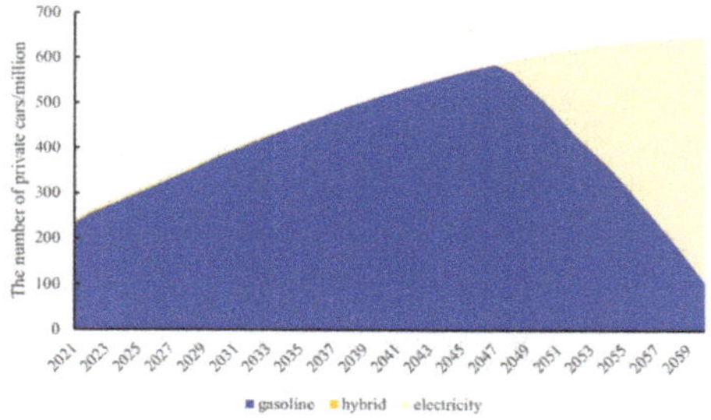

Figure 8. The number of private cars.

4.4. CO_2 Emissions from the Transport Sector

Carbon neutrality is the major scenario in our case study. The only difference between the scenario of BAU and carbon neutrality is that there are no emission constraints in the BAU scenario. CO_2 emissions from the transport sector in China are supposed to reach a peak in 2030 and decrease to 100 million tonnes in 2060, as shown in Figure 9. The results show that although the emission peak year of the transport sector is 2030, CO_2 emissions from freight transport must be reduced from now on. The rate of decline for CO_2 emissions from the freight sector is even from 2020 to 2035 and from 2045 to 2060 and a rapid drop can be observed between 2035 and 2045, thanks to the fossil-fuelled freight vehicles being decommissioned on a large scale. Owing to the cost advantage and the lower CO_2 emissions of natural gas as compared to diesel, gas-fuelled trucks are preferred in the initial stage of optimisation. These newly added gas-fuelled trucks are out of service ten years later, leading to a large supply gap for road freight transport. Facing the challenge of decarbonisation, policymakers have to choose secondary energy-powered vehicles directly. The proportion of hydrogen-powered road transport in Figure 3a increases from 13% to 31% between 2035 and 2045, which can confirm our statement. However, different from the freight transport sector, CO_2 emissions from passenger transport increase first and then decrease. They reach a peak in 2041 at 552 million tonnes, the level falling to 41 million tonnes in 2060, while CO_2 emissions from the freight transport sector are 59 million tonnes in 2060. Notwithstanding CO_2 emissions from the passenger transport sector growing from 453 million tonnes to 540 million tonnes from 2030 to 2045, the total CO_2 emissions from the transport sector still decline by 309 million tonnes due to the drop in CO_2 emissions from the freight sector. Meanwhile, from Figure 9b, two major transitions in the CO_2 emission structure of the transport sector in China can be noticed. The major source of CO_2 emissions will gradually change to natural gas before 2035. With the reduction in natural gas consumption after 2035, CO_2 emissions will be dominated by oil again. What is more the small increase in CO_2 emissions caused by oil between 2030 and 2040 is due to the increase in oil consumption for passenger transport, especially private cars, which can be verified in Figure 10. However, due to the decrease in the use of natural gas, this has not affected the downward trend for total CO_2 emissions.

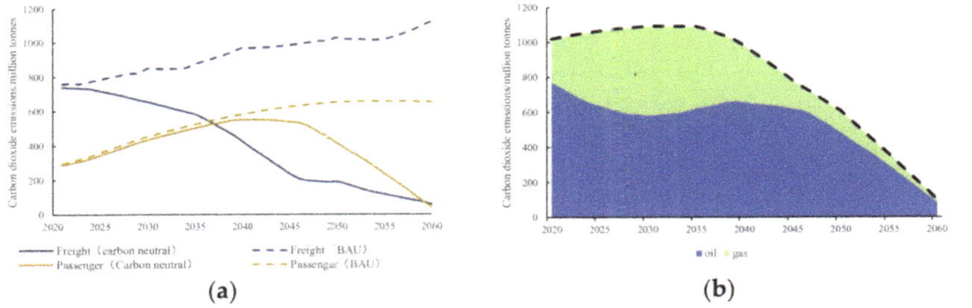

Figure 9. CO_2 emissions from the transport sector. (**a**) CO_2 from freight and passenger transport. (**b**) CO_2 from oil and gas.

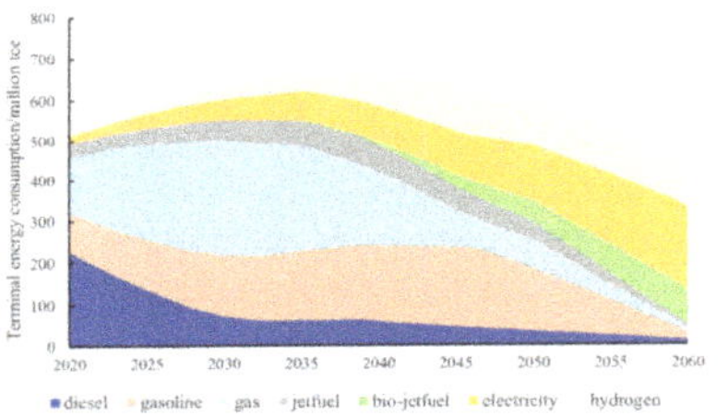

Figure 10. End-use energy consumption in the transport sector in China.

End-use energy consumption in the transport sector in China is presented in Figure 10. Here, we use the electrothermal equivalent method to calculate electricity consumption. It can be seen that there is also a peak in energy consumption, although traffic demand continues to increase. Improved fuel efficiency and modal shift in passenger transport contributes to the decrease in end-use energy consumption. The peak of end-use energy consumption appears in 2037 with 667 million tce. In 2060, end-use energy consumption in the transport sector in China is only 442 million tce, lower than that in 2020. What is more, Figure 10 shows the significant transitional role of natural gas in the decarbonisation of the transport sector in China. The largest amount of natural gas will be used in 2030, reaching 233 billion m^3, which is nearly six times that in 2019. However, only 7 billion m^3 of gas is needed in 2060, which is much lower than the figure for 2019. Moreover, the extensive application of bio-jet-fuel is also noteworthy. Only a handful of routes in China use biomass kerosene, while 91% of the energy consumed by the aviation sector consists of bio-jet-fuel in 2060, based on our optimistic estimate of the cost of bio-jet-fuel. Additionally, hydrogen and electricity use in the transport sector will increase rapidly. Electricity becomes the major energy source for the transport system from 2051 onwards in China; this might be because hydrogen has not been competitive in price compared with electricity. We consider this conclusion to be realistic, for the reason that we regard all hydrogen as being produced by renewable electricity. Power demand and hydrogen demand for the transportation sector are shown in Table 3. If hydrogen is produced by electricity totally and we assume that 45 kWh of electricity can develop a kilogram of hydrogen in 2020 and that 35 kWh of electricity can develop a kilogram of hydrogen in 2060, then 2391 billion kWh is needed for the transport sector. Due to the decline in the cost of charging piles after 2045, electric vehicles have greater cost advantages in the road subsector and occupy part of the original hydrogen-powered vehicle market. Consequently, the demand for hydrogen in the transport sector decreases slightly. In previous work in our group, over 25 trillion kWh of electricity was reckoned to be needed in China in 2060 when all hydrogen is generated by power [50]. Therefore, our results imply that the electricity that the transport sector needs in 2060 is nearly 10% of that needed by the whole of society, which is four times higher than the percentage in 2019. Decarbonisation in the transport sector will put unprecedented pressure on the power system.

Table 3. Power and hydrogen demand in the transport sector in China.

	Power Demand /Billion kWh	Hydrogen Demand /Billion kg	Equivalent Power Demand /Billion kWh
2030	410	6	682
2040	718	16	1361
2050	1101	24	1996
2060	1605	22	2391

4.5. The Costs of Decarbonisation for the Transport Sector

Figure 11 presents the results of the costs of decarbonisation in the transport sector in China. It can be found that the costs of the transport sector are rising, while the proportion goes down consistently. The cost of decarbonisation is about CNY 29.2 trillion in China, nearly three times the cost in 2020, while China only spends CNY 12.7 trillion to merely satisfy the demands of the transport sector. It can also be noticed that there is no big difference in costs between the two scenarios before 2035, meaning that the method of expansion selected in the first 15 years is particularly critical. Choosing an appropriate means of expansion can lay a solid foundation for further decarbonisation while saving costs. Meanwhile, three sharp increases in costs in the scenario of carbon neutrality can be observed in Figure 11. The reasons for these are diverse and they can be explained by reference to Table 4. The first rise between 2035 and 2040 is caused by freight vehicles. Due to the retirement of gas-fuelled trucks in significant quantities and the increase in the number of hydrogen-powered trucks, freight cost increased sharply. The second rise between 2045 and 2050 is caused by infrastructure. More charging piles are needed to support the large number of new electric vehicles in the road subsector. The third rise between 2055 and 2060 is caused by private cars. The cost of private cars consists of two parts in this period. One part is the increased demand for vehicles, and the other part is the demand for vehicles needed as replacements after the retirement of gasoline and electric vehicles, which explains the discrepancy in the costs of private cars before 2055.

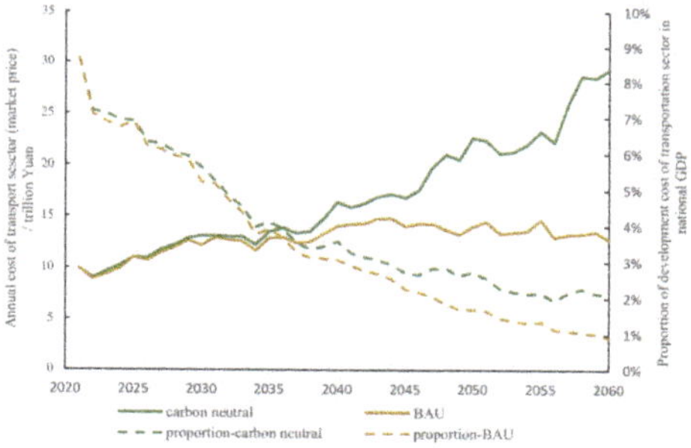

Figure 11. The costs of decarbonisation in the transport sector.

Table 4. Proportion of different costs.

	Private Cars	Freight Vehicles	Passenger Vehicles (Public)	Fuel	Infrastructure
2025	52%	10%	11%	24%	3%
2030	45%	20%	10%	23%	2%
2035	48%	14%	11%	25%	2%
2040	41%	24%	9%	23%	3%
2045	39%	19%	13%	26%	3%
2050	37%	20%	8%	19%	16%
2055	34%	20%	11%	17%	18%
2060	49%	16%	9%	12%	14%

4.6. Sensitivity Analysis

4.6.1. The Necessity of Considering Infrastructure Construction when Modelling

In order to illustrate the importance of considering transport facilities when modelling, two scenarios are compared in Figure 12. The major dissimilarities between the two scenarios concern planning for metros and electric road vehicles. Without considering metro tracks, the development of a subway will be overestimated. The difference between the two scenarios is 47 trillion pkm in 2060, implying that ignoring infrastructure construction leads to 60% more subways being planned. What is more, neglecting facilities also leads to more charging piles in reality. If infrastructure is not considered in the planning period, the number of charging piles required each year will be 45% higher than that on average when considering the facilities in the planning period. Above all, the cost of the whole transport sector each year is 7.3% higher when not considering infrastructure construction on average.

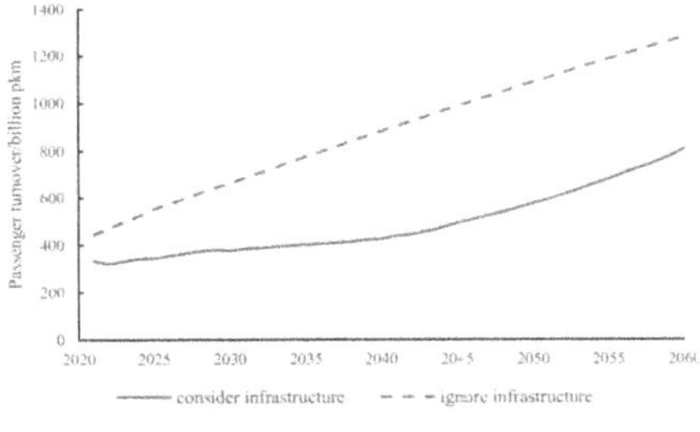

Figure 12. The passenger turnover offered by subways in two scenarios.

4.6.2. Impact of the Proportion of Specific Requirements in Passenger Transport on the Results

In Section 3.7, $FIXPRO_{taxi}$, $FIXPRO_{coach}$ and $FIXPRO_{plane}$ were set to be 0.6, due to their characteristics of being either convenient or fast, while other values were set to 0.4. In this section, diverse values are set to explore the impact of values on annual costs. The results are presented in Figure 13. It can be found that with increases in variable demand, annual cost generally decreases. If there is no variable demand in all passenger transport modes, the annual cost can be 7.5% lower on average, compared with no traffic mode conversion. Even if the transport sector does not carry out decarbonisation, 3.1% of costs can be saved every year if the transport modes can shift freely, compared with the scenario in which modes cannot shift. The sensitivity analysis for the parameter $FIXPRO_{ser}$ is of practical significance. The modification of people's travelling concepts is beneficial for reducing the costs of the transport sector.

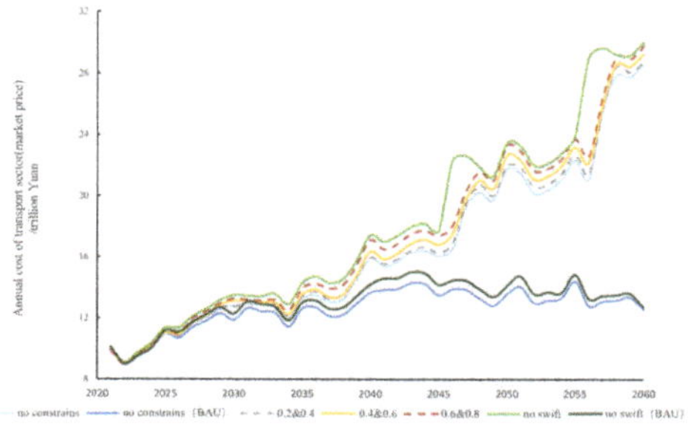

Figure 13. Annual costs of the transport sector with different $FIXPRO_{ser}$ values.

4.6.3. The Necessity of Giving Priority to Decarbonisation in the Freight Sector

The basic scenario discussed in Sections 4.1–4.5 indicates that giving priority to the decarbonisation of freight transport is economically efficient. However, the reality is that passenger transport and freight transport reduce CO_2 emissions simultaneously. In other words, the passenger transport also has decarbonisation requirements each year in China. In this section, the basic scenario is named the freight first scenario, while the other scenario with extra constraints on the passenger sector is regarded as a passenger first scenario. In the passenger first scenario, we simulated a policy that requires the number of electric private cars to increase by 10% every year and the passenger turnover undertaken by other electric vehicles to increase by 5% every year before 2045. At the same time, the passenger turnover of diesel-fuelled trains cannot increase. The annual cost of the transport sector under two different policies is presented in Figure 14. It must be admitted that the costs can be lower in the later stage of the planning period if the passenger first policy is implemented. However, compared with the freight first scenario, the passenger first scenario pays an additional 5% of the cost per year on average.

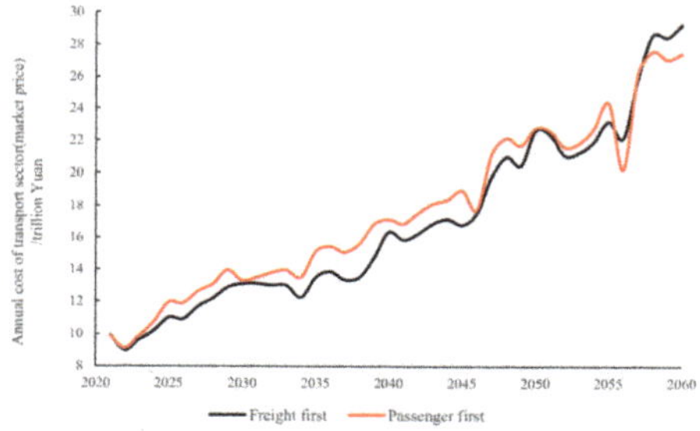

Figure 14. Annual cost of the transport sector under two policies.

5. Policy Recommendations

Four policy recommendations can be summarised based on the analysis from Section 4.

Firstly, make full use of the existing inter-regional transport infrastructure rather than undertake large-scale expansion. Over the past two decades, China's transport infrastructure has developed rapidly, especially rail-related infrastructure.By making full use of existing infrastructure, development costs can be saved and carbon emissions caused by massive infrastructure construction can be reduced.

Secondly, priority should be given to reducing CO_2 emissions from the freight sector rather than from the passenger sector. The freight sector is currently the largest source of carbon emissions in China's transport system. If its CO_2 emissions are not reduced immediately, a greater cost will be paid later. The immediate implementation of emission reduction measures in the freight sector is also in line with the current requirements of the government.

Thirdly, guide passengers to choose more travel modes with low CO_2 emission density through propaganda or other means. Different transport modes have different marginal costs of CO_2 emission reduction. Choosing less carbon-intensive modes to travel could reduce the emission-reduction costs and take full advantage of existing railway infrastructure.

Fourthly, be alert to the rapid growth in power demand caused by the direct emissions reduction of the transport sector, which is a matter for policymakers in the power sector. The reduction in direct carbon emissions in the transport sector relies almost solely on secondary energy, such as electricity, which will lead to a huge increase in electricity consumption in the transport sector. Therefore, power sector planners need to take into account the substantial increase in power consumption in the terminal sector under the carbon-neutrality target.

6. Conclusions

In this study, a long-term multi-regional model is proposed to plan the transport sector, including both freight and passenger transport, which considers modal shift and infrastructure construction. This model can be used in the transport sector, where demand will increase alarmingly during the planning period. China is taken as a case study. Detailed development paths, energy consumption, CO_2 emissions, the composition of cost and sensitivity analysis have been discussed. The main conclusions are as follows:

Firstly, inter-regional infrastructure construction in China is ahead of schedule under the current spatial resolution. Only airports in East China have to be expanded by 2040. By 2060, the number of airport runways should be 138% of the current number. There is no expansion in other infrastructure considered in this paper. Nonetheless, infrastructure such as subway tracks and charging piles is far from sufficient. Ignoring infrastructure construction when modelling leads to an increase of over 7% in costs.

Secondly, oil consumption shows a downwards trend since 2021. The consumption of oil is only 0.4 trillion tce. Natural gas plays a transitional role in the transport sector, reaching a peak in 2030 and falling to 7 bcm in 2060. Electricity and hydrogen will be the chief fuels in the transport sector in the future. The promotion of electric vehicles is influenced by the price of charging piles. Electric cars will replace some hydrogen-powered cars when charging piles are cheap enough. Direct decarbonisation of the transport sector in China depends mainly on the utility of electricity and hydrogen, leading to an increase in power demand. In 2030, it is expected to use 0.4 trillion kWh and 6 billion kg of hydrogen. By 2060, the transport sector is expected to consume 2.4 trillion kWh of electricity and 22 billion kg of hydrogen.

Thirdly, CO_2 emissions from freight transport have now peaked and will fall to 59 million tonnes in 2060. Shipping contributes the most in 2060. The reduction in CO_2 emissions from freight transport occurs relatively quickly from 2035 to 2045, with an average annual decrease of 33 million tonnes. CO_2 emissions from passenger transport reach a peak of 552 million tonnes in 2041 and fall to 41 million tonnes by 2060.

Fourthly, the costs of decarbonisation of the transport sector rise from CNY 9 trillion per year to CNY 29 trillion per year in 2060. Increasing the changeable demand of passengers appropriately can reduce the cost effectively by up to 7%. More importantly, prioritising decarbonisation of the freight sector can also reduce costs by about 5%.

In future research, we suggest more elaborate spatial scales. This paper mainly focuses on inter-regional infrastructure construction due to the fact that in this study it is difficult to identify the distances between cities in one region. However, as urbanisation progresses, the traffic volume between cities in one region will further increase. When the inter-city infrastructure in one region reaches saturation point, the choice of future traffic modes needs to be carefully discussed.

Author Contributions: Writing—original draft preparation, methodology, C.L.; writing—review and editing, P.L.; Conceptualisation, supervision, Z.L. All authors have read and agreed to the published version of the manuscript.

Funding: This research was funded by the National Key Research and Development Program of China (2019YFE0100100) and the Phase IV Collaboration between BP and Tsinghua University.

Institutional Review Board Statement: Not applicable.

Informed Consent Statement: Not applicable.

Data Availability Statement: All data used for this analysis are available from the publicly available sources cited or from the corresponding author upon reasonable request.

Acknowledgments: The authors gratefully acknowledge the support by the National Key Research and Development Program of China and the Phase IV Collaboration between BP and Tsinghua University.

Conflicts of Interest: The authors declare no conflict of interest and that the funders had no role in the design of the study; in the collection, analyses, or interpretation of data; in the writing of the manuscript, or in the decision to publish the results.

Abbreviations

Variables	**Meaning**
$cost$	The total cost of the transport sector during the planning period
$infc_t$	Cost of infrastructure in year t
$buyc_t$	Cost of purchasing new vehicles in year t
$mainc_t$	Cost of maintaining existing vehicles in year t
$fuelc_t$	Cost of fuel in year t
I	Discount rate
$chargec_t$	The sum costs of purchasing and maintaining charging piles in year t
$metroc_t$	The sum costs of constructing and maintaining subway tracks in year t
$railc_t$	The sum costs of constructing and maintaining ordinary and high-speed rails in year t
$airportc_t$	The sum costs of constructing and maintain airports in year t
$portc_t$	The sum costs of constructing and maintaining ports in year t
$nnb_{t,ser,f}$	The number of newly added vehicles of mode ser powered by fuel f in year t
$nb_{t,ser,f}$	The number of vehicles of mode ser powered by fuel f in year t
$ts_{t,r,rr,ser,f}$	Transport turnover from region r to region rr supplied by mode ser powered by fuel f in year t
$totalttb_t$	Total travel time budget in the country in year t
$retire_{t,ser,f}$	The number of newly added vehicles decommissioned before the expiration date of mode ser powered by fuel f in year t
$ycarbon_t$	CO_2 emissions from the transport sector in year t
Parameters	**Meaning**
$BUY_{t,ser,f}$	The cost of purchasing one vehicle of mode ser powered by fuel f in year t
$MAIN_{t,ser,f}$	The cost of maintaining one vehicle of mode ser powered by fuel f in year t
$TRFC_{t,ser,f}$	Fuel consumption per unit transport turnover of mode ser in year t
$TD_{t,r,rr,ser}$	Transport demand of mode ser between region r and rr in year t
$FIXPRO_{ser}$	The proportion of specific requirements in mode ser
$NBD_{t,r}$	The demand for private cars in region r powered by fuel f in year t
$SPEED_{pas,f}$	The speed of passenger transport mode pas powered by fuel f
TTB	Travel time budget of a person in one day
POP_t	The population of the country in year t
$LOAD_{ser,f}$	The maximum turnover a vehicle of mode ser powered by fuel f can serve in a year
$MILEAGE$	The average mileage of private cars in a year
PRO	The minimum ratio of the amount of infrastructure and traffic demand
FC_f	Carbon dioxide emission factor of fuel f
$TRCO2_t$	The upper limit of carbon dioxide emissions from the transport sector

Appendix A

Table A1. Maximum service life of vehicles.

	Gasoline	Diesel	Hybrid	Electricity	Gas	Hydrogen	Jet-Fuel	Bio-Jet-Fuel
Car	15	—	15	10	—	—	—	—
Taxi	15	—	—	8	8	—	—	—
Subway	—	—	—	30	—	—	—	—
Bus	—	10	—	10	10	10	—	—
Coach	—	10	—	10	10	10	—	—
Train	—	100	—	30	—	—	—	—
Plane	—	—	—	15	—	—	15	15
Boat	—	30	—	25	30	30	—	—
Truck	—	10	—	10	10	10	—	—
Railway	—	100	—	30	—	—	—	—
Aircraft	—	—	—	20	—	—	20	20
Vessel	—	35	—	30	35	35	—	—

Table A2. Investment costs of vehicles in 2020 (CNY million).

	Gasoline	Diesel	Hybrid	Electricity	Gas	Hydrogen	Jet-Fuel	Bio-Jet-Fuel
Car	0.2	—	0.27	0.3	—	—	—	—
Taxi	0.2	—	—	0.3	0.2	—	—	—
Subway	—	—	—	40	—	—	—	—
Bus	—	0.4	—	1.7	0.5	2	—	—
Coach	—	0.4	—	1.7	0.5	2	—	—
Train	—	50	—	90 (high-speed 170)	—	—	—	—
Plane	—	—	—	—	—	—	385	385
Boat	—	0.8	—	—	12.6	—	—	—
Truck	—	0.3	—	1	0.4	1.2	—	—
Railway	—	10	—	50	—	—	—	—
Aircraft	—	—	—	—	—	—	467	385
Vessel	—	35	—	—	45.5	—	—	—

Table A3. Scope of the investment costs of infrastructure in 2020 (CNY million).

Metro Rail CNY M/KM	Rail CNY M/KM	Highrail CNY M/KM	Airport CNY M/Runway	Port CNY M/Berth	Charging Pile CNY M/Number
500–1500	68–316	102–900	10,000–20,000	200	0.03

Note: Infrastructure costs in different regions vary for topographical and economic reasons.

Table A4. Costs of fuels in 2020 (CNY/kg).

	Gasoline	Diesel	Electricity	Gas	Hydrogen	Jet-Fuel	Bio-Jet-Fuel
Northeast	9.45	7.56	0.71	5.28	65	7	60
North	9.44	7.77	0.75	4.67	65	7	60
Central	9.45	7.71	0.67	5.23	65	7	60
East	9.39	7.7	0.64	5.89	65	7	60
South	9.45	7.72	0.65	4.50	65	7	60
Northwest	9.30	7.61	0.56	3	65	7	60
Xinjiang	9.34	7.61	0.45	1.86	65	7	60
Inner Mongolia	9.36	7.59	0.45	2.46	65	7	60
Tibet	9.60	7.79	0.66	5.95	65	7	60

Table A5. Modal speed (Km/h).

Car	Taxi	Subway	Bus	Coach	Train	Plane	Boat
55	38	35	18	45	60 (high-speed 220)	900	33

References

1. Tracking Transport 2020. Available online: https://www.iea.org/reports/tracking-transport-2020 (accessed on 13 April 2022).
2. BP Energy Outlook 2022. Available online: https://www.bp.com/content/dam/bp/business-sites/en/global/corporate/pdfs/energy-economics/energy-outlook/bp-energy-outlook-2022.pdf (accessed on 13 April 2022).
3. Tracking Transport 2021. Available online: https://www.iea.org/reports/tracking-transport-2021 (accessed on 13 April 2022).
4. Maduekwe, M.; Akpan, U.; Isihak, S. Road transport energy consumption and vehicular emissions in Lagos, Nigeria: An application of the LEAP model. *Transp. Res. Interdiscip. Perspect.* **2020**, *6*, 100172. [CrossRef]
5. Fan, J.-L.; Wang, J.-X.; Li, F.; Yu, H.; Zhang, X. Energy demand and greenhouse gas emissions of urban passenger transport in the Internet era: A case study of Beijing. *J. Clean. Prod.* **2017**, *165*, 177–189. [CrossRef]
6. Tsita, K.G.; Pilavachi, P.A. Decarbonizing the Greek road transport sector using alternative technologies and fuels. *Therm. Sci. Eng. Prog.* **2017**, *1*, 15–24. [CrossRef]
7. Virley, S. The effect of fuel price increases on road transport CO_2 emissions. *Transp. Policy* **1993**, *1*, 43–48. [CrossRef]
8. Brand, C.; Anable, J.; Tran, M. Accelerating the transformation to a low carbon passenger transport system: The role of car purchase taxes, feebates, road taxes and scrappage incentives in the UK. *Transp. Res. Part A Policy Pract.* **2013**, *49*, 132–148. [CrossRef]
9. Santos, G.; Rembalski, S. Do electric vehicles need subsidies in the UK? *Energy Policy* **2021**, *149*, 111890. [CrossRef]
10. Dhar, S.; Pathak, M.; Shukla, P.R. Electric vehicles and India's low carbon passenger transport: A long-term co-benefits assessment. *J. Clean. Prod.* **2017**, *146*, 139–148. [CrossRef]
11. Larsson, M.; Mohseni, F.; Wallmark, C.; Grönkvist, S.; Alvfors, P. Energy system analysis of the implications of hydrogen fuel cell vehicles in the Swedish road transport system. *Int. J. Hydrog. Energy* **2015**, *40*, 11722–11729. [CrossRef]
12. Li, Y.; Kimura, S. Economic competitiveness and environmental implications of hydrogen energy and fuel cell electric vehicles in ASEAN countries: The current and future scenarios. *Energy Policy* **2021**, *148*, 111980. [CrossRef]
13. Virger, T.; Azimov, U.; Adeniyi, O. Biomass-based fuel blends as an alternative for the future heavy-duty transport: A review. *Renew. Sustain. Energy Rev.* **2022**, *161*, 112391. [CrossRef]
14. Altun, Ş.; Öner, C.; Yaşar, F.; Adin, H. Effect of n-Butanol Blending with a Blend of Diesel and Biodiesel on Performance and Exhaust Emissions of a Diesel Engine. *Ind. Eng. Chem. Res.* **2011**, *50*, 9425–9430. [CrossRef]
15. Fageda, X.; Teixidó, J.J. Pricing carbon in the aviation sector: Evidence from the European emissions trading system. *J. Environ. Econ. Manag.* **2022**, *111*, 102591. [CrossRef]
16. Rathore, H.; Jakhar, S.K. Differential carbon tax policy in aviation: One stone that kills two birds? *J. Clean. Prod.* **2021**, *296*, 126479. [CrossRef]
17. Rehmatulla, N.; Parker, S.; Smith, T.; Stulgis, V. Wind technologies: Opportunities and barriers to a low carbon shipping industry. *Mar. Policy* **2017**, *75*, 217–226. [CrossRef]
18. Stolz, B.; Held, M.; Georges, G.; Boulouchos, K. Techno-economic analysis of renewable fuels for ships carrying bulk cargo in Europe. *Nat. Energy* **2022**, *7*, 203–212. [CrossRef]
19. Helgeson, B.; Peter, J. The role of electricity in decarbonizing European road transport–Development and assessment of an integrated multi-sectoral model. *Appl. Energy* **2020**, *262*, 114365. [CrossRef]
20. Documentation for the TIMES Model: PART I. 2021. Available online: https://iea-etsap.org/index.php/documentation (accessed on 5 April 2022).
21. LEAP: The Low Emissions Analysis Platform. 2022. Available online: https://leap.sei.org (accessed on 5 April 2022).
22. Zhang, H.; Chen, W.; Huang, W. TIMES modelling of transport sector in China and USA: Comparisons from a decarbonization perspective. *Appl. Energy* **2016**, *162*, 1505–1514. [CrossRef]
23. Korkmaz, P.; Schmid, D.; Fahl, U. Incorporating uncertainties towards a sustainable European energy system: A stochastic approach for decarbonization paths focusing on the transport sector. *Energy Strategy Rev.* **2021**, *38*, 100707. [CrossRef]
24. Hong, S.; Chung, Y.; Kim, J.; Chun, D. Analysis on the level of contribution to the national greenhouse gas reduction target in Korean transportation sector using LEAP model. *Renew. Sustain. Energy Rev.* **2016**, *60*, 549–559. [CrossRef]
25. Sadri, A.; Ardehali, M.M.; Amirnekooei, K. General procedure for long-term energy-environmental planning for transportation sector of developing countries with limited data based on LEAP (long-range energy alternative planning) and EnergyPLAN. *Energy* **2014**, *77*, 831–843. [CrossRef]
26. Matthias, V.; Bieser, J.; Mocanu, T.; Pregger, T.; Quante, M.; Ramacher, M.O.P.; Seum, S.; Winkler, C. Modelling road transport emissions in Germany–Current day situation and scenarios for 2040. *Transp. Res. Part D Transp. Environ.* **2020**, *87*, 102536. [CrossRef]
27. Tong, H.; Peng, J.; Zhang, Y.; Fang, T.; Zhang, J.; Men, Z.; Liu, Y.; Wu, L.; Wang, T.; Ren, F.; et al. Environmental benefit analysis of "road-to-rail" policy in China based on a railway tunnel measurement. *J. Clean. Prod.* **2021**, *316*, 128227. [CrossRef]

28. Salvucci, R.; Tattini, J.; Gargiulo, M.; Lehtilä, A.; Karlsson, K. Modelling transport modal shift in TIMES models through elasticities of substitution. *Appl. Energy* **2018**, *232*, 740–751. [CrossRef]
29. Strauss, J.; Li, H.; Cui, J. High-speed Rail's impact on airline demand and air carbon emissions in China. *Transp. Policy* **2021**, *109*, 85–97. [CrossRef]
30. Kurtuluş, E.; Çetin, İ.B. Analysis of modal shift potential towards intermodal transportation in short-distance inland container transport. *Transp. Policy* **2020**, *89*, 24–37. [CrossRef]
31. Zhang, J.; Zhang, S.; Wang, Y.; Bao, S.; Yang, D.; Xu, H.; Wu, R.; Wang, R.; Yan, M.; Wu, Y.; et al. Air quality improvement via modal shift: Assessment of rail-water-port integrated system planning in Shenzhen, China. *Sci. Total Environ.* **2021**, *791*, 148158. [CrossRef]
32. Schwanen, T.; Banister, D.; Anable, J. Scientific research about climate change mitigation in transport: A critical review. *Transp. Res. Part A Policy Pract.* **2011**, *45*, 993–1006. [CrossRef]
33. Hickman, R.; Ashiru, O.; Banister, D. Transitions to low carbon transport futures: Strategic conversations from London and Delhi. *J. Transp. Geogr.* **2011**, *19*, 1553–1562. [CrossRef]
34. Daly, H.E.; Ramea, K.; Chiodi, A.; Yeh, S.; Gargiulo, M.; Gallachóir, B.Ó. Incorporating travel behaviour and travel time into TIMES energy system models. *Appl. Energy* **2014**, *135*, 429–439. [CrossRef]
35. Brand, C.; Tran, M.; Anable, J. The UK transport carbon model: An integrated life cycle approach to explore low carbon futures. *Energy Policy* **2012**, *41*, 107–124. [CrossRef]
36. Isik, M.; Dodder, R.; Kaplan, P.O. Transportation emissions scenarios for New York City under different carbon intensities of electricity and electric vehicle adoption rates. *Nat. Energy* **2021**, *6*, 92–104. [CrossRef]
37. Kany, M.S.; Mathiesen, B.V.; Skov, I.R.; Korberg, A.D.; Thellufsen, J.Z.; Lund, H.; Sorknæs, P.; Chang, M. Energy efficient decarbonisation strategy for the Danish transport sector by 2045. *Smart Energy* **2022**, *5*, 100063. [CrossRef]
38. China. 2022. Available online: https://www.iea.org/countries/china (accessed on 4 April 2022).
39. Tattini, J.; Gargiulo, M.; Karlsson, K. Reaching carbon neutral transport sector in Denmark–Evidence from the incorporation of modal shift into the TIMES energy system modeling framework. *Energy Policy* **2018**, *113*, 571–583. [CrossRef]
40. National Bureau of Statistic. *China Statistical Yearbook 2021*; China Statistics Press: Beijing, China, 2021.
41. ICCSD. *The Comprehension of Carbon Neutral: The Low Carbon Development Action Roadmap of China during 2020–2050*; China Citic Press: Beijing, China, 2021.
42. Ministry of Transport of the People's Republic of China. *China Transportation Statistical Yearbook 2019*; People's Communications Press: Beijing, China, 2019.
43. Ministry of Transport of the People's Republic of China. *China Transportation Statistical Yearbook 2020*; People's Communications Press: Beijing, China, 2020.
44. China Urban Rail Transit Association. *Statistical Yearbook of Urban Rail Transit Mainland China 2020*; China Railway Press: Beijing, China, 2020.
45. Civil Aviation Authority of China. *Statistical Bulletin on the Development of Civil Aviation Industry 2020*; China Civil Aviation Publishing House: Beijing, China, 2020.
46. National Railway Administration. *Railway Statistical Bulletin 2020*; China Railway Press: Beijing, China, 2020.
47. Norouzi, N.; Fani, M.; Ziarani, Z. The fall of oil Age: A scenario planning approach over the last peak oil of human history by 2040. *J. Pet. Sci. Eng.* **2020**, *188*, 106827. [CrossRef]
48. Norouzi, N. Post-COVID-19 and globalization of oil and natural gas trade: Challenges, opportunities, lessons, regulations, and strategies. *Energy Res.* **2021**, *45*, 14338–14356. [CrossRef]
49. Schafer, A.; Victor, D.G. The future mobility of the world population. *Transp. Res. Part A Policy Pract.* **2000**, *34*, 171–205. [CrossRef]
50. Song, S.; Li, T.; Liu, P.; Li, Z. The transition pathway of energy supply systems towards carbon neutrality based on a multi-regional energy infrastructure planning approach: A case study of China. *Energy* **2022**, *238*, 122037. [CrossRef]

processes

Article

Two-Step Intelligent Control for a Green Flexible EV Energy Supply Station Oriented to Dual Carbon Targets

Shanshan Shi [1], Chen Fang [1], Haojing Wang [1], Jianfang Li [2], Yuekai Li [2], Daogang Peng [2] and Huirong Zhao [2,*]

1 State Grid Shanghai Municipal Electric Power Company Electric Power Research Institute, Shanghai 200437, China; shishsh@sh.sgcc.com.cn (S.S.); fangch@sh.sgcc.com.cn (C.F.); haojingw@126.com (H.W.)

2 College of Automation Engineering, Shanghai University of Electric Power, Shanghai 200090, China; jianfangli2021@163.com (J.L.); mo2720337186@163.com (Y.L.); pengdaogang@126.com (D.P.)

* Correspondence: hrzhao@shiep.edu.cn; Tel.: +86-189-3038-0616

Citation: Shi, S.; Fang, C.; Wang, H.; Li, J.; Li, Y.; Peng, D.; Zhao, H. Two-Step Intelligent Control for a Green Flexible EV Energy Supply Station Oriented to Dual Carbon Targets. *Processes* **2021**, *9*, 1918. https://doi.org/10.3390/pr9111918

Academic Editors: Pei Liu, Ming Liu and Xiao Wu

Received: 13 September 2021
Accepted: 22 October 2021
Published: 27 October 2021

Publisher's Note: MDPI stays neutral with regard to jurisdictional claims in published maps and institutional affiliations.

Copyright: © 2021 by the authors. Licensee MDPI, Basel, Switzerland. This article is an open access article distributed under the terms and conditions of the Creative Commons Attribution (CC BY) license (https://creativecommons.org/licenses/by/4.0/).

Abstract: As China proposes to achieve carbon peak by 2030 and carbon neutrality by 2060, as well as the huge pressure on the power grid caused by the load demand of the energy supply stations of electric vehicles (EVs), there is an urgent need to carry out comprehensive energy management and coordinated control for EVs' energy supply stations. Therefore, this paper proposed a two-step intelligent control method known as ISOM-SAIA to solve the problem of the 24 h control and regulation of green/flexible EV energy supply stations, including four subsystems such as a photovoltaic subsystem, an energy storage subsystem, an EV charging subsystem and an EV battery changing subsystem. The proposed control method has two main innovations and contributions. One is that it reduces the computational burden by dividing the multi-dimensional mixed-integer programming problem of simultaneously optimizing the 24 h operation modes and outputs of four subsystems into two sequential tasks: the classification of data-driven operation modes and the rolling optimization of operational outputs. The other is that proper carbon transaction costs and carbon emission constraints are considered to help save costs and reduce carbon emissions. The simulation analysis conducted in this paper indicates that the proposed two-step intelligent control method can help green/flexible EV energy supply stations to optimally allocate energy flows between four subsystems, effectively respond to peak shaving and valley filling of power grid, save energy costs and reduce carbon emissions.

Keywords: electric vehicle; intelligent control; integrated energy system; dual carbon target

1. Introduction

With the increasingly prominent problems of environmental pollution and fossil fuel depletion, countries all over the world have designed different low-carbon development routines. Among them, China proposed a dual-carbon target to achieve carbon peak by 2030 and carbon neutrality by 2060. In China, energy carbon emissions account for about 80% of total carbon emissions. Therefore, the core task of China's dual-carbon target is to promote low-carbon energy transition and revolutions, and the essential measures are to reduce fossil energy consumption and increase the proportion of renewable energy. In the field of transportation, the replacement of fuel vehicles is an inevitable trend for electric vehicles (EVs). In 2020, the stock of EVs in China was 4.5 million, more than twice the stock in 2018 [1]. The development of EVs in China is affected by mileage, cost and policy impact. Small electrical cars, E-vans and electric buses are three typical types of EV. Small electric cars are very much in line with people's daily travel needs, especially in rural markets and third- and fourth-tier cities in China. China's E-vans offer some leading advantages in terms of industrialization and have made great progress in their product development, market expansion and application. For example, the JD company proposes to replace all its trucks with pure electric vehicles in the next five years, and the self-built logistics industry, led by the e-commerce sector, is gradually replacing traditional fuel trucks with pure electric trucks. Therefore, with the guidance of national policies in relation to the satisfaction of end-market demand, E-vans will have a very large development space in the future. Electric buses are the main route of China's new energy vehicles. It is reported that Shenzhen, China took the lead in realizing 100% pure electrification of urban buses in 2017. At this stage, when the on-board power supply, power, driving range, charging speed and charging capacity tend to be mature, popularizing pure electric buses to the whole country is not only an important measure to help save energy and reduce emissions, but also a key step to promote the development of the urban public transport industry.

However, the daily energy supply demand of EVs has obvious peak–valley effects, which increases the difficulty of peak shaving and valley filling within the power grid. There are two options for the energy supply of EVs: charging and battery changing. In China, EVs' charging piles can be seen in both public parking lots and special charging stations, while shorter EVs' battery changing services can only be provided by special battery changing stations. In order to meet the needs of different EV users as well as increasing operational flexibility through low-carbon measures, some EVs' energy supply stations provide charging and battery changing services simultaneously, and are also equipped with photovoltaic systems and energy storage devices. These kinds of multi-functional green/flexible EV energy supply stations are referred to as EV "Photovoltaic-Storage-Charging-Change" stations.

The organic combination of photovoltaic-storage-charging-change systems can help solve the problem of the insufficient power distribution capacity of EVs' energy supply stations, reduce the impact on the power grid, and also increase renewable energy utilization. However, due to the system's complexity, the 24 h daily optimal control and regulation of the "Photovoltaic-Storage-Charging-Change" stations for EVs, considering additional carbon emission constraints, is relatively difficult, which is mainly shown in the following aspects:

1. The 24 h daily control and regulation problem of simultaneously optimizing operation modes and outputs for four subsystems is a mixed-integer programming problem [2], which should solve two issues. One is to determine the operation mode of each subsystem for all 24 h of a day, and the other is to determine the best output of each subsystem.
2. The dynamic characteristics of distributed energy sources including photovoltaic, energy storage, charging and battery changes in the EVs' "Photovoltaic-Storage-Charging-Change" stations are different from each other in terms of time-scale, and are mostly nonlinear and multi-dimensional. This increases the complexity of the optimization control and regulation problem.
3. There are coupling and restriction relationships between different subsystems, which also increases their complexity.
4. Additional carbon transaction costs and carbon emission constraints also increase the complexity.

Existing research studies that are relevant to the aforementioned optimization problem of regulation and control are generally divided into two categories: artificial intelligence algorithms and analytical methods.

On the one hand, artificial intelligence algorithms, including the particle swarm and genetic algorithm, are widely used to solve distributed energy control problems with non-convex characteristics and nonlinearity [3]. Jha, M., et al. proposed a multiport converter for integrating the PV, charging docks, and energy storage device (ESD) with the grid, and proposed an intelligent energy management system by adapting particle swarm optimization for efficient switching between the sources [4]. Improved methods such as the swarm optimization searcher [5], gravitational search algorithm [6], and particle swarm algorithm [7], with time-varying acceleration constant iteration, etc., are also applied to the optimization of distributed energy sources. In addition, deep learning and other intelligent algorithms are used by some researchers to solve the problem. For example, Ndiaye, El

Hadji Mbaye, et al. proposed an intelligent algorithm based on adaptive neuro-fuzzy inference system to control a photovoltaic generator (PVG) connected to a load and a battery, which allows the PV system to function normally by charging and discharging the battery whatever the weather conditions [8]. Qu Kaiping, et al. used the interior point method and the knowledge transfer Q algorithm to form a cascade algorithm to deal with the non-convex energy equation of the distributed energy network [9]. Liu Xiaokang took the generalized heterogeneous multi-agent system as the object, carried out research on two aspects of cooperative control and its application in the DC microgrid, and designed an optimal power control strategy based on reinforcement learning [10]. MyungJae Shin, et al. proposed a multi-agent-based energy management deep reinforcement learning method, established a distributed power generation model for EV charging stations containing photovoltaic systems and energy storage systems, and obtained a targeted power load dispatching plan to reduce operating costs [11].

One the other hand, two types of analytical methods are involved in this research field: the stepwise decomposition solution and the unified solution. The stepwise decomposition solution mainly optimizes and solves each independent subsystem, and then couples the information of each system, and finally iteratively solves the optimal solution that satisfies the conditions. The unified solution starts from the overall model, and uses linearization-related techniques to approximate the original optimization problem to a mixed integer programming or linear programming problem, which mainly includes methods such as 0-1 variables and convex relaxation techniques. Given that EV energy supply stations are connected to the power grid, some researchers used hierarchical control, improved droop control [12] or sequence arrangement control strategies [13], distributed algorithms, consensus algorithms [14], and heuristic algorithms [15] to guide the charging load of EV energy supply stations. Researchers who were focused on system frequency stability realized overall load variance and charging cost reduction in [16,17], and realized peak shaving and valley filling in [18,19]. Meng Chao, et al. aimed at the collaborative control problem of nonlinear multi-agent systems. They proved that followers can approach the leader's output infinitely, not only meeting the predetermined performance but also keeping the error within the specified range with high accuracy [20]. In the EV to grid(V2G) area, combined with the charging load model and the charging optimization model, some researchers formulated multi-agent charging management model [21], and built a dynamic multi-agent collaboration framework [22,23]. With the idea of aggregation, a model was constructed through dynamic programming algorithms [24] and probability statistics [25] to ensure an optimal balance between charging costs and grid revenue. Other researchers also focus on the configuration of multiple energy sources. In [26], Vavilapalli, S., et al. proposed a Power Balancing Control (PBC) method for the configuration of a grid energy storage system for photo voltaic (PV) applications with three different power sources to operate the system in three different modes of operation, and conducted a Hardware-In-the-Loop (HIL) simulation to prove the effectiveness of the proposed PBC algorithm.

To sum up, there have been many researches contributing to the optimization of configuration, energy management and real-time control of integrated energy systems with multiple energy sources. Generally, most studies focus on theoretical research, and the effectiveness of the proposed methods in these studies are mainly verified by simulations. In this paper, a real pilot green/flexible EV "Photovoltaic-Storage-Charging-Change" energy supply station in Shanghai, China is taken as the research object. To solve the realistic 24 h daily optimal control and regulation problem of the green/flexible EV "Photovoltaic-Storage-Charging-Change" station as well as supporting the dual carbon target, a two-step intelligent control method referred to as ISOM-SAIA is proposed in this paper. The proposed ISOM-SAIA algorithm is a combination of a data-driven intelligent classification method for improving a self-organizing map (SOM) proposed in [27] and a rolling optimization method for combining the simulated annealing method (SA) [28,29] and the immune algorithm (IA) [30–32]. It classifies the operation modes of all subsystems, for each hour, by ISOM in the first step, and then calculates the optimal operation outputs of all subsystems for each hour over a day by SAIA in the second step. The performance of the peak shaving and valley filling of the power grid, as well as the energy costs and carbon emission reduction of the proposed two-step ISOM-SAIA control method, are verified by simulation studies.

The paper is organized as follows. Section 2 presents the system architecture and control model description of EVs' "Photovoltaic-Storage-Charging-Change" stations. Section 3 gives the formation law of the proposed two-step intelligent control method for EVs' "Photovoltaic-Storage-Charging-Change" stations, in terms of its orientation to the dual carbon target. In Section 4, the proposed two-step intelligent control is illustrated by a numerical case study. The conclusions are given in Section 5.

2. "Photovoltaic-Storage-Charging-Change" System Architecture and Control Model Description

2.1. "Photovoltaic-Storage-Charging-Change" System Architecture

The system architecture of an EV "Photovoltaic-Storage-Charging-Change" station discussed in this paper is shown in Figure 1. It mainly includes four parts: the photovoltaic system, the energy storage system, the charging system, and the battery changing system. Detailed information of these parts is given as follows:

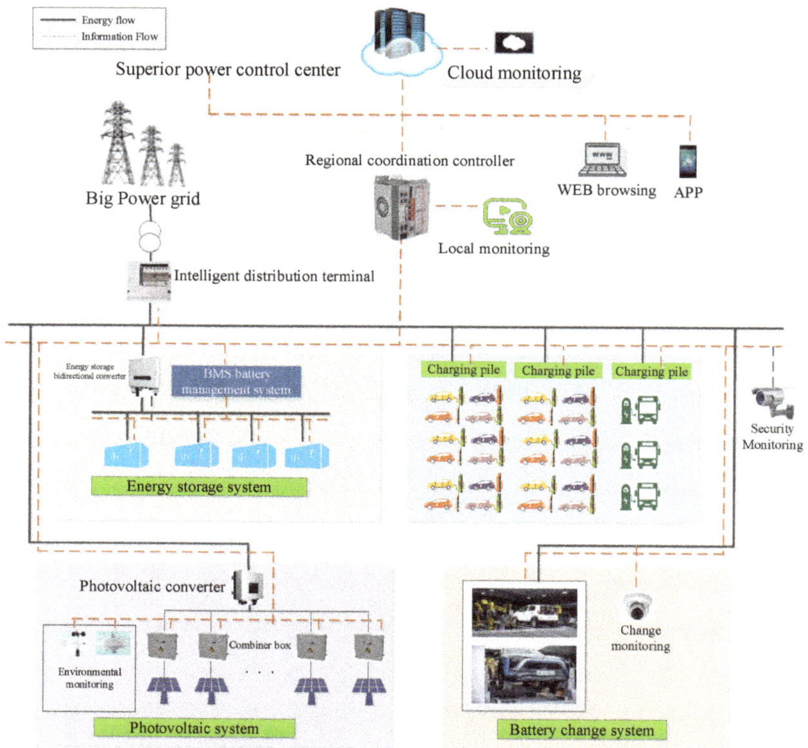

Figure 1. "Photovoltaic-Storage-Charging-Change" integrated system architecture.

(a) Photovoltaic system

The photovoltaic system mainly includes photovoltaic modules, photovoltaic inverters, and other major equipment. The electricity generated by photovoltaics can be transmitted to the energy storage system, charging piles, and the battery changing system, as well as to the main power grid. In this paper, the photovoltaic power generation model only considers the impact of light intensity and ambient temperature, as shown in Formulas (1) and (2) [33].

$$P_{pv}(i) = P_{ste}\frac{G(i)}{G_{ste}}(1 + k(T_{pt}(i) - T_{ste}(i))) \tag{1}$$

$$T_{pt}(i) = T_{et}(i) + 30 \times \frac{G(i)}{1000} \tag{2}$$

where $P_{pv}(i)$—photovoltaic system power generation at time i, kW; P_{ste}—rated output power of a single photovoltaic panel in a standard test environment, kW; T_{ste}—rated temperature of a single photovoltaic panel in a standard test environment, °C; $G(i)$—light intensity at time i, kW/m^2; G_{ste}—the light intensity under the standard test environment, which is 1 kW/m^2; $T_{pt}(i)$—photovoltaic system temperature at time i, °C; $T_{et}(i)$—environmental temperature at time i, °C.

(b) Energy storage system

The energy storage system includes devices such as energy storage converters and Lithium iron phosphate battery systems. It is mainly responsible for absorbing photovoltaic power generation as well as providing a backup energy source to satisfy peak energy demands. The energy storage system model, as shown in Formula (3), only considers system states such as current charging and discharging modes, and the residual power of each item of energy storage equipment, while ignoring the internal power electronic control. The limitation of the charging and discharging processes of the lithium iron phosphate battery pack in the energy storage system is that its maximum charging and discharging power cannot exceed its maximum rate.

$$P_{dis}(i) = P_{disc}(i)\omega_c(i)\eta_{disc} + P_{disd}(i)\omega_d(i)\eta_{disd} \tag{3}$$

where $P_{dis}(i)$—total charging and discharging power of the energy storage system at the time i, kW; $P_{disc}(i)$—charging power of the energy storage system at the time i, kW; $P_{disd}(i)$—discharging power of the energy storage system at the time i, kW; $\omega_c(i)$—charge state of the energy storage system at the time i, if charging, $\omega_c(i) = 1$; $\omega_d(i)$—discharge state of the energy storage system at the time i, if discharging, $\omega_d(i) = -1$.

(c) Charging system

Charging piles in the charging system are used as energy interactions between EVs and the power grid. Under normal conditions. EVs take power from grid through charging piles. When the power grid is short of power, EVs can also be used as an adjustable load to transmit electric energy to the power grid through V2G technology. The remaining electricity and charging demand of EV at any time of the day are represented by Formula (4) [34].

$$S_{EV}(i+1) = S_{EV}(i) + \lambda_{cp}P_{EV}(i)\Delta t - P_{EV,dcp}(i)\Delta t \tag{4}$$

where $S_{EV}(i+1)$—remaining EV battery power at time i + 1, %; $S_{EV}(i)$—remaining EV battery power at time i, %; λ_{cp}—charging efficiency of the EV; Δt—time interval, min; $P_{EV,dcp}(i)$—the power consumption of the EV at the time i, kW.

(d) Change system

Battery changing is a supplementary option for EV charging. At present, the shortest charging time by high-power fast charging piles for EVs is about 30 min to 2 h, which is too long for general EV users. However, an automatic battery changing process includes removing the low-power battery from EVs and loading a fully charged one. The whole process only takes 3 to 5 min based on the operation experiences of a pilot station in Shanghai, China. Therefore, battery changing is an important supplementary option for EV users, in addition to charging.

In the battery changing system, there is a certain amount of fully charged batteries in stock to ensure that EV users can exchange fully charged batteries at any time. The low-power batteries replaced from EVs are charged on the battery rack and can be reused for changing after it is fully charged. Therefore, the batteries in the battery changing system can be used as backup energy storage to relieve energy supply tension. The model of the battery changing system is described in Formulas (5)–(8):

$$P_{es}(i) = \sum_{n=1}^{156} P_{n,es}(i)\lambda_{ces}\omega_{n,es}(i) \quad (5)$$

$$S_{N,es}(i) = Q_{es0} - P_{EV,dcp}(i)\Delta t \quad (6)$$

$$S_{es}(i) = \sum_{N=0}^{N_{EV}} S_{N,es}(i),\ N = 0,1,2,\ldots,N_{EV} \quad (7)$$

$$S_{tes}(i+1) = S_{tes}(i) + P_{es}(i)\Delta T - S_{es}(i) \quad (8)$$

where $P_{es}(i)$—total charging power of the battery changing system at time i, kW; n—the number of chargers in the battery changing system; $P_{n,es}(i)$—the charging power of the nth charger at time i, kW; λ_{ces}—the charging efficiency of the charging compartment to the battery to be replaced; $\omega_{n,es}(i)$—the state of the nth charger at time i, charging is 1 and discharging is -1, otherwise it is 0; N—the number of EVs that need to be replaced; $S_{N,es}(i)$—the EV's electric quantity that is replaced at time i; Q_{es0}—initial power; $S_{es}(i)$—total power of the EV that is replaced at time i; $S_{tes}(i+1)$—total amount of change electricity required by the EV at time i + 1; $S_{tes}(i)$—total amount of change electricity required by the EV at time i; ΔT—time interval, min.

2.2. Control Model of "Photovoltaic-Storage-Charging-Change" Integrated System

The "Photovoltaic-Storage-Charging-Change" integrated system is a comprehensive green/flexible EV energy supply system that integrates the photovoltaic subsystem, the energy storage subsystem, the charging subsystem, and the battery changing subsystem. The 24 h daily control objective of this integrated system is to meet the EVs' energy demands by coordinating different subsystems at the lowest cost. The control model of the "Photovoltaic-Storage-Charging-Change" integrated system includes regular operation constraints, carbon emissions constraints, as well as the optimization objective function.

A. Regular operation constraints

In the operation of the "Photovoltaic-Storage-Charging-Change" integrated system, the internal constraints of each subsystem need to be met.

(a) Photovoltaic output constraints:

$$P_{pv,min} \le P_{pv}(i) \le P_{pv,max} \quad (9)$$

(b) Energy storage system constraints [35]:

$$20\%S_{dis,max} \le S_{dis}(i) \le 90\%S_{dis,max} \quad (10)$$

$$S_{dis}(i+1) = S_{dis}(i) - \frac{\eta_{dis}P_{dis}(i)}{S_{dis,max}} \quad (11)$$

$$\begin{cases} 0 \le P_{disc}(i) \le \gamma_{disc}S_{diss} \\ -\gamma_{disc}S_{diss} \le P_{disd}(i) \le 0 \end{cases} \quad (12)$$

(c) Charging system constraints [36]:

$$S_{EV,min} \le S_{EV}(i) \le S_{EV,max} \quad (13)$$

$$\Delta P_{EV}(i) = (1 - \lambda_{cp}\alpha_{EV})P_{EV}(i-1) \quad (14)$$

$$\eta_{min}P'_{EV}(i) \le P_{EV}(i) \le \eta_{max}P'_{EV}(i) \quad (15)$$

$$\sum_{i=1}^{T} P'_{EV}(i) = \sum_{i=1}^{T} P_{EV}(i) \quad (16)$$

(d) Changing system constraints:

$$\eta_{ces,min}P_{es,max} \leq \lambda_{ces}P_{n,es}(i)\Delta T \leq \eta_{ces,max}P_{es,max} \tag{17}$$

where $P_{pv}(i)$—the power generation of the photovoltaic system at the time i, kW; $P_{disd}(i)$—the discharge power of the energy storage system at the time i, kW; $P_{disc}(t)$—charging power of the energy storage system at the time i, kW; η_{disc}—charge efficiency of the energy storage system; η_{disd}—discharge efficiency of the energy storage system; $S_{dis}(i)$—capacity of the energy storage system at the time i, kWh; $S_{dis,max}$—maximum capacity of the energy storage system, kWh; S_{diss}—rated capacity of the lithium iron phosphate battery pack of the energy storage system, kWh; γ_{disc}—maximum allowable charge rate of the lithium iron phosphate battery pack of the energy storage system; γ_{disd}—maximum discharge rate of the lithium iron phosphate battery pack of the energy storage system; $S_{EV,min}$—the lower limit of battery storage energy, %; $S_{EV,max}$—the upper limit of battery storage energy, %; α_{EV}—power adjustment coefficient; η_{min}—minimum charging margin in disordered charging mode; η_{max}—maximum charging margin in disordered charging mode; $P_{EV}(i)$—power value of orderly charging of EVs at the time i, kW; $P'_{EV}(i)$—power value of disorderly charging of EVs at the time i, kW. $\eta_{ces,min}$—the minimum margin for replaced battery in the charging compartment of the battery changing system; $P_{es,max}$—maximum charging power of the charging compartment, kW; $\eta_{ces,max}$—the maximum margin for replaced battery in the charging compartment of the battery changing system.

B. Carbon transaction costs and carbon emission constrains

In the operation of the "Photovoltaic-Storage-Charging-Change" integrated system, carbon transaction costs are especially considered in this paper as shown in Formula (18) [37].

$$C_e(i) = \sum \sigma_i(Q_{et} - Q_{qt}) \tag{18}$$

where $C_e(i)$ is the carbon transaction cost, σ_i is the carbon transaction price of the sub system at time i, Q_{iet} is subsystem carbon emissions at time i, Q_{iqt} is subsystem's carbon emission allowance at time i.

On the basis of Formulas (9)–(17), the maximum carbon emission quota constraints of the "Photovoltaic-Storage-Charging-Change" integrated system are considered in this paper, as shown in Formulas (19) and (20) [38]:

$$C_{CO2} = \sum_{t_i}^{t_j} P_G\gamma_{CO2}\pi_{CO2} \tag{19}$$

$$C_{CO2} \leq C_{max} \tag{20}$$

where γ_{CO2}, π_{CO2} are the average proportion of carbon-containing power sources in the distribution network and the carbon emission coefficient per kWh.

C. Optimization objective function

The optimization objective function considers energy costs, carbon transaction costs, and the net balance of energy supply and demand, as shown in Formulas (21)–(24):

$$minF = \lambda_1F_1 + \lambda_2F_2 \tag{21}$$

$$F_1 = \sum_{t_i}^{t_j}[C_n(P(n,i)) \quad P_{RES}(i) + C_w(P(n,i)) + C_{excit}(i) + C_{dc}(i) + C_e(i)] \tag{22}$$

$$F_2 = \frac{1}{T}\sum_{t_i}^{t_j}[P_{RES}(i)]^2 \tag{23}$$

$$P_{RES}(i) = \left| \begin{array}{l} P_G(i) + \mu_{pv}P_{pv}(i)N_{pv} + \\ \mu_{disd}F_{disd}(i)N_{dis} - \mu_{cp}P_{cp}(i)N_{cp} - \\ P_{load}(i) - E_{load}(i) - \mu_{disc}P_{disc}(i)N_{dis} - \mu_{es}P_{es}(i)N_{es} \end{array} \right| \tag{24}$$

where $P_{RES}(i)$—net power of the integrated system of "Photovoltaic-Storage-Charging-Change" at the time i, kW; $P_G(i)$—power provided by the distribution network to the system at time i, kW; μ_{pv}—transmission efficiency of photovoltaic system, %; μ_{disd}—the discharge efficiency of the storage system, %; μ_{cp}—the transmission efficiency of the charging system, %; $P_{cp}(i)$—the power generated by the charging system at the time i, kW; N_{pv}—the number of power generation units in the photovoltaic system; N_{disd}—the number of energy storage units in the energy storage system; N_{cp}—the number of AC/DC charging piles in the charging system; N_{es}—the number of batteries in the battery changing compartment; μ_{disc}—charging efficiency of the energy storage system, %; μ_{es}—charging efficiency of the battery in the battery exchange system, %; $P_{es}(i)$—the charging power of the battery at the time i, kW; $P_{load}(i)$—total load in the integrated urban "Photovoltaic-Storage-Charging-Change" system at the time i, kW; $E_{load}(i)$—the conventional power load of the charging pile in the integration of urban "Photovoltaic-Storage-Charging-Change" at the time i, kW; $C_n(P(n,i))$—the power generation cost of the nth type of power generation unit; $C_w(P(n,i))$—the maintenance cost of the nth type of power generation unit; $C_{excit}(i)$—the cost for motivating users to participate in the dispatch; $C_{dc}(i)$—the cost for the power station to purchase electricity from the grid; λ_1, λ_2—the weighting factors, $\lambda_1 + \lambda_2 = 1$.

In order to overcome the possible error of subjective experience in weighting, the entropy weight method is applied to the weight in this paper. For a detailed description of the entropy weight method, please refer to the literature [39].

3. Two-Step Intelligent Control of "Photovoltaic-Storage-Charging-Change" System Based on ISOM-SAIA

3.1. Overall Architecture of Two-Step Intelligent Control ISOM-SAIA

The overall architecture of two-step intelligent control (ISOM-SAIA) for the integrated EV "Photovoltaic-Storage-Charging-Change" system is shown in Figure 2. The tasks of the two-step intelligent control are to classify the operation modes of all subsystems for each hour using a data-driven intelligent classification method known as the improved self-organizing mapping neural network (ISOM) in the first step, and then calculate the optimal operation outputs of all subsystems for each hour over a day using a rolling optimization method, called simulated annealing immune control (SAIA), in the second step.

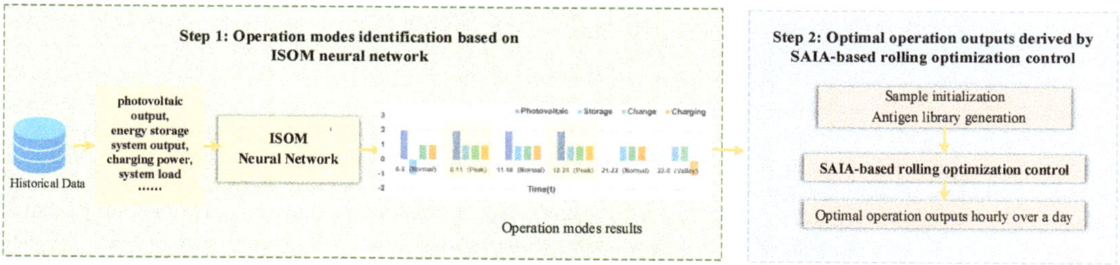

Figure 2. Overall architecture of two-step intelligent control based on ISOM-SAIA.

3.2. Step 1: Classification of Operation Modes Based on ISOM

The first step of the two-step intelligent control for the integrated EV "Photovoltaic-Storage-Charging-Change" system is to classify the operation modes of all subsystems in each hour of a day based on a data-driven intelligent clustering method known as the improved self-organizing map neural network (ISOM). The ISOM neural network consists of three layers: the input layer, the competition layer and the output layer. The nodes in all layers are connected by weights: $\omega_{ij}(i = 1, \ldots, n; j = 1, \ldots, m)$. The input data of ISOM neural network include the system time, EV charging quantity, energy storage SOC and changed battery SOC, etc. The outputs of the ISOM neural network are the matching operation modes (charging, discharging, optional, not working) of all subsystems in three special time periods of the day (peak periods, flat periods, valley periods). As shown in Figure 3, the ISOM neural network classification procedure is given as follows:

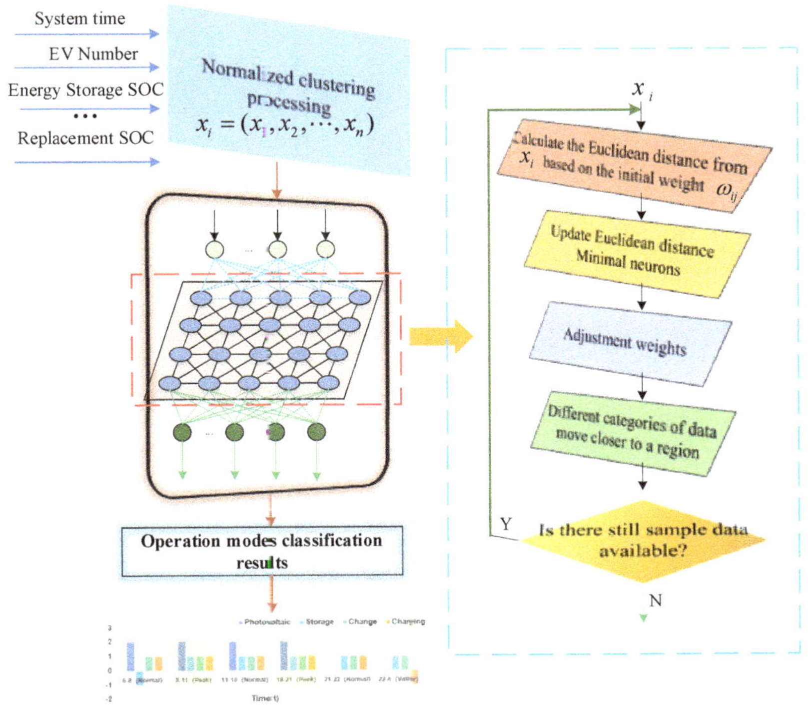

Figure 3. ISOM-based operation mode classification process.

(a) Perform normalized clustering processing on input data sample $x_i = (x_1, x_2, \cdots, x_n)$. The state variables, represented by x_i, are the system time, EV charging quantity, energy storage system, battery SOC, etc.;

(b) Initialize the connection weights of ISOM neural network and locate it in the middle area of the input data, set the time count $t = 0$, determine the learning rate, and the learning radius r;

(c) Calculate the Euclidean distance between the input sample processed in step (a) and the neurons in the competition layer according to Equation (25);

$$d_j = \sum_{i=1}^{n} \left(x_i^k - \omega_{ij}\right)^2 \tag{25}$$

(d) Select the neuron of the competition layer with the smallest Euclidean distance as the winning neuron to replace the original neuron;

(e) Adjust the weights between the input layer and the competition layer, and the weights between the competition layer and output layer according to Equations (26) and (27);

$$\omega_{ij} = \omega_{ij} + \xi_1 (x_i - \omega_{ij}) \tag{26}$$

$$\omega_{jk} = \omega_{jk} + \xi_2 \left(y_k - \omega_{jk}\right) \tag{27}$$

where y_k is the category of different control modes (k = 1, 2, 3, . . .).

(f) If there are still input data samples, then $t = t + 1$, and return to step (c), otherwise go to (g).

(g) Output the matching operation modes (power output, power input, optional, and not working) of all subsystems in three special time periods of the day (peak periods, flat periods, valley periods).

3.3. Step 2: SAIA-Based Rolling Optimization Control

The second step of the two-step intelligent control for the integrated EV "Photovoltaic-Storage-Charging-Change" system is to derive the optimal operation outputs of all subsystems in each hour of a day with a rolling optimization process based on a simulated annealing immune algorithm (SAIA) combined with a simulated annealing (SA) and immune algorithm (IA). Figure 4 shows the flow chart of SAIA-based rolling optimization control procedure, which has three main procedures—(a), (b) and (c)—as follows.

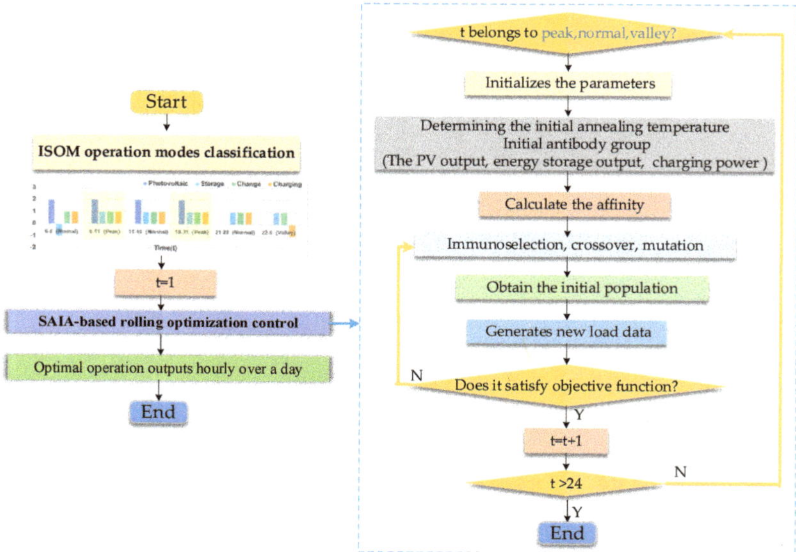

Figure 4. SAIA-based rolling optimization control procedure.

(a) Based on the classification results from Step 1 in Section 3.2, the operation modes of four subsystems in each time period of peak, flat, and valley are obtained.
(b) Starting from t = 1, check which special time period t belongs to, and derive the optimal operation output of all subsystems using the SAIA algorithm based on the objective function and constraints given in Formulas (9)–(24). The SAIA optimization calculation process is given from (1) to (8) as follows:

(1) Let the total population be N, the number of memory cell antibodies be N_m, the crossover probability be P_c, the mutation probability be P_m, the temperature cooling coefficient be C, the number of memory cells be g, and the initial annealing temperature be T;

(2) The system time, EV energy supply loads, and system operating status are used as antigens to initialize the parameters;

(3) Randomly generate $N - N_m$ antibodies, mainly using photovoltaics, energy storage, charging piles, etc. as the initial antibody group objects, and form the initial population together with N_m antibodies in the memory unit; if the memory unit is empty, randomly generate N antibodies to form the initial population, make $k = 0$;

(4) Calculate the actual load value of each antibody P_i (photovoltaic system output, energy storage system output, electric vehicle charging power, rechargeable battery charging power, etc.), according to Formulas (3) and (4) to determine the current temperature of

the system operation requirements and control output Affinity $TF(P_i)$ between each other, save N_m antibodies in the memory unit P_g;

$$TF(P_i) = \frac{e^{-(f(P_i)-f(P_g))/t}}{\sum_{i=1}^{N} e^{-(f(P_i)-f(P_g))/t}} \tag{28}$$

(5) Implement the immune selection, crossover, and mutation operations to determine the type of control strategy;

(6) Perform the simulated annealing operation of fixed-step sampling on the parameters of energy equipment, such as the photovoltaic, energy storage, charging and swapping parameters, in the population, and new antibodies are generated. If the result satisfies Formulas (3)–(5), it is updated and the new load value is calculated; Otherwise, it remains unchanged;

$$\min[1, \exp(-\Delta/T_k)] \geq \text{random}[0, 1] \tag{29}$$

where T_k—annealing temperature in k.

(7) Determine whether the algorithm convergence criterion is satisfied. If it is satisfied, perform the de-temperature operation, output the control strategy result, and end the algorithm; otherwise, return to step (5);

(8) Let $k = k + 1$, input the Formulas (3)–(6) to the de-temperature operation.

$$T_k = CgT_{k-1}, C \in (0, 1) \tag{30}$$

(c) If t < 24 h, $t = t + 1$; otherwise, the algorithm ends and derives the optimal operation modes of all subsystems in each hour of a day.

4. Simulation Analysis

4.1. Data Source and System Parameters of "Photovoltaic-Storage-Charging-Change" System

The data source of the simulation analysis in this section comes from a demonstration project of an EV "Photovoltaic-Storage-Charging-Change" station in Shanghai, China. It is equipped with four subsystems: photovoltaic, energy storage, charging, and battery changing. In addition, it provides both fast charging and battery changing services for EVs. The installed capacity, power generation costs, and maintenance costs of each subsystem are shown in Table 1.

Table 1. Parameters of each power generation unit.

Type	Installed Capacity (kWh)	Power Generation Cost (Yuan kWh)	Maintenance Cost (Yuan kWh)
Photovoltaic	10	0.41	0.0401
Energy Storage	1000	0.68	0.0843
Charging System	4000	0.56	0.0512
Battery Changing System	2340	0.68	0.0753

The typical daily EV charging and battery changing load curves of the green/flexible EV energy supply station in summer and winter are shown in Figure 5.

4.2. Tests for Two-Step Intelligent Control of "Photovoltaic-Storage-Charging-Change" System

4.2.1. Tests for Step 1: Classification of Daily Operation Modes Based on ISOM Neural Network

In this section, the EVs' daily energy demands and the system parameters of the EVs' "Photovoltaic-Storage-Charging-Change" system, as shown in Section 4.1, are used in the tests of the first step of two-step intelligent control method proposed in this paper. Each subsystem of the EVs' "Photovoltaic-Storage-Charging-Change" system has four operation modes. They are, respectively, the power output state (indicated by 2), the power input state (indicated by −1), the power output or input state (indicated by 1), and the not-working state (indicated by 0).

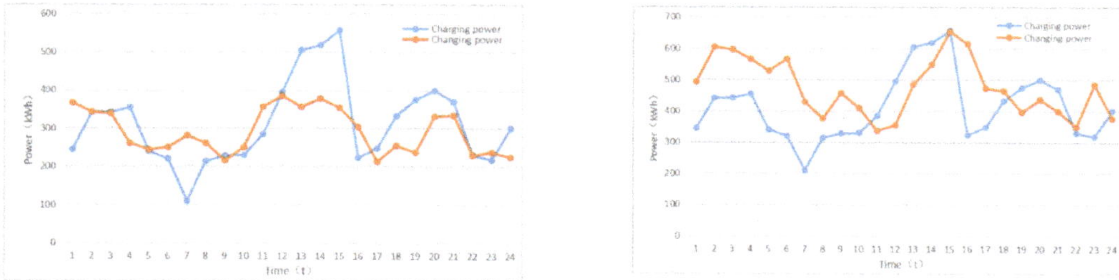

(a) (b)

Figure 5. (**a**) EV energy demand curve of charging and battery changing on a typical day in summer; (**b**) EV energy demand curve of charging and battery changing on a typical day in winter.

Based on the method for the classification of ISOM-based operation modes given in Section 3.2, the hourly operation modes of all subsystems in the "Photovoltaic-Storage-Charging-Change" system are derived and shown in Figure 6. From Figure 6, the operation modes of all subsystem during different time periods are basically clear: the photovoltaic subsystem can only output electric power during the daytime; the energy storage system can output power to relieve the high EV energy demand during some peak periods, and can input power during normal and valley periods; the battery changing subsystem is optional for power output or input during the entirety of the day; the charging subsystem is optimal for power output or input during most hours of the day, and can only input power during several normal and valley periods.

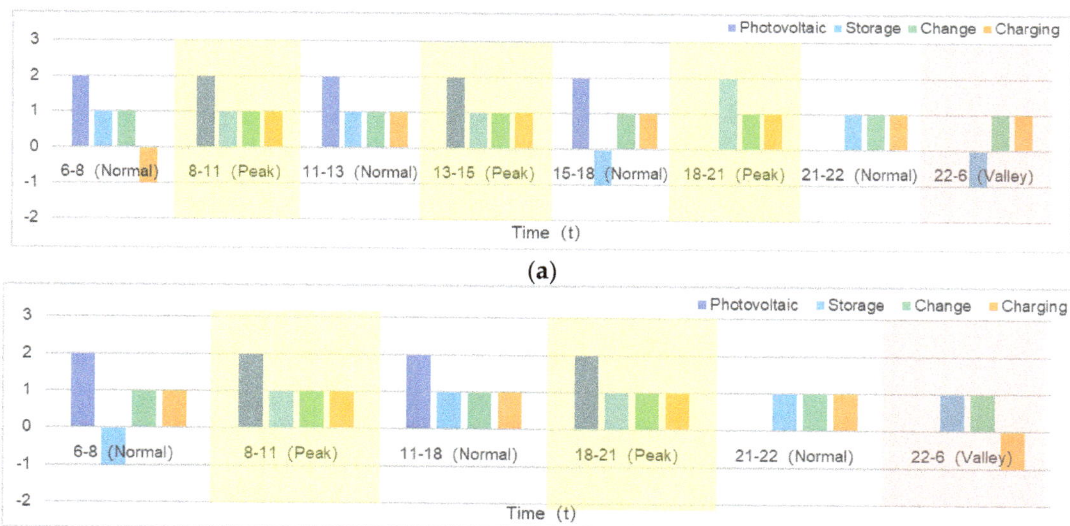

(b)

Figure 6. Daily operation mode classification results based on ISOM neural network: (**a**) in summer; (**b**) in winter.

4.2.2. Tests for Step 2: Optimal Daily Operation Outputs by SAIA-Based Rolling Optimization

In this section, the second step of the two-step intelligent control method proposed in Section 3.3 of this paper is tested based on the operation mode classification result derived from Section 4.2.1.

The optimal operation outputs of each subsystem and the total power of the integrated system purchased from grid for all hours of a typical summer day and a typical winter day are derived via the SAIA-based rolling optimization control method, and are shown in Figure 7. In order to test the performance of the proposed two-step intelligent control method in terms of the peak shaving and valley filling of a power grid, as well as energy costs and carbon emission reduction, the following simulation analysis is carried out.

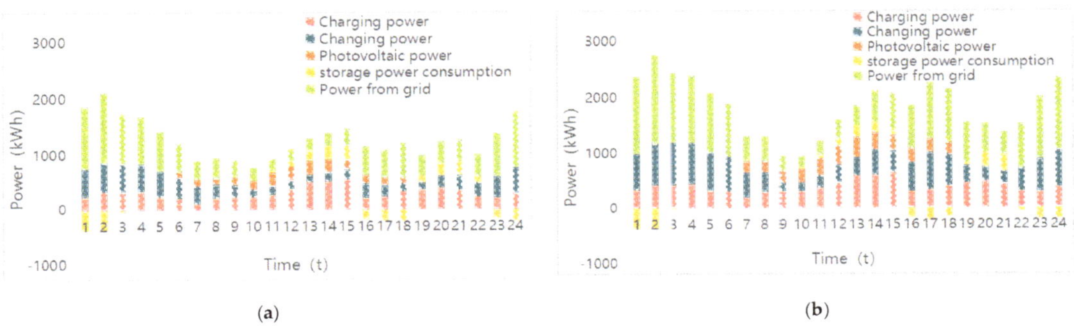

Figure 7. Results of coordinated regulation of various systems on a typical day: (**a**) in summer; (**b**) in winter.

A. Simulation analysis of power grid peak shaving and valley filling performance

In Figures 8 and 9, the operation outputs of the battery changing subsystem and the energy storage subsystem derived by the proposed two-step intelligent control ISOM-SAIA are compared with original operation records. Both subsystems contribute to the peak shaving and valley filling of the grid. Figure 10 provides more detailed statistics on the performance of peak shaving and valley filling of power grid: the peak load reduction can reach 2387.57 kWh and the valley filling is 2132.68 kWh on a typical day in summer; the peak load reduction can reach 1663.18 kWh and the valley filling 2346.73 kWh on a typical day in winter.

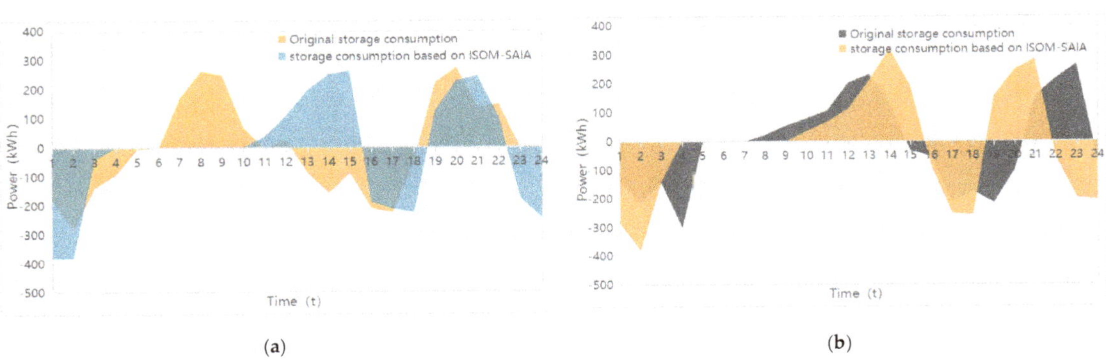

Figure 8. Comparison of power outputs of energy storage subsystem on a typical day: (**a**) in summer; (**b**) in winter.

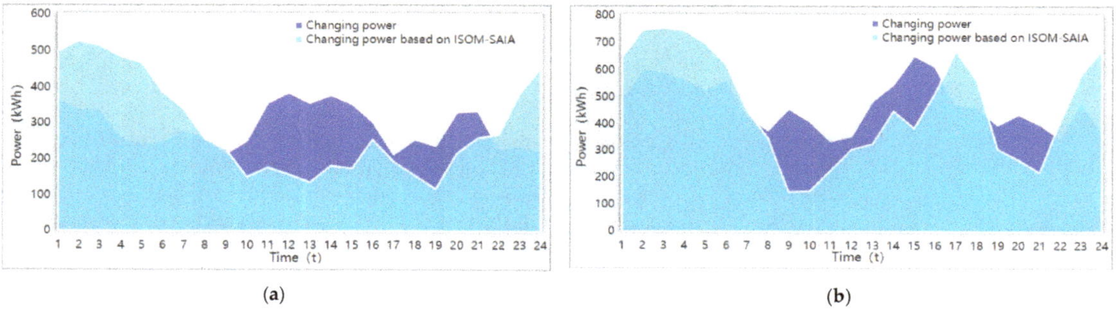

Figure 9. Comparison of power consumption of battery changing subsystem on a typical day: (**a**) in summer; (**b**) in winter.

Figure 10. Comparison of the results of peak shaving and valley filling: (**a**) in summer; (**b**) in winter.

B. Simulation analysis of carbon transaction costs and carbon emission constraints

In order to test the performance of the carbon transaction cost and carbon emission constraints (Formulas (18)–(20)) considered in the proposed ISOM-SAIA two-step intelligent control method, Figure 11 compares the power purchased from power grid of the EV "Photovoltaic-Storage-Charging-Change" station on a typical day with and without carbon constraints. It indicates that after considering the carbon transaction costs and carbon emission constraints, the total electricity purchased by the EV energy supply station from the grid is reduced from the original 15,738.32 kWh to 12,629.64 kWh on a typical summer day, and it drops from 23,215.62 kWh to 18,428.3 kWh on a typical winter day, while meeting the same daily EV energy demand. It can save 17,634.68 yuan on a typical summer day, and 10,453.82 yuan on a typical winter day.

Due to the high proportion of fossil-fuel fired power generation in the power grid, the carbon emissions per kilowatt hour of power purchased from grid are higher than those of photovoltaic power generation in the EV energy supply station. Therefore, when meeting the same EV energy demand, reducing the proportion of the grid-source power in the EV energy supply station can reduce both the energy costs and carbon emissions. Therefore, the carbon transaction costs and carbon emission constraints considered in the proposed two-step intelligent control method can help to achieve both economic and environmental benefits.

In addition, to further study the sensitivity of the maximum carbon emission limit, the overall operation benefits of the EV "Photovoltaic-Storage-Charging-Change" station, under different carbon emission limits, are analyzed and shown in Figure 12. It indicates that different carbon emission limits can have a significant effect on the benefits of the system. Therefore, it is necessary to select an appropriate carbon emission limit when implementing the proposed two-step intelligent control ISOM-SAIA for the EV "Photovoltaic-Storage-Charging-Change" station to ensure both high environmental and economic benefits.

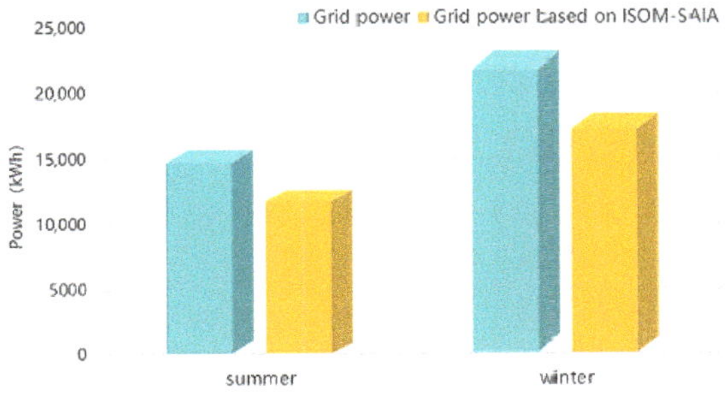

Figure 11. Power purchased from grid on a typical day.

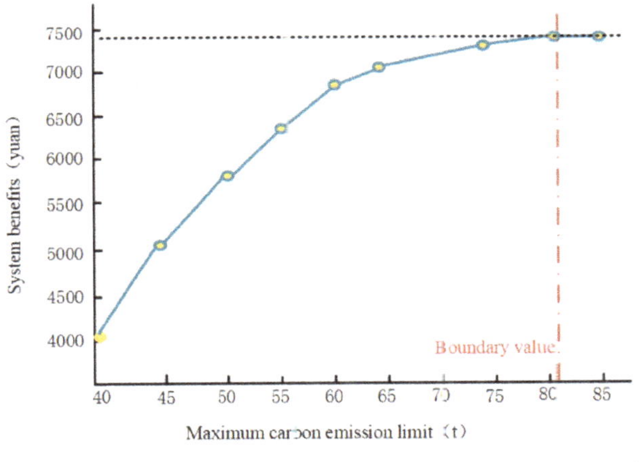

Figure 12. System benefits under different maximum carbon emission limits.

5. Conclusions

A two-step intelligent control method, ISOM-SAIA, is proposed in this paper to solve the problem of the 24 h control and regulation of a green/flexible EV energy supply station including four subsystems such as a photovoltaic subsystem, an energy storage subsystem, an EV charging subsystem, and an EV battery changing subsystem, under China's Dual Carbon Target. The innovations and contributions of this paper are listed as follows:

(1) The two-step intelligent control method, ISOM-SAIA, proposed in this paper effectively reduces the computation burden of solving the multi-dimensional mixed-integer programming problem of the simultaneous optimization of 24 h operation modes and outputs of four subsystems in the green/flexible EV energy supply station into two steps: step 1 for the data-driven classification of operation modes, and step 2 for the rolling optimization of operation outputs.

(2) Proper carbon transaction costs and carbon emission constraints are considered in the proposed control method, and the simulation results indicate that the added carbon transaction costs and carbon emission constraints can help to save costs and reduce carbon emissions for the green/flexible EV energy supply station.

In conclusion, the two-step intelligent control method, ISOM-SAIA, proposed in this paper is a useful control method for the 24 h control and regulation of integrated green/flexible EV energy supply stations. It can help the stations to optimally allocate energy flows between four subsystems, effectively respond to the peak shaving and valley filling of the power grid, save energy costs, and reduce carbon emissions.

Author Contributions: Conceptualization, S.S. and C.F.; Methodology, C.F., H.W. and H.Z.; Resources, S.S., D.P. and H.Z.; Software, C.F., H.W. and Y.L.; Validation, C.F., J.L., Y.L. and D.P.; Writing—original draft, S.S., J.L. and H.Z.; Writing—review and editing, S.S. and H.Z. All authors have read and agreed to the published version of the manuscript.

Funding: This work is supported by the Science and Technology Project of the State Grid Corporation of China (No. 52094019007G), and the Shanghai Rising-Star Program (No. 20QB1400300).

Institutional Review Board Statement: Not applicable.

Informed Consent Statement: Not applicable.

Conflicts of Interest: The authors declare no conflict of interest.

References

1. IEA. *Global EV Outlook 2021*; IEA: Paris, France, 2021. Available online: https://www.iea.org/reports/global-ev-outlook-2021 (accessed on 31 July 2021).
2. Xu, L.; Zhang, J.; Bao, Z.; Cao, Y. A comparative study between IPSO and MIP for co-ordinated scheduling of electricity and heat within a microgrid. *Trans. Inst. Meas. Control* **2013**, *35*, 444–456. [CrossRef]
3. Yang, T.; Zhao, L.; Wang, C. Review on application of artificial intelligence in power system and integrated energy system. *Autom. Electr. Power Syst.* **2019**, *43*, 2–14.
4. Jha, M.; Blaabjerg, F.; Khan, M.A.; Bharath Kurukuru, V.S.; Haque, A. Intelligent Control of Converter for Electric Vehicles Charging Station. *Energies* **2019**, *12*, 2334. [CrossRef]
5. Beigvand, S.D.; Abdi, H.; La Scala, M. Combined heat and power economic dispatch problem using gravitational search algorithm. *Electr. Power Syst. Res.* **2016**, *33*, 160–172. [CrossRef]
6. Jiang, X.; Jing, Z.; Li, Y.; Wu, Q.; Tang, W. Modelling and operation optimization of an integrated energy based direct district water-heating system. *Energy* **2014**, *64*, 375–388. [CrossRef]
7. Meng, A.; Mei, P.; Lu, H. Crisscross optimization algorithm for combined heat and power economic dispatch. *Power Syst. Prot. Control* **2016**, *44*, 90–97.
8. Ndiaye, E.H.M.; Ndiaye, A.; Faye, M.; Gueye, S. Intelligent Control of a Photovoltaic Generator for Charging and Discharging Battery Using Adaptive Neuro-Fuzzy Inference System. *Int. J. Photoenergy* **2020**, *2020*, 8649868. [CrossRef]
9. Qu, K.; Zhang, X.; Yu, T.; Han, C.J. Knowledge transfer based Q-learning algorithm for optimal dispatch of multi-energy system. *Autom. Electr. Power Syst.* **2017**, *41*, 18–25.
10. Liu, X. *Heterogeneous Multi-Agent Cooperative Control and Its Application in DC Microgrid*; Huazhong University of Science and Technology: Wuhan, China, 2019.
11. Meng, Y.; Cao, Y.; Zeng, J. Design and Implementation of Intelligent Management System for Electric Vehicle Charging Station. *Instrum. Technol.* **2016**, *33*, 22–24.
12. Zhang, Y.; Luo, H.; Wang, Y. Research on Hybrid Energy Storage Power Distribution and Bus Voltage Stabilization of Photovoltaic DC Microgrid Based on Fuzzy-Sag Control. *Electr. Meas. Instrum.* **2020**, 1–9. Available online: http://kns.cnki.net/kcms/detail/23.1202.TH.20200806.1637.014.html (accessed on 26 March 2021).
13. Qu, F.; Zhao, J.; Cai, Z.; Hu, C.; Dai, S.; Sun, Q. Cooperative optimization control strategy of electric vehicle and temperature-controlled load virtual power plant. *J. Electr. Power Syst. Autom.* **2021**, *33*, 48–56.
14. Tang, M.; Qu, X.; Yao, R.; Zhang, Y.; Chen, W.; Jia, K. Multi-photovoltaic Coordinated Control Strategy in DC Distribution Network Based on Discrete Consensus Algorithm. *Autom. Electr. Power Syst.* **2020**, *44*, 89–95.
15. Blasius, E.; Federau, E.; Janik, P.; Leonowicz, Z. Heuristic Storage System Sizing for Optimal Operation of Electric Vehicles Powered by Photovoltaic Charging Station. *Int. J. Photoenergy* **2016**, *2016*, 3980284. [CrossRef]
16. Zhang, X.; Li, L.; Fu, Y. Stability Analysis and Regional Cooperative Control of Controllable Inertia Optical Storage Interconnection System. *High Volt. Technol.* **2021**, *47*, 1694–1705.
17. Jiang, Y. Simulation of electric vehicle charging load cooperative control based on microgrid. *Comput. Simul.* **2020**, *37*, 87–90.
18. Ramadhani, U.H.; Shepero, M.; Munkhammar, J.; Widén, J.; Etherden, N. Review of probabilistic load flow approaches for power distribution systems with photovoltaic generation and electric vehicle charging. *Int. J. Electr. Power Energy Syst.* **2020**, *120*, 106003 [CrossRef]
19. Shimazu, T.; Takahashi, A.; Shimofuji, K.; Funabiki, S.; Nagata, T. Reactive power control of power conditioning systems to avoid voltage deviation in high-voltage distribution systems caused by both of photovoltaic generation and electric vehicle charging. *Electr. Eng. Jpn.* **2018**, *138*, 416–422.
20. Meng, C.; Jiang, Y.; Dai, J. Study on the scheduled performance control of a class of nonlinear multi-agent systems. *J. Nanchang Hangkong Univ.* **2019**, *33*, 1–6.

21. Xu, S.; Yan, Z.; Feng, D.; Zhang, L. Electric vehicle charging collaborative control strategy based on multi-agent. *Electr. Power Autom. Equip.* **2014**, *34*, 7–13.
22. Hao, R.; Ai, Q.; Zhu, Y. Energy Internet collaborative optimization control based on multi-agent consistency. *Autom. Electr. Power Syst.* **2017**, *41*, 10–17.
23. Zhou, Y.; Wang, K.; Li, G.; Han, B.; Liu, Z. Distributed hierarchical control strategy of microgrid based on multi-agent consensus algorithm. *Autom. Electr. Power Syst.* **2017**, *41*, 142–149.
24. Mesbahi, T.; Bartholomeüs, P.; Rizoug, N.; Sadoun, R.; Khenfri, F.; Le Moigne, P. Advanced Model of Hybrid Energy Storage System Integrating Lithium-Ion Battery and Supercapacitor for Electric Vehicle Applications. *IEEE Trans. Ind. Electron.* **2021**, *68*, 3962–3972. [CrossRef]
25. Shin, M.; Choi, D.; Kim, J. Cooperative Management for PV/ESS-Enabled Electric Vehicle Charging Stations: A Multiagent Deep Reinforcement Learning Approach. *IEEE Trans. Ind. Inform.* **2020**, *16*, 3493–3503. [CrossRef]
26. Vavilapalli, S.; Padmanaban, S.; Subramaniam, U.; Mihet-Popa, L. Power Balancing Control for Grid Energy Storage System in Photovoltaic Applications-Real Time Digital Simulation Implementation. *Energies* **2017**, *10*, 928. [CrossRef]
27. Kohonen, T. Self-organized formation of topologically correct feature maps. *Biol. Cybern.* **1982**, *43*, 59–69. [CrossRef]
28. Musharavati, F.; Hamouda, A.S.M. Enhanced simulated-annealing-based algorithms and their applications to process planning in reconfigurable manufacturing systems. *Adv. Eng. Softw.* **2012**, *45*, 80–90. [CrossRef]
29. Trung, N.T.; Anh, D.T. Comparing Three Improved Variants of Simulated Annealing for Optimizing Dorm Room Assignments. In Proceedings of the IEEE-RIVF International Conference on Computing and Communication Technologies, Danang, Vietnam, 13–17 July 2009.
30. Zhang, Y.; Liu, C. An Improved Evolution Algorithm of Immune Detectors for Network Data Analysis. In Proceedings of the International Conference on Intelligent Computing and Human-Computer Interaction (ICHCI), Sanya, China, 4–6 December 2020.
31. Jiang, J.; Song, C.; Ping, H.; Zhang, C. Convergence Analysis of Self-adaptive Immune Particle Swarm Optimization Algorithm. In Proceedings of the 15th International Symposium on Neural Networks (ISNN), Minsk, Belarus, 25–28 June 2018.
32. Shi, J.; Su, Y.D.; Xie, M. Research on Application of IGA (Immune Genetic Algorithm) to the Solution of Course-Timetabling Problem. In Proceedings of the 4th International Conference on Computer Science and Education, Nanning, China, 25–28 July 2009.
33. Yang, Y. *Research on Control Strategy of Power Electronic Transformer with Access of Photovoltaic-Storage System and Charging Station*; North China Electric Power University: Beijing, China, 2020.
34. Ni, C.; Liu, X. Battery Capacity Optimal Configuration and Economical Analysis of Energy Storage Photovoltaic System. *Zhejiang Electr. Power* **2019**, *38*, 1–10.
35. Chen, T.; Xu, X.; Yan, Z.; Zhu, Y. Optimal operation based on deep reinforcement learning for energy storage system in photovoltaic-storage charging station. *Electric Power Autom. Equip.* **2021**, *41*, 1–9.
36. Shang, Z.; Wang, G. Research on Day-ahead Optimal Economic Dispatching Strategy for Micro-energy-grid and Analysis of Electric Energy Storage Strategy. *Electr. Power Sci. Eng.* **2020**, *36*, 9–16.
37. Tang, W.; Zhang, Y.; Xuan, D.; Jiang, D. Low-carbon Park Economic Operation Scheduling Strategy Taking into Account the Tiered Carbon Trading Mechanism. *Distrib. Util.* **2021**, *38*, 10–18.
38. Mei, G.; Gong, J.; Zhang, Y. Scheduling Strategy for Multi-energy Complementary Virtual Power Plant Considering the Correlation Between Wind and Solar Output and Carbon Emission Quota. *Proc. CSU-EPSA* **2021**, *33*, 62–69.
39. Hananeh, F.; Alireza, Z.; Shahram, J. The role of demand response in single and multi-objective wind-thermal generation scheduling: A stochastic programming. *Energy* **2014**, *64*, 853–867.

processes MDPI

Article

Life Cycle Energy Consumption and Greenhouse Gas Emissions Analysis of Primary and Recycled Aluminum in China

Tianduo Peng [1], Lei Ren [2,3,4], Ershun Du [1], Xunmin Ou [2,3,4,*] and Xiaoyu Yan [5,*]

1 Institute of Climate Change and Sustainable Development, Tsinghua University, Beijing 100084, China
2 Institute of Energy, Environment, Economy (3E), Tsinghua University, Beijing 100084, China
3 Tsinghua University—Zhang Jiagang Joint Institute for Hydrogen Energy and Lithium-Ion Battery Technology, Tsinghua University, Beijing 100084, China
4 Tsinghua-Rio Tinto Joint Research Centre for Resources, Energy and Sustainable Development, Laboratory for Low Carbon Energy, Tsinghua University, Beijing 100084, China
5 Environment and Sustainability Institute, University of Exeter, Penryn, Cornwall TR10 9FE, UK
* Correspondence: ouxm@mail.tsinghua.edu.cn (X.O.); Xiaoyu.Yan@exeter.ac.uk (X.Y.)

Abstract: Aluminum production is a major energy consumer and important source of greenhouse gas (GHG) emissions globally. Estimation of the energy consumption and GHG emissions caused by aluminum production in China has attracted widespread attention because China produces more than half of the global aluminum. This paper conducted life cycle (LC) energy consumption and GHG emissions analysis of primary and recycled aluminum in China for the year 2020, considering the provincial differences on both the scale of self-generated electricity consumed in primary aluminum production and the generation source of grid electricity. Potentials for energy saving and GHG emissions reductions were also investigated. The results indicate that there are 157,207 MJ of primary fossil energy (PE) consumption and 15,947 kg CO_{2-eq} of GHG emissions per ton of primary aluminum ingot production in China, with the LC GHG emissions as high as 1.5–3.5 times that of developed economies. The LC PE consumption and GHG emissions of recycled aluminum are very low, only 7.5% and 5.3% that of primary aluminum, respectively. Provincial-level results indicate that the LC PE and GHG emissions intensities of primary aluminum in the main production areas are generally higher while those of recycled aluminum are lower in the main production areas. LC PE consumption and GHG emissions can be significantly reduced by decreasing electricity consumption, self-generated electricity management, low-carbon grid electricity development, and industrial relocation. Based on this study, policy suggestions for China's aluminum industry are proposed. Recycled aluminum industry development, restriction of self-generated electricity, low-carbon electricity utilization, and industrial relocation should be promoted as they are highly helpful for reducing the LC PE consumption and GHG emissions of the aluminum industry. In addition, it is recommended that the central government considers the differences among provinces when designing and implementing policies.

Keywords: primary and recycled aluminum; life cycle analysis; energy consumption; greenhouse gas emissions

Citation: Peng, T.; Ren, L.; Du, E.; Ou, X.; Yan, X. Life Cycle Energy Consumption and Greenhouse Gas Emissions Analysis of Primary and Recycled Aluminum in China. *Processes* **2022**, *10*, 2299. https://doi.org/10.3390/pr10112299

Academic Editors: Ming Liu and Xiao Wu

Received: 9 October 2022
Accepted: 31 October 2022
Published: 5 November 2022

Publisher's Note: MDPI stays neutral with regard to jurisdictional claims in published maps and institutional affiliations.

Copyright: © 2022 by the authors. Licensee MDPI, Basel, Switzerland. This article is an open access article distributed under the terms and conditions of the Creative Commons Attribution (CC BY) license (https://creativecommons.org/licenses/by/4.0/).

1. Introduction

China has the largest aluminum production capacity in the world. In 2020, 37.08 million tons of primary aluminum and 6.0 million tons recycled aluminum were produced in China, sharing more than half of the world's aluminum output [1,2]. As an energy- and GHG emissions-intensive industry, aluminum production puts tremendous pressure on China's energy conservation and greenhouse gas (GHG) emissions reduction. Aluminum smelting is one of the largest energy-consuming sectors of China's non-ferrous metal industry, which accounted for about 6.5% of China's total electricity consumption in 2020 [3,4]. GHG emissions from primary aluminum ingot production were as high as 426 mt (megaton) $CO_{2\text{-eq}}$ in 2020, accounting for approximately 5% of China's total GHG emissions [5]. China is now making great efforts to promote energy efficiency, capacity utilization, and the adoption of advanced technologies in the aluminum industry to realize the potential of energy saving and GHG emissions reduction towards its carbon peaking and carbon neutrality target. However, the production of aluminum is a complex and multi-process industrial activity. Therefore, it is necessary to assess the GHG footprint of the entire aluminum production process using the life cycle (LC) approach, instead of focusing solely on the smelting process.

Many academic and industrial institutions have carried out research on the GHG emissions of aluminum from the perspective of LC [6–13]. For example, the International Aluminum Institute (IAI) established a general process-based calculating tool to estimate the energy input, material flow, and environmental emissions from primary aluminum production worldwide. The LC inventory data and environmental metrics of the global primary aluminum industry in 2019 were published recently in which China's quantitative results were of poor quality due to the lack of statistics data [6]. Argonne National Laboratory developed the annually updated GREET model to investigate the LC energy input, water consumption, and air emissions of multiple materials used in vehicle manufacturing, with the ability to evaluate the performance of American aluminum products [7]. The Aluminum Association (TAA) released a report on the environmental footprint of aluminum products in North America based on operational data for the year of 2016 [8]. Aluminum for Future Generation (AFG) and the European Aluminum Association (EAA) [9] jointly built the LC inventory data for aluminum production and transformation processes in Europe according to the data for the year 2010. Liu et al. [10] reviewed the status and application of the life cycle assessment method in the aluminum industry, providing the strength and weakness of the method to address sustainability in the aluminum industry.

The energy consumption and GHG emissions of China's aluminum industry have also received extensive attention from researchers and have been widely analyzed in previous studies [14–23], in which LC analysis is an important perspective. Gao et al. [18] assessed the LC GHG emissions of primary aluminum in China in comparison with the global average level and estimated the reduction potentials in detail. Du et al. [19] evaluated the reduction potentials for LC GHG emissions and energy use from aluminum-intensive vehicles in China and found that the reduction potentials were insignificant because of the high LC intensities of energy consumption and GHG emissions. Ding et al. [20] quantitatively analyzed the energy consumption and GHG emissions caused by primary aluminum and recycled aluminum production in China in 2008 and compared this with 2003. Guo et al. [21] calculated the national average LC energy consumption and GHG emissions for primary aluminum production in China in the years of 2003, 2006, 2008, and 2010, and conducted correlation analysis between GHG emissions and key factors. Liu et al. [22] investigated the features, trajectories, and driving forces of the energy-related GHG emissions of China's primary and recycled aluminum industry from the LC analysis perspective in the period 2004–2013.

Electricity generation has a significant impact on energy consumption and GHG emissions for aluminum production. However, both the scale of self-generated electricity associated with primary aluminum production and the generation mix of grid electricity have significant regional variations in China. Therefore, its necessary to carry out provincial LC analysis on primary and recycled aluminum production considering the individual or regional actual power consumption to make the estimation more accurate and policy recommendations more effective. Several studies have assessed the provincial variations in the LC GHG emissions of China's aluminum industry [24–26]. For example, Hao et al. [24] explored the national and provincial average-level GHG emissions intensity of China's primary aluminum production in 2013 and 2020, considering the impact of provincial disparity in the electricity generation system based directly on GHG emissions factors of each process from other studies. Zhang et al. [25] assessed the LC environmental footprint of primary aluminum and recycled aluminum production in China based on a bottom-up approach with statistical data, and simply analyzed the provincial aluminum production density and environmental impacts in different provinces.

However, significant shortcomings still exist in the literature: (1) Most of these previous studies only focused on primary aluminum but neglected recycled aluminum; (2) very few studies considered the substantial impacts of self-generated electricity on primary aluminum production at both the national and provincial levels; (3) the characterization of the aluminum production process was not sufficiently comprehensive and detailed, with a failure to examine the potentials for energy saving and GHG emissions reduction of aluminum production in some studies; and (4) key data and assumptions need to be updated periodically based on the latest information and technologies.

This study aimed to fill the gaps by performing a comprehensive LC analysis of China's aluminum industry in 2020 based on a process-based approach. It focused on the following innovative aspects: (1) Both primary and recycled aluminum were studied using the same method and tool; (2) detailed investigation of the national and provincial self-generated electricity and grid mix was conducted to reflect the real electricity system of Chinese aluminum production; (3) the system boundary was extended to cover the upstream end-use energy production, main raw and auxiliary materials, and all related processes; (4) up-to-date data on the energy input and material flows were used; and (5) potentials for energy saving and GHG reductions were estimated and comparative analysis with international levels was conducted.

The structure of this paper is organized as follows: Section 2 explains the system boundary and research methodology and introduces the key data acquisition process and basic assumptions. Section 3 details the results and discussions. Section 4 concludes this paper and proposes policy implications.

2. Methodology

2.1. System Boundary

Figure 1 shows the research boundary of this study and the material flow of aluminum production. This study covers the main processes of primary and recycled aluminum production. The former includes bauxite mining, bauxite transport, alumina production, anode production, fluoride salt production, aluminum electrolysis, and ingot casting, and the latter contains aluminum scrap collection/transportation, pre-treatment, smelting, and ingot casting. Fuel inputs in aluminum production processes cover all key end-use energy types, including coal, natural gas (NG), oil, gasoline, diesel, electricity, and coke. Upstream production stages of process fuel are within the system boundary, which means that primary fossil energy (PE) consumption and GHG emissions occurring in resource exploitation, resource transportation, fuel production, fuel transportation, and distribution are covered. It should be pointed out that facilities construction and maintenance are excluded in this study. The process fuel consumed in all these stages is sourced in the form of three main types of PE: raw coal, raw NG, and petroleum. Three key types of GHGs, including CO_2, CH_4, and N_2O, and the perfluorocarbons (CF_4 and C_2F_6) emitted from the anode effect in the electrolysis process are taken into consideration, and they are measured in CO_2 equivalents ($CO_{2\text{-eq}}$) according to the global warming potential (GWP) factor.

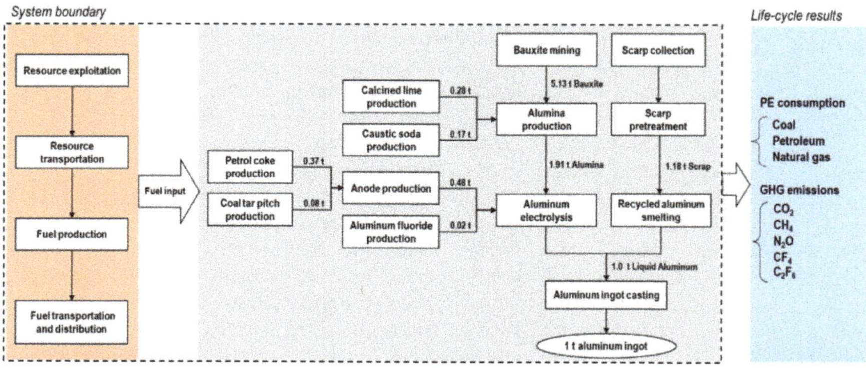

Figure 1. System boundary and material flow of primary and recycled aluminum production in China.

2.2. Calculation Method

The total direct energy consumption per ton of aluminum produced is the sum of the fuel consumption of each process, shown as Equation (1):

$$EN_{direct} = \sum_i \sum_j EN_{direct,i,j} \quad (1)$$

where EN_{direct} represents the total fuel consumed per ton of aluminum (MJ/t Al ingot); $EN_{direct,i,j}$ denotes the fuel type j consumed in the stage/process i (MJ/t Al ingot). The LC PE consumption is obtained with Equation (2):

$$EN_{LC} = \sum_i \sum_j EN_{direct,i,j} \cdot EN_j \quad (2)$$

where EN_{LC} is the LC PE consumption intensity of aluminum (MJ/t Al ingot); EN_j is the LC PE intensity of fuel type j (MJ/MJ fuel obtained and utilized). For electricity, EM_j can be defined as a weighted average intensity of all electricity generation sources using Equation (3):

$$EN_{electricity} = \frac{1}{1-\eta} \sum_k \left(EN_{electricity,k} \cdot SH_k \right) \quad (3)$$

where $EN_{electricity}$ is the average PE intensity of the electricity grid (MJ/MJ electricity generation and supply); η is the transmission loss of the electricity grid (%), $EN_{electricity,k}$ is the LC PE intensity of the electricity source k (MJ/MJ electricity generation and supply); and SH_k is the share of the electricity source k in the total electricity generation of the grid (%). Here, k represents coal, oil, NG, hydro, nuclear, solar, wind, biomass, and others, respectively.

The LC GHG emissions of primary aluminum are generated from four parts: (1) the combustion of fossil fuels; (2) CO_2 emitted from anode material consumption; (3) CO_2 emissions from the calcination of limestone; and (4) perfluorocarbons (PFCs) containing tetrafluoromethane (CF_4) and hexafluoroethane (C_2F_6) from the anode effect in the electrolysis process. Equations (4)–(8) show the calculation principles for the total LC GHG emissions per ton aluminum:

$$EM_{LC} = EM_{LC,En} + EM_{AC} + EM_{AE} + EM_{Limestone} \quad (4)$$

$$EM_{LC,En} = \sum_i \sum_j \left(EN_{direct,i,j} \cdot EM_j / 1000 \right) \quad (5)$$

$$EM_{AC} = \frac{44}{12} P_A \cdot (1 - S_A - A_A) \quad (6)$$

$$EM_{AE} = 6500 EM_{CF_4} + 9200 EM_{C_2F_6} \quad (7)$$

$$EM_{Limestone} = P_{Limestone} \cdot EF_{Limestone} \quad (8)$$

where EM_{LC} is the LC GHG emissions intensity of aluminum (kg $CO_{2\text{-eq}}$/t Al ingot); $EM_{LC,En}$ represents the GHG emissions associated with fuel utilization (kg $CO_{2\text{-eq}}$/Al ingot); EM_{AC}, $EM_{Limestone}$, and EM_{AE} denote the GHG emissions intensities of anode consumption, calcining limestone, and the anode effect (kg $CO_{2\text{-eq}}$/t Al ingot), respectively; P_A is the carbon anode consumption per ton of primary aluminum (kg/ Al ingot); S_A is the average sulfur content of the carbon anode (%); A_A is the ash content of the carbon anode (%); EM_{CF_4} and $EM_{C_2F_6}$ are the CF_4 and C_2F_6 emissions per ton of primary aluminum (kg/t Al ingot); $P_{Limestone}$ is the limestone consumption per ton of primary aluminum (kg/t Al ingot); $EF_{Limestone}$ is the emission factor for calcining limestone (kg $CO_{2\text{-eq}}$/kg limestone); and EM_j is the LC GHG emissions intensity of fuel type j (kg $CO_{2\text{-eq}}$/MJ fuel obtained and utilized). In the case of electricity, EM_j can be defined as follows:

$$EM_{electricity} = \frac{1}{1-\eta} \sum_k \left(EM_{electricity,k} \cdot SH_k \right) \quad (9)$$

where $EM_{electricity}$ is the average GHG emissions intensity of the electricity grid (g $CO_{2\text{-eq}}$/MJ electricity generation and supply); $EM_{electricity,k}$ is the LC GHG emissions intensity of the electricity source k (g $CO_{2\text{-eq}}$/MJ electricity generation and supply).

GHG generated during recycled aluminum production comes from fuel utilization, which can also be estimated by Equation (5).

2.3. Key Data and Assumptions

2.3.1. Basic Process Data Collection

The material flow during LC of aluminum production is shown in Figure 1. The details of the basic data on the import rate of raw materials, fuel input in each process, and fuel structure of each process/material are based on the latest public official and industry statistics and reports, which reflect the average technological level of China's aluminum industry. There are three ways to collect these data: (1) The energy and feedstock input data was mainly obtained from the yearbooks of the non-ferrous metal sector [3] and the energy sector [4,27], and the LCA database of the International Aluminum Institution (IAI) [2,28]; (2) for the data of intermediate input products such as anode production, fluoride salt, etc., refer to commercial databases and literature research [7,29]; (3) for some processes or intermediate inputs for which it was difficult to obtain the data in the base year (2020) due to limited statistics and research, this study made corrections and assumptions on the basis of data from previous years [22,30–32]. China's bauxite resources are becoming increasingly scarce, and the limited domestic bauxite reserves are unable to meet the huge domestic demand, 54.6% of which needed to be imported from abroad in 2020, about 25.5% of aluminum scrap came from imports in 2020, and alumina was 4.9% [33]. It should be noted that the production process of imported materials occurs abroad, so this part is no longer included in China's analysis, but the transportation process from the exporter to China were taken into consideration in this study.

2.3.2. Data on Transportation

Data relating to the transportation of raw materials and intermediate products for aluminum production are shown in Table 1. Imported materials and scrap are assumed to be first transported to the port by ocean ships and then shipped to smelting plants by rail. The maritime transport distance is the value obtained by weighting the distance between the geographical location of the importing countries and the main ports of China according to the import volumes [3,34]. The domestic transport distance of imported bauxite and alumina refers to the average transport distance of metal ore while the data of aluminum scrap refers to that of non-ferrous metals. The average transport distance of domestic aluminum scrap is assumed to be 500 km by road to reach the smelting facilities from the main scrap distribution center [20,35]. Other materials, including caustic soda, calcined lime, fluoride salt, petrol coke, and coal bar pitch, are assumed to be directly transported by road from plants to the electrolytic aluminum refineries for use, with a distance of 176 km, which is the average distance of China's road freight transportation [35]. Table 2 presents China's energy intensity and fuel structure of multiple transport modes in 2020.

Table 1. Transportation parameters of the aluminum industry in China in 2020 [1,3,34,35].

Item	Transport Mode
Bauxite	Ocean shipping: 54.6% (15,755 km); Railway: 100% (670 km)
Alumina	Ocean shipping: 4.9% (8600 km); Railway: 100% (670 km)
Aluminum scrap	Ocean shipping: 25.5% (8033 km); Railway: 25.5% (500 km); Road: 70% (500 km)
Other	Road: 100% (176 km)

Note: The sum of the proportions of individual transport modes may exceed 100% because of intermodal transport.

Table 2. Energy intensity and fuel structure of various transport modes.

Item	Energy Intensity	Fuel Structure	Date Source
Ocean shipping [1]	0.076 MJ/t · km	Fuel oil (100%)	[36]
Railway [2]	0.129 MJ/t · km	Coal (2.5%), Diesel (23.0%), electricity (65%), others (9.0%)	[37]
Road	0.498 MJ/t · km	Diesel (72%), gasoline (28%)	[36,38]

Note: [1] The fuel consumed in ocean shipping is fuel oil while other fuel types are not considered. [2] The fuel structure of railway was established according to a field survey and the literature [39].

2.3.3. Data on Electricity

As a high-energy consumption industry, the demand for electricity in primary aluminum production is huge. In order to obtain a stable electricity supply and reduce the cost of electricity, most electrolytic aluminum enterprises have built their own thermal power plants and purchase coal for electricity generation in addition to purchasing electricity from the national/regional grid. The electricity mix of the electrolytic aluminum industry does not necessarily match the grid mix found in aluminum-producing regions/nations due to the high proportion of self-generated electricity. According to AFG [40], about 30% of the electricity consumed in the electrolytic aluminum industry globally comes from a self-generated thermal power plant, with the rest being purchased as grid electricity. This ratio is higher in China, with about 65% estimated by market surveys [41,42]. The nationwide sources of electricity generation for electrolytic aluminum in China were assessed based on this data and China's grid mix in 2020 [27], as shown in Table 3. Coal power still dominates in the electricity mix of the electrolytic aluminum industry, the percentage of which is as high as 87.1%. The average transmission loss was 5.6% for grid electricity in China in 2020 [27]. This is negligible for self-generated electricity because the power plants are usually located close to electrolytic aluminum production facilities, avoiding long-distance transmission of electricity. In this study, we assumed that the grid electricity purchased is consumed in recycled aluminum production and the processes of primary aluminum production other than electrolysis. It should be noted that the diversity of electricity generation of primary aluminum production among provinces in China leads to significant regional differences in the results of the LC analysis, which is further discussed in Section 3.3.

Table 3. Nationwide sources of electricity generation for the grid mix and electrolytic aluminum mix.

Item	Coal	Oil	NG	Hydro	Nuclear	Solar	Other
Grid mix (%)	63.2	0.1	3.2	17.0	4.7	11.1	0.7
Electricity for aluminum electrolysis (%)	87.1	0.0	1.1	6.0	1.6	3.9	0.2

2.3.4. Data on Life Cycle Intensities of Each Process Fuel

Table 4 presents China's LC PE and GHG intensity of each process fuel in 2020. The definition of LC PE intensity is PE consumed to obtain and utilize 1 MJ fuel over the whole LC stages, including resource exploitation, resource transportation, fuel production, fuel transportation and distribution, and fuel combustion. LC GHG emissions intensity is the total GHG emitted to obtain and utilize 1 MJ fuel. Here, we update the data based on the methodology from our previous study [38] and national energy statistics [4].

Table 4. LC PE and GHG emissions intensity of each fuel.

Item	EN_j	$EN_{j,Coal}$	$EN_{j,NG}$	$EN_{j,Oil}$	EM_j
Unit	MJ/MJ	MJ/MJ	MJ/MJ	MJ/MJ	$gCO_{2,e}$/MJ
Raw coal	1.068	1.065	0.001	0.002	98.0
Raw NG	1.135	0.035	1.052	0.048	67.0
Crude oil	1.093	0.024	0.036	1.033	78.8
Clean coal	1.083	1.067	0.002	0.014	99.1
Refined NG	1.139	0.036	1.056	0.048	68.8
Diesel	1.251	0.058	0.047	1.146	91.6
Gasoline	1.260	0.060	0.047	1.153	89.4
Fuel oil	1.190	0.046	0.042	1.102	90.1
Grid power	1.910	1.817	0.064	0.029	172.7
Coke [1]	2.172	1.236	0.002	0.015	105.9
Coal power	2.844	2.802	0.004	0.038	260.3
NG power	2.561	0.014	2.534	0.013	149.9
Oil power	3.899	0.157	0.145	3.597	295.4
Nuclear power	0.063	0.052	0.005	0.006	6.5
Other power [2]	0	0	0	0	5.0

Note: [1] Data on coker refers to the research results from Research Center for Eco-Environmental Sciences, Chinese Academy of Sciences [43]; however, it should be noted that that the data was obtained based on the 2010 situation. In this paper, we assume it did not change in the short term. [2] The fossil energy consumption of other electricity sources with non-fossil energy as the raw material (such as hydro, biomass, and wind) are very low and often negligible [38].

2.3.5. Data on GHG Emissions from Anode Consumption, Calcining Limestone, and Anode Effect

The parameters of anode consumption, calcining limestone, and anode effect shown in Table 5 were obtained from the NDRC guideline [32]. The GHG emissions factor of anode consumption and calcining limestone are determined by the physical properties of the material, resulting in similar values among all studies. CF_4 and C_2F_6 are emitted in the primary aluminum reduction process because of the anode effect, and their emission factors are closely related to the type and technological level of the electrolytic bath. Currently, China's electrolytic aluminum production mainly adopts world advanced Point-center Feed Prebake (PFPB) technology. Therefore, the value referenced by the NDRC guideline is lower than that provided by IAI [28]. In addition, the updated GWP factors from the Intergovernmental Panel on Climate Change (IPCC) were employed in this study: 6500 for CF_4 and 9200 for C_2F_6 [44].

Table 5. The parameters of anode consumption, calcining limestone, and anode effect [28,32,44].

Item	Unit	Value
Average sulfur content of carbon anode	%	2
Average ash content of carbon anode	%	0.4
Emission factor for calcining limestone	kg CO_2/kg limestone	0.405
CF_4 emission factor of anode effect	kg /t Al ingot	0.034
C_2F_6 emission factor of anode effect	kg /t Al ingot	0.0034

3. Results and Discussions

3.1. Life Cycle Results of Primary Aluminum

Figure 2 shows the LC results of primary aluminum. In total, 77,547 MJ of direct energy and 157,207 MJ of LC PE are consumed to produce one ton of primary aluminum ingot, with the proportion of aluminum electrolysis exceeding more than 60% and 76%, respectively. The total LC PE consumption is 2.3 times as much as the total direct energy input. The difference between LC PE consumption and direct energy input at the aluminum electrolysis stage is apparently higher than that at other stages because the main fuel consumed in the electrolysis process is electricity and the energy conversion efficiency of transforming PE into electricity is lower than in other fuel types. As Figure 2b shows, coal contributes the most in the LC PE consumption and accounts for 90.53% of the total due to China's coal-dominated power structure. In the transportation of bauxite, the LC PE consumption associated with the transportation of imported bauxite accounts for about 95% due to China's massive dependence on overseas sources and the long ocean shipping distance.

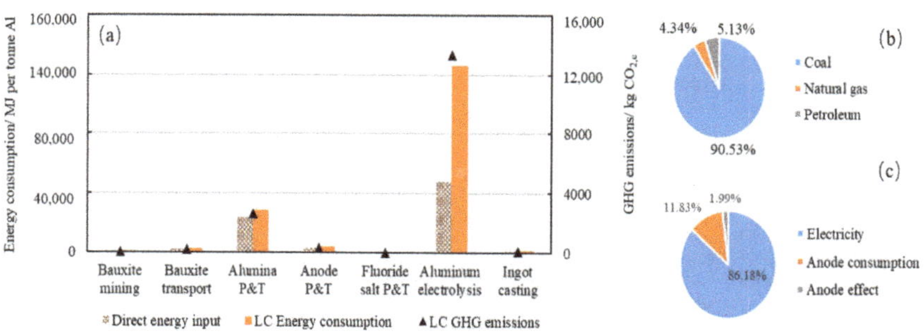

Figure 2. LC results of primary aluminum in mainland China in 2020: (**a**) Direct energy input, LC PE consumption, and GHG emissions of each stage/process; (**b**) Proportion of each primary fossil energy; (**c**) Breakdown of GHG emissions of aluminum electrolysis.

The process of primary aluminum production will generate 15,947 kg $CO_{2\text{-eq}}$ GHG emissions from the LC perspective, in which the aluminum electrolysis shares the most at 80.0% (12,704 kg $CO_{2\text{-eq}}$), presented as Figure 2c. The overall emission structure is similar to that of energy consumption. The GHG emissions from electricity consumption contribute the most in the total GHG emissions of aluminum electrolysis, with a share of 86.18%, followed by that of anode consumption, which is estimated to be 1503 kg $CO_{2\text{-eq}}$, 11.83% of the total GHG emissions in this stage. At 2400 kg $CO_{2\text{-eq}}$, the GHG emissions from alumina P&T are the second largest emissions source, with a share of 15.05%. In total, 113 kg $CO_{2\text{-eq}}$ of GHG are generated from calcium carbonate decomposition, with a proportion of 4.73% of the alumina P&T stage, while the rest is generated by fuel combustion.

3.2. Life Cycle Results of Recycled Aluminum

The result of each process/stage in descending order is aluminum smelting, scrap pretreatment, ingot casting, and scrap collection/transportation, shown as Figure 3. In total,

8612 MJ of direct energy is consumed for per ton of recycled aluminum ingot production (only 11.1% of primary aluminum ingot), which are dominated by aluminum smelting with a share of 53.19%. In total, 11,730 MJ of LE PE is consumed per ton of recycled aluminum production, only 7.46% of primary aluminum, with the contribution from each process similar to those of direct energy input. Up to 25.5% of the scrap aluminum is imported from abroad, which leads to a large amount of energy input in the ocean shipping of imported scarp aluminum, which accounts for 36% of the whole transport stage. Compared to primary aluminum production, the structure of LC PE consumption changes significantly (shown as the sub-graph of Figure 3): coal accounts for 35.74% and is no longer the dominant energy while NG contributes the most, with a share of 54.31%.

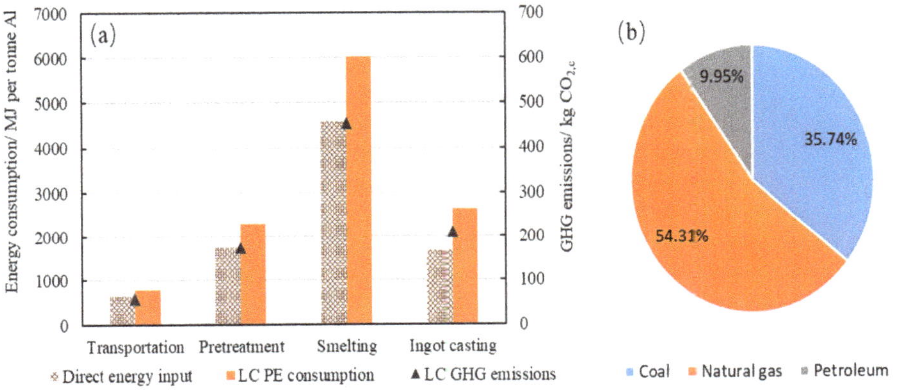

Figure 3. LC results of recycled aluminum in mainland China in 2020 (**a**); and the proportion of each primary fossil energy (**b**).

The process of recycled aluminum production will generate 845 kg $CO_{2\text{-eq}}$ GHG emissions, only 5.40% of those of primary aluminum ingot. The GHG emissions from the process of collection/transportation, pretreatment, smelting, and ingot casting are 57, 161, 424, and 203 kg $CO_{2\text{-eq}}$, representing 6.80%, 19.03%, 50.12%, and 24.05% of the total emissions, respectively.

Compared with primary aluminum, recycled aluminum shows obvious advantages in energy consumption and GHG emissions. It should be pointed out that the fuel of smelting furnaces in modern recycled aluminum factories is mostly fuel oil or NG. However, there are few official statistics on the nationwide average energy data, which makes it difficult to explore the fuel structure of recycled aluminum. In this study, we adopted the data of NG from [25]. If the smelting furnace is fueled by fuel oil, the LC PE and GHG emissions will increase due to the fact that the LC intensities of fuel oil are higher than those of NG.

3.3. Regional Disparity of LC Results

This part shows the provincial differences among 31 provinces in mainland China. Hong Kong, Macao, and Taiwan in China are excluded in this study due to the lack of sufficient data.

The spatial distribution of primary and recycled aluminum production in mainland China in 2019 are shown in Figure 4a,b, respectively (data of aluminum production in 2020 has not been published). The production capacity of China's primary aluminum is highly concentrated, mainly distributed in 17 provinces in the north and central west regions. Shandong and Xinjiang are the two most important primary aluminum production provinces in China currently, accounting for more than 40% of the national production [3]. Recycled aluminum production is distributed in all provinces of mainland China except Jilin, Guangxi, Hainan, Tibet, and Ningxia. The southeastern coastal provinces are major production areas, particularly the three coastal provinces of Jiangsu, Shandong, and Guangdong, accounting for more than 35% of the national production [3].

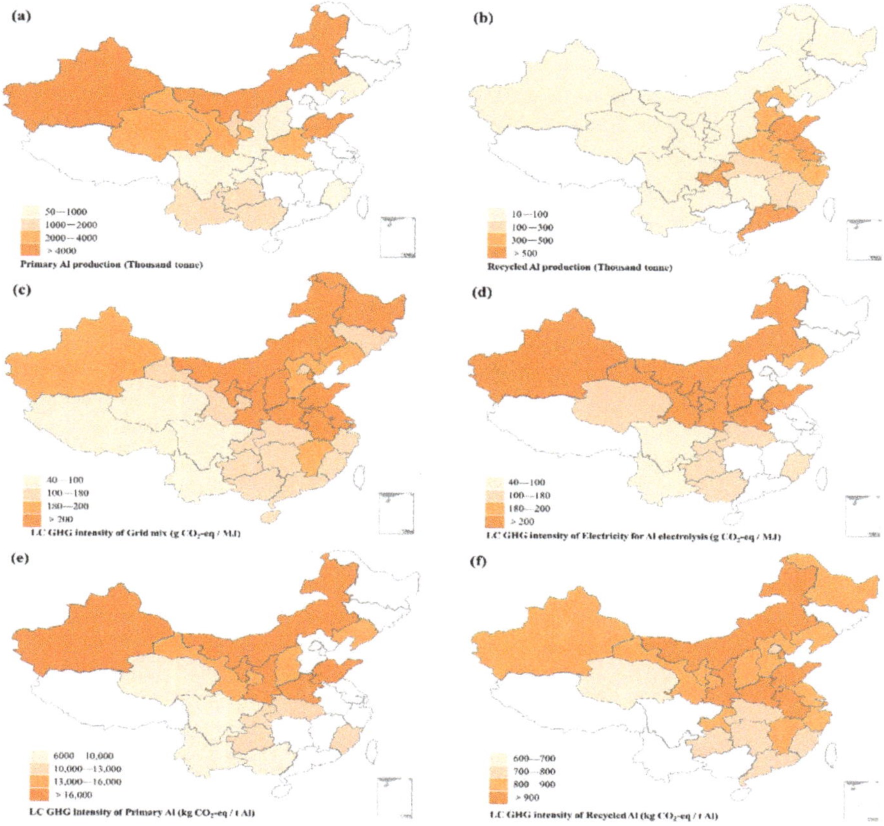

Figure 4. Provincial aluminum industry data in mainland China: (**a**) Primary aluminum and (**b**) recycled aluminum production in 2019; (**c**) LC GHG emissions intensity of the grid mix; (**d**) LC GHG emissions intensity of electricity for aluminum electrolysis; (**e**) GHG emissions intensity of primary aluminum; and (**f**) GHG emissions intensity of recycled aluminum in 2020.

Some previous studies [24,45,46] estimated the grid mix and LC GHG emissions intensity at the provincial level in China, considering the inter-province power transmissions for the provincial grid mix. Here, we recalculated the LC emissions intensity of electricity (see Figure 4c) based on their estimations [46]. The LC intensity shows a downward trend from northeast to southwest. Shannxi, Shandong, and Inner Mongolia show the highest intensity with more than 250 g $CO_{2\text{-eq}}$/MJ, which is about 1.1 times that of the nationwide average level. Due to the differences in the regional policy and self-generated electricity cost, the proportion of self-generated electricity in the electrolytic aluminum industry varies from province to province. Based on the industry survey and literature review, we obtained the provincial self-generation rate (as shown in Table A1). Combining the provincial grid mix, GHG emissions intensity of different power generation sources, and proportion of self-generated electricity, we estimated the provincial GHG emissions intensity of electricity for aluminum electrolysis, as shown in Figure 4d. The distribution of GHG emissions intensity basically coincides with primary aluminum production, particularly those in

Shandong, Xinjiang, and Inner Mongolia, as these provinces have an intensity of more than 255 g $CO_{2\text{-eq}}$/MJ.

We assume that, apart from provincial electricity diversity, no other differences exist among provinces, such as production technology, energy efficiency, fuel input and mix, and other factors. For primary aluminum, there is a geographical mismatch between the raw materials production processes (such as bauxite mining, alumina production, and anode) and electrolytic aluminum production process. As a result, raw materials in many provinces need to be purchased across provinces or imported from other countries. However, there are no reliable data on the interprovincial flows of raw materials, making it impossible to analyze the LC supply chains in certain provinces. Therefore, we only consider the provincial electricity mix disparity in the aluminum electrolysis process, and still used the nationwide average data for other processes.

The LC GHG emissions per ton of primary aluminum ingot are shown in Figure 4e and Table A1. Generally, the LC GHG emissions intensity of primary aluminum is positively correlated with the production volume and shows significant provincial disparity. The provinces with larger production volumes, such as Shandong and Xinjiang, also have higher emissions intensities. The five provinces (Inner Mongolia, Shandong, Shaanxi, Xinjiang, and Ningxia) with GHG intensities exceeding the national average intensity account for 59.2% of China's total production. The GHG emissions intensity of the southwest provinces is lower than that of the northern provinces. For instance, the GHG emissions intensities of Yunnan and Sichuan are 39.2% and 37.2% that of Shandong (17,540 kg $CO_{2\text{-eq}}$), respectively. Provinces with low GHG emissions intensities share two characteristics: (1) relatively high proportions of renewable electricity in the grid mix; and (2) low proportions of self-generated electricity for primary production. Figure 4f and Table A1 show the LC GHG emissions per ton of recycled aluminum ingot. Generally, the LC GHG emissions intensities show a negative correlation with the spatial distribution of recycled aluminum production, indicating that the provinces with lower emissions intensities tend to have higher production volumes. As shown in Figure 4f, the LC GHG emissions intensities of the southeastern coastal provinces are lower than those of the northern provinces. Shannxi shows the highest GHG emissions intensity with 927 kg $CO_{2\text{-eq}}$, about 1.1 times that of the national average level and 1 6 times that of Yunnan. This is mainly attributed to the high proportion of coal power in northern China's grid mix.

3.4. *Potential for Energy Conservation and GHG Emissions Reduction*

3.4.1. Decrease in Electricity Consumption of Electrolytic Aluminum

The overall electricity consumption per ton of electrolytic aluminum in China has been steadily declining in recent years due to the improvement in aluminum smelting technology, reaching 13,244 kWh/t Al in 2020 [1]. This is below the global average level (see Figure 5). According to the specification of the aluminum industry [30], the electricity consumption of newly built and reformed electrolytic aluminum liquid must be less than 12,750 kWh/t Al liquid, and the overall electricity consumption of aluminum ingot must be less than 13,200 kWh/t Al ingot. For existing facilities, if the electricity consumption decreases by 100 kWh, the LC PE and GHG emissions per ton of primary aluminum ingot will decrease by 906 MJ and 86 kg $CO_{2\text{-eq}}$, respectively.

3.4.2. Self-Generated Electricity Management

Self-generated electricity plays an important role in China's primary aluminum industry. Compared with many other countries, China has a high proportion of self-generated electricity in the electrolytic aluminum industry. China has implemented relevant policies to strengthen the management of coal-fired power plants for self-generation in order to control the scale of their development and improve their energy efficiency. A 10% decrease in the proportion of self-generated electricity for the electrolytic aluminum industry will reduce the LC PE consumption and GHG emissions by 4504 MJ and 274 kg $CO_{2\text{-eq}}$, respectively. The LC PE and GHG emissions intensities under different proportions of self-generated electricity for the electrolytic aluminum industry are presented in Figure 6. When all the electricity consumed is purchased from the grid, the LC PE consumption and GHG emissions per ton of aluminum ingot are 127,931 MJ and 13,203 kg $CO_{2\text{-eq}}$, a decrease of 18.6% and 16.3% compared to those of 65% self-generation in 2020, respectively.

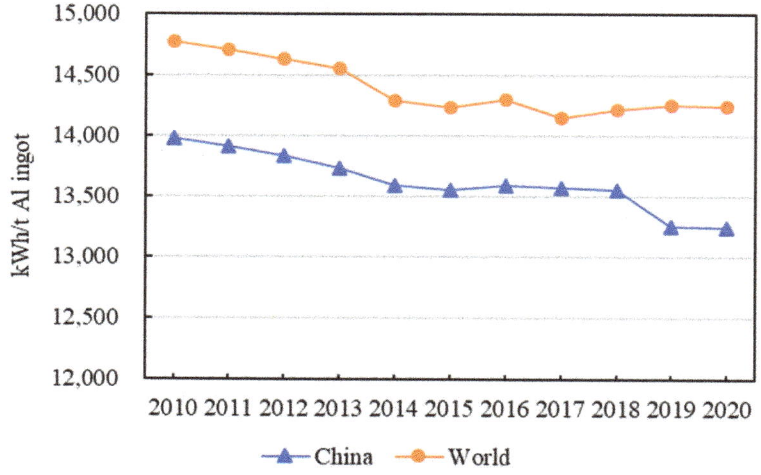

Figure 5. Overall electricity consumption per ton of primary aluminum ingot 16.

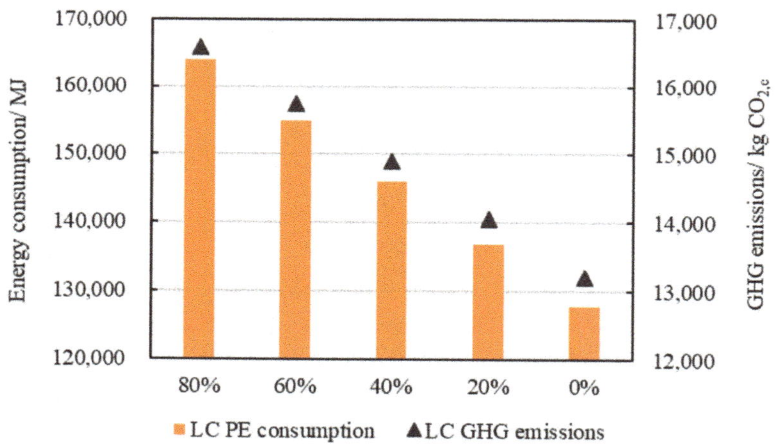

Figure 6. LC PE and GHG emissions intensities under different proportions of self-generated electricity in the electrolytic aluminum industry.

The LC PE and GHG emissions intensities of coal power used in this study were estimated by the method from Peng et al. [38] based on the national average net coal consumption rate at 305 g of coal equivalent/kWh in 2020 [4]. However, the net coal consumption of self-generated power plants is generally higher than this value as the capacities of coal-fired units in self-generation power plants are smaller than those of large-scale thermal power plants [41]. A decrease in the net coal consumption of self-generation power plants by 10 g of coal equivalent/kWh will reduce the LC PE consumption and GHG emissions by 3747 MJ and 425 kg $CO_{2\text{-eq}}$, respectively.

3.4.3. Low-Carbon Electricity Development

China's electricity grid is currently dominated by coal power, with a proportion of 63.2% in 2020, resulting in high LC PE and GHG emissions intensities. China has introduced many measures to vigorously promote the development of renewable electricity and the proportion of coal power in the electricity grid is expected to gradually decrease in the future. Scenario analysis was conducted to estimate the LC PE and GHG emissions intensities of primary and recycled aluminum with different proportions of coal power in the electricity mix (50%, 40%, and 30%). Here, we assume that the proportion of self-generated electricity in the electrolytic aluminum industry is 50% and is the same under different scenarios. For primary aluminum, the LC PE intensities in the 50%, 40%, and 30% coal power scenarios drop by 7.06%, 11.64%, and 16.21% compared with the baseline (2020) while the LC GHG emissions intensities decrease by 7.07%, 11.12%, and 15.18%, respectively. The LC PE intensities for recycled aluminum are reduced by 7.22%, 11.90%, and 16.58% under the 50%, 40%, and 30% coal power scenarios while the LC GHG emissions intensities decline by 11.80%, 17.60%, and 23.41%, respectively. Therefore, decarbonization of grid electricity can play an important role in promoting energy conservation and GHG emissions reduction of the aluminum industry.

3.4.4. Industrial Relocation

The LC GHG emissions intensity is high in provinces with concentrated production of primary aluminum, resulting in the high GHG emissions of China's aluminum production. The southwest provinces are rich in hydropower and therefore have natural advantages in hosting the electrolytic aluminum industry. The appropriate relocation of the primary aluminum industry from north China to the southwest will help reduce PE consumption and GHG emissions. For example, relocating half of Shandong's production to Qinghai and Sichuan will reduce the production-weighted average LC GHG emissions intensity of Chinese aluminum by 6.42% and 15.31%, respectively. The relocation of the primary aluminum industry shows great potential in GHG emissions reduction. However, other key factors such as the consumer market, production cost, and environmental protection should be taken into account holistically in any industrial relocation. This is a topic worth exploring in detail in future studies.

3.5. *Comparative Analysis with Similar Studies*

Table 6 presents a comparative analysis with other studies focusing on China and other regions. The results for China show significant differences among studies, most of which are lower than those of this study mainly because: (1) we extended the system boundary to include both fuel acquisition and aluminum production, taking into consideration fuel conversion efficiencies that are usually neglected by other studies; and (2) in primary aluminum production, the proportion of thermal power in the electricity mix for electrolytic aluminum is higher than that of the grid electricity mix due to coal-fired self-generation, leading to higher LC PE and GHG emissions intensities.

The results for China tend to be worse than those for the U.S. and Europe, which are mainly attributed to the coal-dominated energy structure. The LC GHG emissions intensity of primary aluminum for China is 1.5–3.5 times that of other regions, respectively. The U.S. is taken as an example for analysis. The power consumption for China's per ton aluminum production has been almost close to advanced economies [7]; however, its LC PE consumption and GHG emissions are about two times as much as that in the U.S. This is mainly attributed to two reasons: (1) the fuels consumed for primary aluminum production in the U.S. are mainly fuel oil and natural gas in addition to electricity while this is dominated by coal in China; and (2) coal shares a proportion of more than 60% in China's grid mix [27] while this figure is 14.3% in the U.S. [7].

Table 6. LC results from different studies.

Item	Year	Scope	LC PE Intensity (MJ/t Al Ingot)		LC GHG Emissions Intensity (kg $CO_{2\text{-eq}}$/t Al Ingot)	
			Primary	Recycled	Primary	Recycled
This study	2020	China	157,207	11730	15,947	845
Hao et al., 2016 [24]	2013	China			16,500	
Hao et al., 2016 [24]	2020	China			14,300	
Ding et al., 2012 [20]	2010	China			17,000	715
Ding et al., 2021 [26]	2017	China			14,500	930
Zhang et al., 2016 [25]	2012	China	167,847	7875	15,800	722
He et al., 2020 [23]	2016	China			19,500	350
Anna M K 2014 [11]	2013	Europe			9503	
Anna M K 2014 [11]	2013	Iceland			5560	
IAI 2022 [6]	2018	Global			16,100	
World Aluminum 2018 [12]	2015	Global	166,000		18,000	
Nunez et al. 2016 [13]	2010	Global	163,000		16,500	
Nunez et al. 2016 [13]	2010	RoW *	109,000		10,800	
TAA 2013 [8]	2010	North America			8973	670
GREET 2021 [7]	2020	the U.S.	115,780		7282	

Note: * RoW: the world excluding China (rest of world).

3.6. Limitation and Future Work

Although this study has carried out several extensive works compared to previous similar studies, it still faces three limitations: (1) the average data of the energy efficiency and fuel consumption industry used in the current study cannot reflect the more specific technical characteristics of China's aluminum industry, especially the situation of advanced enterprises; (2) this study failed to assess the GHG emissions contribution of the entire aluminum industry to China's carbon emissions; and (3) it is necessary to explore the low-carbon transition pathway of the aluminum industry in the context of China's carbon peaking and carbon neutrality targets.

4. Concluding Remarks

This study estimated China's national average LC PE consumption and GHG emissions of the aluminum industry in 2020 based on the process-based LC analysis method and proposed several policy recommendations for China's aluminum industry. In view of the different GHG emissions factors of the electricity mix for the aluminum industry, the disparity of the GHG emissions intensity of aluminum production among provinces was estimated. In addition, comparative analysis with other studies was conducted. The following specific conclusions are drawn from this study:

(1) LC PE consumption and GHG emissions per ton of primary aluminum ingot in 2020 were 157,207 MJ and 15,947 $CO_{2\text{-eq}}$, respectively, with the aluminum electrolysis stage being the largest contributor. There are multiple sources of GHG emissions in the LC chain of aluminum production, among which fuel generation and utilization contribute the most, accounting for 88.28% of the total emissions.

(2) China's recycled aluminum shows significant advantages in energy consumption and GHG emissions in comparison with primary aluminum. The LC PE consumption and GHG emissions per ton of recycled aluminum ingot in 2020 were 11,730 MJ and 845 $CO_{2\text{-eq}}$, only 7.46% and 5.30% of those for primary aluminum ingot, respectively. Promotion of the recycled aluminum industry will contribute to energy saving and emission reduction in China's aluminum industry.

(3) Obvious provincial disparity exists in the LC results of China's aluminum industry. In general, the LC PE and GHG emissions intensities of primary aluminum are higher in the main production areas while those of recycled aluminum are lower in the main production areas.

(4) China's aluminum industry has significant potential for energy saving and GHG emissions reduction by decreasing electrolysis electricity consumption, improving the electricity structure (particularly the share of self-generated electricity), reducing the net coal consumption rate of self-generated electricity, and optimizing the geographical distribution of the production capacity.
(5) GHG emissions of China's primary aluminum production are 1.5–3.5 times that of developed economies mainly because China's electricity source and process fuels are both dominated by coal.

The following policy suggestions for China's aluminum industry are made based on this study:

(1) To boost the utilization of renewable electricity in the aluminum industry to improve the performances of the electricity mix in LC PE consumption and GHG emissions.
(2) To restrict new capacities and phase out inefficient capacities of self-generated electricity to limit the scale and reduce the overall net coal consumption rate.
(3) To improve the domestic aluminum scrap recovery system to provide sufficient and stable raw materials for the development of the recycled aluminum industry.
(4) To promote the relocation of the primary aluminum production capacity from the north to the south-west to optimize the spatial distribution of the primary aluminum industry.
(5) To consider the differences among provinces when relevant policies are implemented by the central government in order to make policies more effective nationwide.

Author Contributions: Conceptualization, T.P., X.O. and X.Y.; methodology, T.P. and X.O.; software, T.P., L.R. and X.O.; validation, E.D. and X.O.; formal analysis, E.D. and L.R.; investigation, T.P. and X.O.; data curation, L.R. and E.D.; writing—original draft preparation, T.P. and X.O.; writing—review and editing, T.P., L.R. and X.O.; visualization, L.R. and E.D.; supervision, X.O. and X.Y.; project administration, X.O.; funding acquisition, X.O. All authors have read and agreed to the published version of the manuscript.

Funding: This research was funded by National Natural Science Foundation of China, grant number: 72174103.

Acknowledgments: The authors gratefully acknowledge the financial support from the Rio Tinto Group in the context of the Tsinghua-Rio Tinto Joint Research Center for Resources, Energy and Sustainable Development. Tsinghua University received some funding for this work from the Energy Demand changes Induced by Technological and Social Innovations (EDITS) project, which is part of the initiative coordinated by the Research Institute of Innovative Technology for the Earth (RITE) and International Institute for Applied Systems Analysis (IIASA) (and funded by Ministry of Economy, Trade, and Industry, Japan).

Conflicts of Interest: The authors declare no conflict of interest

Abbreviations/Nomenclature

AFG	Aluminum for Future Generation
Al ingot	aluminum ingot
CF_4	tetrafluoromethane
C_2F_6	hexafluoroethane
$CO_{2\text{-eq}}$	CO_2 equivalents
GHG	greenhouse gas
GWP	global warming potential
IAI	International Aluminum Institute
IPCC	Intergovernmental Panel on Climate Change
LC	life-cycle
mt	megaton

NG	natural gas
NDRC	National Development and Reform Commission
PE	primary fossil energy
PFCs	perfluorocarbons
PFPB	Point-center Feed Prebake
P&T	production and transportation
RoW	the world excluding China (rest of world)
t	tonne
TAA	The Aluminum Association
A_A	the average ash content of carbon anode
EM	greenhouse gas emissions intensity
EM_{LC}	life-cycle greenhouse gas emissions intensity
EN_{direct}	direct energy consumption
EN_{LC}	life-cycle primary fossil energy intensity
P	specific material consumption per tonne of primary aluminum
SH_k	the share of electricity source in the electricity generation of the grid
S_A	the average sulfur content of carbon anode
η	transmission loss of the electricity grid
i	aluminum production process
j	process fuel type
k	electricity source

Appendix A

Table A1. Provincial and nationwide GHG intensity of electricity and aluminum production.

	Grid Mix	Electricity for Al Electrolysis		LC GHG Emissions Intensity (kg $CO_{2\text{-eq}}$/t Al)	
	g $CO_{2\text{-eq}}$/MJ Electricity	Share of Self-Generated Power	LC GHG Intensity (g $CO_{2\text{-eq}}$/MJ Electricity)	Primary Aluminum	Recycled Aluminum
Beijing	142.3	- -	- -	- -	753.9
Tianjin	227.0	- -	- -	- -	921.6
Hebei	198.4	- -	- -	- -	865.1
Shanxi	216.0	0%	216.0	15,443.6	899.9
Inner Mongolia	217.7	89%	255.6	17358.4	923.6
Liaoning	188.5	0%	188.5	14020.1	845.4
Jilin	178.3	- -	- -	- -	0.0
Heilongjiang	206.8	- -	- -	- -	881.7
Shanghai	222.7	- -	- -	- -	913.1
Jiangsu	201.4	- -	- -	- -	871.0
Zhejiang	167.0	- -	- -	- -	802.8
Anhui	222.1	- -	- -	- -	911.9
Fujian	135.8	0%	135.8	11290.4	741.1
Jiangxi	194.7	- -	- -	- -	857.6
Shandong	216.9	98%	259.4	17539.7	924.5
Henan	219.0	60%	243.8	16793.0	919.2

Table A1. *Cont.*

	Grid Mix	Electricity for Al Electrolysis		LC GHG Emissions Intensity ($kg\ CO_{2-eq}/t$ Al)	
	$g\ CO_{2-eq}/MJ$ Electricity	Share of Self-Generated Power	LC GHG Intensity ($g\ CO_{2-eq}/MJ$ Electricity)	Primary Aluminum	Recycled Aluminum
Hubei	122.5	0%	122.5	10604.3	714.9
Hunan	144.8	- -	- -	- -	758.9
Guangdong	160.6	- -	- -	- -	790.2
Guangxi	121.2	30%	162.9	8026.8	0.0
Hainan	136.2	- -	- -	- -	- -
Chongqing	166.6	0%	166.6	12886.6	802.1
Sichuan	43.9	0%	43.9	6533.5	559.2
Guizhou	154.3	0%	154.3	12249.5	777.8
Yunnan	50.4	0%	50.4	6872.5	572.2
Tibet	40.6	- -	- -	- -	- -
Shaanxi	218.4	100%	260.3	17586.9	927.1
Gansu	154.8	70%	228.7	15835.5	818.6
Qinghai	53.3	23%	100.9	9317.3	603.6
Ningxia	206.0	50%	233.1	16233.6	894.7
Xinjiang	199.4	94%	256.6	17343.4	897.9
Nationwide	172.7	65%	229.0	15,947	844.9

Note: "- -" represents there is no primary/recycled aluminum industry capacity in the province.

References

1. China Nonferrous Metals Industry Association. China Nonferrous Metals Industry 2020 Economic Operation Report. 2021. Available online: http://lwzb.stats.gov.cn/pub/lwzb/tzgg/202205/W020220511403032411542.pdf (accessed on 12 October 2021). (In Chinese)
2. International Aluminum Institute (IAI). Statistics of Global Primary Aluminum Production. 2021. Available online: http://www.world-aluminium.org/statistics/ (accessed on 10 November 2021).
3. Editorial Board of the Yearbook of Nonferrous Metals Industry of China. *The Yearbook of Nonferrous Metals Industry of China 2020*; China Nonferrous Metals Industry Association: Beijing, China, 2021. (In Chinese)
4. Bureau of Statistics of China. *China Energy Statistical Yearbook 2021*; China Statistics Press: Beijing, China, 2022.
5. Hu, B. The change of electrolytic aluminum industry under carbon neutrality target. *China Nonferrous Met News*. 2021. Available online: https://acin.org.cn/6772.html (accessed on 19 May 2021). (In Chinese)
6. International Aluminium Institute (IAI). 2019 Life Cycle Inventory (LCI) Survey Global Summary. 2022. Available online: https://international-aluminium.org/resource/life-cycle-inventory-lci-data-and-environmental-metrics/ (accessed on 16 July 2022).
7. Argonne National Laboratory (ANL). The Greenhouse Gases, Regulated Emissions, and Energy Use in Transportation Model (GREET). 2021. Available online: https://greet.es.anl.gov/greet_excel_model.models (accessed on 12 August 2022).
8. The Aluminum Association (TAA). The Environmental Footprint of Semi-Finished Aluminum Products in North America. 2022. Available online: https://www.aluminum.org/sites/default/files/2022-01/2022_Executive-Summary_Semi-Fab_LCA.pdf (accessed on 16 May 2022).
9. European Aluminum Association. Life Cycle Inventory Data for Aluminum Production and Transformation Processes in Europe. 2013. Available online: http://ec.europa.eu/DocsRoom/documents/13762/attachments/5/translations/en/renditions/pdf (accessed on 14 November 2021).
10. Liu, G.; Müller, D.B. Addressing sustainability in the aluminum industry: A critical review of life cycle assessments. *J. Clean. Prod.* **2012**, *35*, 108–117. [CrossRef]
11. Anna, M.K. *A Cradle-To-Gate Life Cycle Assessment of Primary Aluminum Production at Norðurál*; University of Iceland: Reykjavik, Iceland, 2014.

12. World Aluminum. Aluminum Carbon Footprint Technical Support Document. 2018. Available online: http://www.world-aluminium.org/media/filer_public/2018/02/15/carbon_footprint_technical_support_document_v1_published.pdf (accessed on 12 September 2019).
13. Nunez, P.; Jones, S. Cradle to gate: Life cycle impact of primary aluminum production. *Int. J. Life Cycle. Assess.* **2016**, *21*, 1594–1604. [CrossRef]
14. Yue, Q.; Wang, H.; Gao, C.; Du, T.; Liu, L.; Lu, Z. Resources saving and emissions reduction of the aluminum industry in China. *Resour. Conserv. Recycl.* **2015**, *104*, 68–75. [CrossRef]
15. Lin, B.; Xu, L. Energy conservation of electrolytic aluminum industry in China. *Renew Sustain. Energy Rev.* **2015**, *43*, 676–686. [CrossRef]
16. Li, M.; Mi, Z.; Coffman, D.M.; Wei, Y.M. Assessing the policy impacts on non-ferrous metals industry's CO_2 reduction: Evidence from China. *J. Clean. Prod.* **2018**, *192*, 252–261. [CrossRef]
17. Li, Q.; Zhang, W.; Li, H.; He, P. CO_2 emission trends of China's primary aluminum industry: A scenario analysis using a system dynamics model. *Energy Policy* **2017**, *105*, 225–235. [CrossRef]
18. Gao, F.; Nie, Z.; Wang, Z.; Li, H.; Gong, X.; Zuo, T. Greenhouse gas emissions and reduction potential of primary aluminum production in China. *Sci. China Ser. E Technol. Sci.* **2009**, *52*, 2161–2166. [CrossRef]
19. Du, J.; Han, W.; Peng, Y.; Gu, C. Potential for reducing GHG emissions and energy consumption from implementing the aluminum intensive vehicle fleet in China. *Energy* **2010**, *35*, 4671–4678. [CrossRef]
20. Ding, N.; Gao, F.; Wang, Z.; Gong, X. Comparative analysis of primary aluminum and recycled aluminum on energy consumption and greenhouse gas emission. *Chin. J. Nonferr. Met.* **2012**, *22*, 2908–2915.
21. Guo, J.; Gao, F.; Wang, Z.H.; Liu, Y.; Hu, J.Y.; Li, Z. Correlation Analysis between Emission and Impact Factors of Greenhouse Gases Caused by Production of Primary Aluminum. *Mater. Sci. Forum* **2014**, *787*, 135–143. [CrossRef]
22. Liu, Z.; Geng, Y.; Adams, M.; Dong, L.; Sun, L.; Zhao, J.; Dong, H.; Wu, J.; Tian, X. Uncovering driving forces on greenhouse gas emissions in China' aluminum industry from the perspective of life cycle analysis. *Appl. Energy* **2016**, *166*, 253–263. [CrossRef]
23. He, X.; Kim, H.C.; Wallington, T.J.; Zhang, S.; Shen, W.; De Kleine, R.; Wu, Y. Cradle-to-gate greenhouse gas burdens for aluminum and steel production and cradle-to-grave GHG benefits of vehicle lightweighting in China. *Resour. Conserv. Rec.* **2020**, *152*, 104497. [CrossRef]
24. Hao, H.; Geng, Y.; Hang, W. GHG emissions from primary aluminum production in China: Regional disparity and policy implications. *Appl. Energy* **2016**, *166*, 264–272. [CrossRef]
25. Zhang, Y.; Sun, M.; Hong, J.; Han, X.; He, J.; Shi, W.; Li, X. Environmental footprint of aluminum production in China. *J. Clean. Prod.* **2016**, *133*, 1242–1251. [CrossRef]
26. Ding, N.; Liu, N.; Lu, B.; Yang, J. Life cycle greenhouse gas emissions of aluminum based on regional industrial transfer in China. *J. Ind. Ecol.* **2021**, *25*, 1657–1672. [CrossRef]
27. China Electricity Council (CEC). *National Power Industry Statistics 2020*; China Electricity Council: Beijing, China, 2021; Available online: https://www.cec.org.cn/upload/1/editor/1640595481946.pdf (accessed on 12 July 2022).
28. IAI. Global Data on Perfluorocarbon (PFC) Emissions. 2018. Available online: http://www.world-aluminium.org/statistics/perfluorocarbon-pfc-emissions/#bubble (accessed on 26 May 2019).
29. Integrated Knowledge for our Environment (iKE). Chinese Life Cycle Database. Available online: http://www.ike-global.com/products-2/chinese-lca-database-clcd (accessed on 26 May 2019).
30. Ministry of Industry and Information Technology(MIIT). Standard Regulations for Aluminum Industry. 2013. Available online: http://www.gov.cn/gzdt/2013-07/24/content_2454273.htm (accessed on 25 May 2019). (In Chinese)
31. Zhang, Z. Energy saving of carbon anode enterprises. *Carbon Technol.* **2012**, *31*, B29–B32.
32. National Development and Reform Commission (NDRC). Guideline for accounting GHG emissions from the aluminum smelter. Beijing. 2013. Available online: http://qhs.ndrc.gov.cn/zcfg/201311/W020131202589518801729.pdf (accessed on 25 May 2019). (In Chinese)
33. General Administration of Customs of China. Import statistics of aluminum ore and concentrate in 2020. Available online: http://www.customs.gov.cn/eportal/ui?msgDataId=78ec17e885704f22b552e7cfe4e4002b (accessed on 12 July 2021).
34. International Trade Statistics Database. Available online: https://comtrade.un.org/data/ (accessed on 21 July 2022).
35. National Bureau of Statistics of China. National Data on the Average Distance of Cargo Transportation. 2020. Available online: http://data.stats.gov.cn/easyquery.htm?cn=C01 (accessed on 25 May 2021). (In Chinese)
36. Ministry of Transport. Statistical Bulletin of Transportation Development. 2020. Available online: http://www.gov.cn/xinwen/2020-05/12/content_5510817.htm (accessed on 15 May 2020). (In Chinese)
37. National Railway Administration. Statistical Bulletin of the Railway in 2020. Available online: http://www.nra.gov.cn/xwzx/zlzx/hytj/202204/P020220902306794538609.pdf (accessed on 25 April 2021). (In Chinese)
38. Peng, T.; Zhou, S.; Yuan, Z.; Ou, X. Life Cycle Greenhouse Gas Analysis of Multiple Vehicle Fuel Pathways in China. *Sustainability* **2017**, *9*, 2183. [CrossRef]
39. State Railway Administration. *Completion of National Railway Energy Consumption and Conservation Indicators (2015–2019)*; State Railway Administration: Beijing, China, 2020. (In Chinese)
40. Aluminum for Future Generation (AFG). Energy & Water Consumption for Primary Aluminum Production. Available online: http://primary.world-aluminium.org/aluminium-facts/energy-water/ (accessed on 16 May 2019).

41. China Bond Rating Co., Ltd. Guide to Regional and Enterprise Competitiveness of Electrolytic Aluminum in 14 Provinces of China. 2017. Available online: http://finance.sina.com.cn/stock/hkstock/hkstocknews/2017-05-24/doc-ifyfkkme0318771.shtml (accessed on 16 May 2019).
42. Antaike. The Proportion of Self-Generated Power Capacity in China's Electrolytic Aluminum Industry Shows a Downward Trend. 2020. Available online: https://news.cnal.com/2020/12-02/1606872807561727.shtml (accessed on 20 January 2021). (In Chinese)
43. Ding, N.; Yang, J. Life cycle inventory Analysis of Fossil Energy in China. *China Environ. Sci.* **2015**, *35*, 1592–1600.
44. IPCC. *Climate Change 2014: Mitigation of Climate Change Chapter 10: Industry*; Cambridge University Press: Cambridge, UK, 2014.
45. IAI. Life Cycle Model of Chinese Grid Power and Application to the Aluminum Industry. 2017. Available online: http://www.world-aluminium.org/publications/ (accessed on 10 November 2021).
46. Zhuo, Z.; Du, E.; Zhang, N.; Nielsen, C.P.; Lu, X.; Xiao, J.; Wu, J.; Kang, C. Cost increase in the electricity supply to achieve carbon neutrality in China. *Nat. Commun.* **2022**, *13*, 1–13. [CrossRef] [PubMed]

processes

MDPI

Article

Optimization of Cascade Cooling System Based on Lithium Bromide Refrigeration in the Polysilicon Industry

Shutong Yang, Youlei Wang and Yufei Wang *

School of Chemical Engineering and Environment, China University of Petroleum, Beijing 102249, China; yst0720@sohu.com (S.Y.); youleiwang@yeah.net (Y.W.)
* Correspondence: wangyufei@cup.edu.cn

Abstract: Cascade cooling systems containing different cooling methods (e.g., air cooling, water cooling, refrigerating) are used to satisfy the cooling process of hot streams with large temperature spans. An effective cooling system can significantly save energy and costs. In a cascade cooling system, the heat load distribution between different cooling methods has great impacts on the capital cost and operation cost of the system, but the relative optimization method is not well established. In this work, a cascade cooling system containing waste heat recovery, air cooling, water cooling, absorption refrigeration, and compression refrigeration is proposed. The objective is to find the optimal heat load distribution between different cooling methods with the minimum total annual cost. Aspen Plus and MATLAB were combined to solve the established mathematical optimization model, and the genetic algorithm (GA) in MATLAB was adopted to solve the model. A case study in a polysilicon enterprise was used to illustrate the feasibility and economy of the cascade cooling system. Compared to the base case, which only includes air cooling, water cooling, and compression refrigeration, the cascade cooling system can reduce the total annual cost by USD 931,025·y^{-1} and save 7,800,820 kWh of electricity per year. It also can recover 3139 kW of low-grade waste heat, and generate and replace a cooling capacity of 2404 kW.

Keywords: cascade cooling system; waste heat recovery; LiBr-H_2O absorption refrigeration; heat load distribution; temperature breakpoints

Citation: Yang, S.; Wang, Y.; Wang, Y. Optimization of Cascade Cooling System Based on Lithium Bromide Refrigeration in the Polysilicon Industry. *Processes* **2021**, *9*, 1681. https://doi.org/10.3390/pr9091681

Academic Editors: Pei Liu, Ming Liu and Xiao Wu

Received: 31 August 2021
Accepted: 16 September 2021
Published: 18 September 2021

Publisher's Note: MDPI stays neutral with regard to jurisdictional claims in published maps and institutional affiliations.

Copyright: © 2021 by the authors. Licensee MDPI, Basel, Switzerland. This article is an open access article distributed under the terms and conditions of the Creative Commons Attribution (CC BY) license (https://creativecommons.org/licenses/by/4.0/).

1. Introduction

In the process industry, cooling systems are a key element that takes away waste heat or cools down streams to a target temperature. Three main cooling methods are usually applied in industry, i.e., air cooling, water cooling, and refrigeration.

Air cooling takes ambient air as the cooling medium to cool the process stream. In general, air cooling has a higher capital cost, but it can save water and does not suffer from severe fouling problems in comparison with water cooling. Air cooling has been studied by many researchers. Doodman et al. [1] proposed an optimization model for air cooler designing. Manassaldi et al. [2] presented a disjunctive mathematical model for the optimal design of air coolers, which minimize the total annual cost considering both minimum heat transfer area and minimum fan power consumption. The research above studied the design and selection of air coolers. Other scholars studied the environmental effects on air cooler performance, including the impact of ambient air temperature and heat load variation [3], the freezing of air coolers [4], and fouling effects [5].

Water cooling uses water as the cooling medium. Cooling water systems are widely used due to their high heat transfer efficiency and relatively low cost. Studies on cooling water systems have been continued for many decades. Kim and Smith [6] proposed a pinch-based method to combine a cooling tower with a water cooler network and studied the interactions between the two parts. Reusing cooling water for different coolers makes cooling tower performance better. Panjeshahi and Ataei [7] extended the design of cooling water systems, intending to minimize costs and to maximize resource conservation based on the pinch approach. Castro et al. [8] also focused on decreasing operation costs in cooling water systems through minimizing the water flowrate. However, all research previously conducted mainly focuses on minimizing the water flowrate. In order to meet the heat transfer demand, more contact areas are required. The MINLP model proposed by Ponce-Ortega [9] is different from the previous models minimizing the energy consumption; it takes the capital cost for coolers and utility costs into consideration simultaneously.

Absorption refrigeration technology uses a binary solution composed of refrigerant and absorbent as working fluids for refrigeration. Compared with traditional compression refrigeration technology, absorption refrigeration technology can be driven by waste heat, rather than power. Over the past few decades, a number of studies were conducted on absorption refrigeration. Srikhirin et al. [10] reviewed the research on working pairs before 2001. They pointed out that there are more than 40 refrigerant compounds and 200 absorbent compounds available, but the most widely used working pairs are still LiBr-H_2O and NH_3-H_2O. Sun et al. [11] divided the working fluids of absorption refrigeration into NH_3 series, H_2O series, ethanol series, halogenated hydrocarbon series, and other series according to the different refrigerants, and listed the characteristics of various working pairs and their corresponding research literature. Another important study of absorption refrigeration is the performance of the cycle, that is, the coefficient of performance (COP). Karamangil et al. [12] developed visualization software and applied it to the simulation of cycle performance. The results showed that the operating temperature of generator, absorber, evaporator and condenser would all affect the COP of the cycle. Kaynakli and Kilic [13] also studied the influence of different operating parameters on system performance, such as an increase of the temperature of the generator and evaporator, decreases in the heat load of absorber and generator, and increases in the COP of the cycle. Absorption refrigeration is an attractive way for low-grade waste heat recovery. Ebrahimi et al. [14] discussed the technical and economic problems of using absorption refrigeration to recover waste heat from servers in data centers. The recovery process of Liquefied Natural Gas (LNG) requires low temperature cooling, which is generally provided by vapor compression refrigeration. If the absorption refrigeration system is driven by waste heat generated by the generating gas turbine, it can meet necessary cooling demands while reducing the overall energy consumption [15]. Zhang et al. [16] introduced a waste heat recovery, refrigeration, and application system to improve the energy utilization performance of industrial parks. Salmi et al. [17] proposed a steady-state thermodynamic model of absorption refrigeration cycles (ARCs) with water–LiBr and ammonia–water working pairs for ships. By using different waste heat sources and ARCs, the optimal generation temperatures under evaporation temperature were determined. Yang et al. [18] proposed the cascade utilization of 90–150 °C low-grade waste heat by LiBr/H_2O absorption refrigeration cycle and transcritical CO_2 cycle. The cascade system provides a potential way to generate electricity and refrigeration capacity using low-grade waste heat.

Compression refrigeration technology takes advantage of the phase change of the refrigerant with a low boiling point, to achieve refrigeration. Most of the research on compression refrigeration systems is related to the refrigerants. Dalkilic and Wongwise [19] studied the theoretical performance of traditional vapor-compression refrigeration using various rations of refrigerant mixtures based on HFC134a, HFC152a, HFC32, HC290, HC1270, HC600, and HC600a. Although HC series refrigerants are highly flammable, they are used in many applications because they do not affect the ozone layer and global warming. An exergy analysis of compression refrigeration system was also carried out to evaluate the economic performance of the system. Ahamed et al. [20] found that exergy is related to evaporation temperature, condensation temperature, degree of supercooling and compressor pressure and also depends on the ambient temperature. A large number of studies showed that the main exergy loss occurred in the compressor in the compression refrigeration system.

In the cooling system, the hot stream's temperature influences the temperature of the refrigerating medium (such as chilled water), and the evaporating temperature of the refrigeration system. The evaporating temperature of the refrigeration system has a certain influence on the efficiency of the system. A theoretical analysis of the compression refrigeration system with different refrigerants was performed. The investigation showed that with an increase in the evaporating temperature, the COP of the compression refrigeration system increases [21]. Selbas et al. [22] optimized the subcooled and superheated vapor compression refrigeration cycle through thermo-economy. the results indicated that the heat exchangers capital cost of the system increases with the evaporating temperature. Kaushik and Arora [23] studied the energy and exergic analysis of a lithium bromide absorption refrigeration system. The results showed that an increase in the generator temperature increases the COP and exergic efficiency up to an optimal generator temperature. An increase in the evaporator temperature increases the COP but reduces exergetic efficiency.

Air cooling, water cooling, absorption refrigeration, and compression refrigeration have been studied over a long period of time. Among them, air cooling and water cooling are combined to take away waste heat, and absorption refrigeration and compression refrigeration are applied to cool down the stream to a sub-ambient target temperature. In the polysilicon industry, some hot streams have a large cooling temperature span from over 100 °C to sub-ambient temperature. In a conventional design for such a situation, air/water cooling is used to cool down hot streams to near ambient temperature, and compression refrigeration is used to cool down streams to target temperature. In this situation, absorption refrigeration can be used to firstly recover low-grade waste heat to reduce the duty of air/water cooling, and then the generated cooling duty can be used to reduce the duty of compression refrigeration. However, such cascade cooling systems have not been widely studied.

Therefore, in this work, a cascade cooling system containing waste heat recovery, air cooling, water cooling, absorption refrigeration, and compression refrigeration is proposed. Figure 1 is employed to show the structure of the cascade cooling system. Thout(i,1), Thout(i,2), Thout(i,3) and Thout(i,4) are the outlet temperatures of the hot streams of the waste heat recovery exchanger, air cooler, water cooler, and absorption refrigeration cooler, respectively. They are also the temperature breakpoints between the five heat exchangers. The breakpoints indicate the heat load distribution of the five heat exchangers and influence the capital cost and operation cost of heat exchangers. To obtain the optimal heat load distribution and the temperature breakpoints of hot streams, an optimization method combining Aspen Plus and a GA is proposed. Finally, a case study in a polysilicon enterprise was optimized in this work.

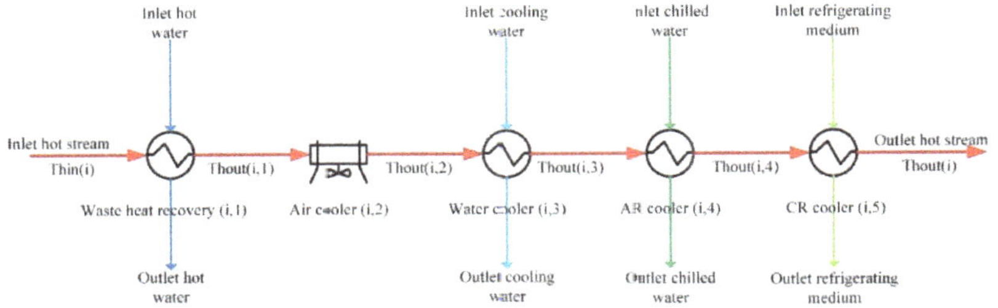

Figure 1. Structure of the cascade cooling system.

2. Problem Statement

Hot streams of a hydrochlorination plant in a polysilicon enterprise have high supply temperatures (more than 130 °C). These streams were cooled to target temperature by air cooler, water cooler, and compression refrigeration, originally. There are two problems with the original cooling process. Firstly, without waste heat recovery, the cooling duty of air/water cooling is large. Secondly, large amounts of energy will be consumed if the streams are cooled by compression refrigeration directly after water cooling. In view of these two problems, the cascade cooling system containing waste heat recovery, air cooling, water cooling, absorption refrigeration, and compression refrigeration was adopted.

The objective of this work is to determine the optimal heat load distribution of coolers and the optimal breakpoints of temperatures between waste heat recovery, air cooling, water cooling, absorption refrigeration, and compression refrigeration in terms of minimum total annual cost (TAC). The mathematical optimization model is established and solved, and the optimal design is based on several assumptions:

The specific heat capacities and film transfer coefficients of hot water, air, cooling water, chilled water, and ethylene glycol are constant. The film transfer coefficients of hot streams are constant. Due to the large temperature span of the hot stream, the specific heat capacity of the hot stream is not constant.

The heat loss during the heat exchange and the transportation of streams is ignored.

3. Model Formulation

3.1. Heat Exchanger Formulation

In Equation (1), Q is the heat load of the heat exchanger which is obtained through the simulation in Aspen; i and j represent the hot streams and cooling medium; and j = 1–5, respectively, represent hot water, air, cooling water, chilled water, and ethylene glycol. M and m are the mass flow rate of hot stream and cooling medium; I_{in} and I_{out} are the inlet and outlet specific enthalpy of hot stream; Cp_m is the specific heat capacity of the cooling medium; T_{in} and T_{out} are the inlet and outlet temperature of hot stream; tin and $tout$ are the inlet and outlet temperature of the cooling medium.

$$Q(i,j) = M(i,j) \cdot (Iin(i,j) - Iout(i,j)) = m(i,j) \cdot Cp_{\mathrm{m}}(i,j) \cdot (tout(i,j) - tin(i,j)) \quad (1)$$

For heat exchangers, dt_{in} and dt_{out} represent the temperature differences on both sides of the heat exchanger, and temperature differences should be greater than the minimum temperature approach difference preset for the heat exchanger [24], as shown in Equations (2) and (3).

$$dt_{\mathrm{in}}(i,j) = T_{in}(i,j) - t_{out}(i,j) \geq \Delta T_{\mathrm{min}}(i,j) \quad (2)$$

$$dt_{\mathrm{out}}(i,j) = T_{out}(i,j) - t_{in}(i,j) \geq \Delta T_{\mathrm{min}}(i,j) \quad (3)$$

The area of the heat exchanger can be calculated by Equation (4), in which hin and $hout$ are film transfer coefficients of both sides of the heat exchanger.

$$A(i,j) = \frac{Q(i,j)}{K(i,j) \cdot \Delta T_{\mathrm{m}}(i,j)} = \frac{Q(i,j)}{\frac{dt_{\mathrm{in}}(i,j) - dt_{\mathrm{out}}(i,j)}{\ln(dt_{\mathrm{in}}(i,j)/dt_{\mathrm{out}}(i,j))}} \cdot \left(\frac{1}{h_{in}} + \frac{1}{h_{out}}\right) \quad (4)$$

Equation (5) shows the capital cost of the heat exchanger; a, b, and c are the cost parameters. Then total capital cost of the heat exchangers in the cascade cooling system is shown as Equation (6).

$$CC_{\mathrm{HEX}}(i,j) = a + b \cdot A(i,j)^c \quad (5)$$

$$CC_{\mathrm{HEX}} = \sum CC_{\mathrm{HEX}}(i,j) = \sum (a + b \cdot A(i,j)^c) \quad (6)$$

3.2. Pump Formulation and Pipe Formulation

Pumps are used to transport the cooling medium used in the cascade cooling system, such as hot water (HW), cooling water (CW), chilled water (CHW), and ethylene glycol (EG). The diameter of the pipe can be calculated by Equation (7), where ρ is the density of the cooling medium, and u is the flow velocity. It is assumed that the density of the cooling medium does not change with temperature.

$$D_{\mathrm{in}}(j) = \sqrt{\frac{4m(j)}{\pi \cdot \rho(j) \cdot u(j)}} \quad (7)$$

According to Equation (1), the mass flow rate of cooling medium can be calculated through Equation (8).

$$m(j) = \sum_{i=1}^{n} \frac{Q(i,j)}{Cp_{\mathrm{m}}(i,j) \cdot (t_{out}(i,j) - t_{in}(i,j))} \tag{8}$$

The capital cost and operation cost of the pump can be calculated through Equations (9) and (10), where α, β, γ are the parameters of the pump capital cost; Pe is the cost of the unit electricity; Hy is the annual operation time; and η_{Pump} is the pump efficiency [25].

$$CC_{\mathrm{Pump}} = \sum_{j \neq 2}^{5} (\alpha + \beta \cdot (\frac{m(j) \cdot \Delta P(j)}{\rho(i)})^{\gamma}) \tag{9}$$

$$OC_{\mathrm{Pump}} = \sum_{j \neq 2}^{5} (Pe \cdot Hy \cdot \frac{m(j) \cdot \Delta P(j)}{1000 \cdot \rho(j) \cdot \eta_{\mathrm{Pump}}}) \tag{10}$$

In the above equation, ΔP is the pressure drop of the pump, which generally contains pipeline pressure drop and heat exchanger pressure drop, as shown in Equation (11).

$$\Delta P(j) = \Delta P_{\mathrm{PIPE}}(j) + \Delta P_{\mathrm{HEX}}(j) \tag{11}$$

The pressure drop of the pipeline is calculated by Equations (12)–(14), in which Re and f are Reynolds (Re) number and Fanning friction coefficient, μ is viscosity, and L is transmission distance. Fanning friction (the pipe is a hydraulically rough pipe) can be calculated through Equation (13) [26], and it is assumed that the Re number is between 3000 and 3×10^6.

$$\mathrm{Re} = \frac{\rho \cdot D_{\mathrm{in}} \cdot u}{\mu} \tag{12}$$

$$f = 0.0055 + \frac{0.5}{\mathrm{Re}^{0.32}} \tag{13}$$

$$\Delta P_{\mathrm{PIPE}} = 4 \cdot f \cdot \frac{L \cdot \rho \cdot u^2}{2 \cdot D_{\mathrm{in}}} \tag{14}$$

The pressure drop of the heat exchanger can be divided into tube and shell side pressure drop. Since the calculation of the shell side pressure drop is very complicated, and usually, the cooling medium flows in tube side and the hot stream flows in the shell side, only the pressure drop of the tube side is considered in this paper. The pressure drop is related to the constant K_{t}, the heat transfer area A, and the film heat transfer coefficient h_{t}. The constant K_{t} is related to the flow rate of the cooling medium m_{t}, viscosity μ_{t}, density ρ_{t}, heat conductivity κ_{t}, specific heat capacity Cp_{t}, and inner diameter Dt_{in} and outer diameter Dt_{out} of tube in the heat exchanger [27]. In this paper, except for the air cooler, other heat exchangers choose the same geometric shape and type, which is a counterflow single shell and single tube heat exchanger.

$$\Delta P_{\mathrm{HEX}} = \Delta P_{\mathrm{t}} = K_{\mathrm{t}} \cdot A \cdot h_{\mathrm{t}}^{3.5} \tag{15}$$

$$K_{\mathrm{t}} = \frac{Dt_{\mathrm{in}}^{1/2} \cdot \mu_{\mathrm{t}}^{11/6}}{0.023^{2.5} \cdot \rho_{\mathrm{t}} \cdot \kappa_{\mathrm{t}}^{7/3} \cdot Cp_{\mathrm{t}}^{7/6} \cdot m_{\mathrm{t}}} \cdot \frac{Dt_{\mathrm{in}}}{Dt_{\mathrm{out}}} \cdot (\frac{\mu_{\mathrm{t}}}{\mu_{\mathrm{r}}})^{-0.63} \tag{16}$$

In the calculation of the heat exchanger pressure drop, it is necessary to take the series and parallel structure of the heat exchanger into account. In general, the total pressure drop of the series structure is the sum of the pressure drop of all heat exchangers. The maximum pressure drop of the branch stream is taken as the pressure drop of the parallel structure [28]. In this paper, the heat exchanger network of the cooling medium and hot stream is in parallel structure.

In addition to considering the capital and operation cost of the pump for the cooling medium, the investment cost of the pipeline should also be calculated. The outer diameter D_{out} of the pipe is calculated from the inner diameter by Equation (17) [29], the constants in the equation are the model parameters for the calculation. Then the capital cost of the pipeline can be calculated by Equations (18)–(20) [29]. In the equation that follows, Wt is the weight of the pipe per unit length, $Pcul$ is the investment cost of the pipe per unit length. In this paper, 80# steel pipes are selected, and A_1, A_2, A_3, and A_4 are the related cost parameters [29].

$$D_{\text{out}}(j) = 1.101 \cdot D_{\text{in}}(j) + 0.006349 \tag{17}$$

$$Wt(j) = 1330 \cdot D_{\text{in}}^2(j) + 75.18 \cdot D_{\text{in}}(j) + 0.9268 \tag{18}$$

$$Pcul(j) = A_1 \cdot Wt(j) + A_2 \cdot D_{\text{out}}^2(j) + A_3 + A_4 \cdot D_{\text{out}}(j) \tag{19}$$

$$CC_{\text{PIPE}} = \sum_{j \neq 2} (L \cdot Pcul(j)) \tag{20}$$

3.3. Air Cooler Formulation

The ambient temperature influences the face velocity of the air cooler. T_{ambient} is the inlet air temperature of the air cooler. V_{F} and V_{NF} are the face velocity and actual face velocity of air cooler [30].

$$V_{\text{NF}} = V_{\text{F}} \cdot \frac{293}{273 + T_{\text{ambient}}} \tag{21}$$

Equation (22) is used to calculate the outside film heat transfer coefficient of the air cooler, which depends on its actual face velocity. In this work, it is assumed that the air flows through triangular pitch banks of finned tubes.

$$h_{\text{a}} = 218.9 \cdot V_{\text{NF}}^{0.718} \tag{22}$$

The energy consumption of the air cooler is related to the fan pressure drop. The air cooler fan pressure drop is related to the air mass flowrate and the number of bundles. Equations (23)–(25) were employed to calculate the fan pressure drop, where G represents the mass velocity rate, G_{max} is the maximum mass velocity when the air flows through the narrow part of air cooler, N_b is the number of bundles in air cooler, and f_{friction} is the friction factor.

$$G = V_{\text{F}} \cdot \rho_{\text{air}} \tag{23}$$

$$G_{\text{max}} = 2G \tag{24}$$

$$\Delta p_{\text{air}} = 9.8 \cdot f_{\text{friction}} \frac{N_{\text{b}} \cdot G_{\text{max}}^2}{2g\rho_{\text{air}}} \tag{25}$$

Equation (26) represents the power consumption of the air cooler fan, V_{air} is the volumetric flow rate of air, and η_{fan} is the fan efficiency. Then the operation cost of air coolers can be calculated by Equation (27).

$$P_{\text{fan}}(i,j) = \frac{\Delta p_{\text{air}} \cdot V_{\text{air}}(i,j)}{\eta_{\text{fan}}} \tag{26}$$

$$OC_{\text{AC}} = \sum_{j=2} P_{\text{fan}}(i,j) \cdot Pe \cdot Hy \tag{27}$$

3.4. Absorption Refrigeration Cycle Formulation

The absorption refrigeration cycle (ARC) used in the cascade cooling system is a traditional single-effect LiBr/H_2O absorption refrigeration cycle. It consists of a generator, condenser, throttle valve, evaporator, absorber, solution heat exchanger, and pump. Figure 2 shows the schematic diagram of the single-effect lithium bromide absorption refrigeration cycle [31].

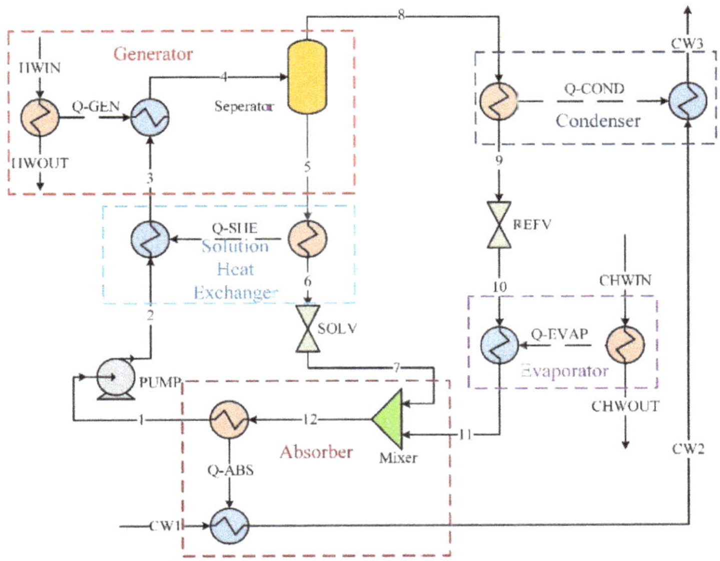

Figure 2. Schematic diagram of single-effect LiBr/H_2O absorption refrigeration cycle.

This absorption refrigeration cycle is simulated in Aspen Plus, and the ELECNRTL Equation of State is selected for the electrolyte system [32]. The working fluid of this cycle is lithium bromide aqueous solution, whose original state (stream 1) is 36.62 °C, 800 Pa and the concentration is 0.58 (mass fraction). The absorbent is LiBr solution, and the refrigerant is pure water. The LiBr solution is heated by heat source (hot water from waste heat recovery) in the generator, and the refrigerant (water) is evaporated to steam and then condensed to saturated liquid in the evaporator and condenser. At the same time, the LiBr solution in the evaporator becomes a strong solution and flows into the absorber through a throttle. The saturated refrigerant liquid steps down to evaporating temperature by a throttle, then evaporates in the evaporator, resulting in a refrigeration effect. Then the saturated steam from the evaporator flows into the absorber and mixes with the strong LiBr solution. The strong solution absorbs the steam and becomes the weak solution as used in the beginning. Since the pressure of the generator is higher than that of the absorber, a pump is used to send the diluted solution back to the generator for the whole process to circulate.

Equation (28) expresses the calculation of the *COP* for the absorption refrigeration cycle, in which $Q_{\text{AR-EAVP}}$ and $Q_{\text{AR-GEN}}$ are the heat load of the evaporator and generator, respectively. The heat load of the generator should be the same as the total heat load of the waste heat recovery in the cascade cooling system. The heat load of the evaporator should also be the same as the total heat load of the absorption refrigeration in the cascade cooling system.

$$COP_{\text{AR}} = \frac{\text{Cooling capacity generated inevaporator}}{\text{Heat absorbed in generator}} = \frac{Q_{\text{AR-EVAP}}}{Q_{\text{AR-GEN}}} \tag{28}$$

For simplicity of calculation, the capital cost of the absorption refrigeration machine is calculated by the sum of the capital costs of the heat exchangers and the working fluid pump. The operation cost is the operation cost of working fluid pump. The heat exchangers are the generator, condenser, evaporator, absorber, and solution heat exchanger. The area of these heat exchangers can be calculated by Equation (4). For the ARC model used in MATLAB, they can be found in reference [33]. Because all the models are the same, they are not listed in this paper.

Equations (31) and (32) are employed to calculate the capital cost and operation cost of the absorption refrigeration cycle. The calculation of the capital cost of the heat exchangers is the same as in Equation (5), and the capital and operation costs of pumps for working fluid and cooling water are same as Equations (9) and (10), while the mass flow rate of the working fluid M_{LiBr} and cooling water $M_{\text{AR-CW}}$ can be calculated by Equations (31) and (32).

$$CC_{\text{AR}} = CC_{\text{AR-GEN}} + CC_{\text{AR-COND}} + CC_{\text{AR-EVAP}} + CC_{\text{AR-ABS}} + CC_{\text{AR-SHE}} + CC_{\text{Pump-LiBr}} \quad (29)$$

$$OC_{\text{AR}} = OC_{\text{Pump-LiBr}} \quad (30)$$

$$M_{\text{LiBr}} = \frac{Q_{\text{AR-GEN}}}{q_{\text{AR-GEN}}} \quad (31)$$

$$M_{\text{AR-CW}} = \frac{Q_{\text{AR-ABS}}}{Cp_{\text{CW}} \cdot (T_{\text{CW2}} - T_{\text{CW1}})} \quad (32)$$

In Equation (31), $q_{\text{AR-GEN}}$ is the heat load of the generator when the working fluid flow rate is 1 kg/s. In Equation (32), $Q_{\text{AR-ABS}}$ is the heat load of the absorber. T_{CW1} and T_{CW2} are the inlet and outlet temperature of the cooling water in the absorber. T_{CW2} can be calculated by Equation (33), in which T_{CW3} is the outlet temperature of the cooling water in the condenser, and $Q_{\text{AR-COND}}$ is the heat load of condenser.

$$\frac{T_{\text{CW2}} - T_{\text{CW1}}}{T_{\text{CW3}} - T_{\text{CW2}}} = \frac{Q_{\text{AR-ABS}}}{Q_{\text{AR-COND}}} \quad (33)$$

3.5. Compression Refrigeration Cycle Formulation

Traditional vapor compression refrigeration (VCR) is composed of an evaporator, a compressor, a condenser, and an expansion valve. This cycle is modeled in the Aspen Plus software, as shown in Figure 3. The REFPROP property method is adopted for simulation. R134a is selected as the refrigerant. For a compressor, the actual compression process is non-isentropic, and the parameter η_{isen} is used to describe the degree of non-isentropicity [34]. η_{isen} for the compressor is set as 0.7 during the simulation.

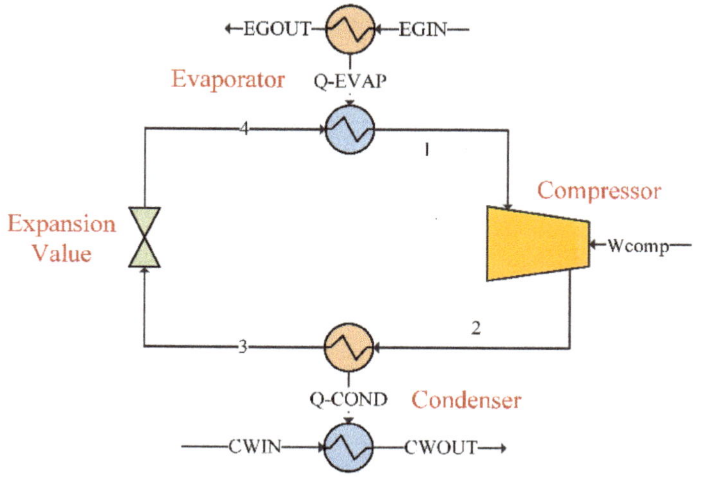

Figure 3. Schematic diagram of the vapor compression refrigeration cycle.

The *COP* of a vapor compression refrigeration cycle is an important system performance indicator. It represents the refrigeration effect per unit of compressor work required and is expressed by Equation (34) [35].

$$COP_{\text{CR}} = \frac{\text{Refrigerating effect (Heat load of evaporator)}}{\text{Comprssion work of the compressor}} = \frac{Q_{\text{CR-EVAP}}}{W_{\text{Comp}}} \quad (34)$$

In Equation (34), $Q_{\text{CR-EVAP}}$ is the refrigeration effect which is equal to the total heat load of compression refrigeration coolers in the hot stream cooling process, W_{Comp} is the non-isentropic work of compressor. Operation cost of compressor can be calculated by Equations (35) and (36) [20]. In Equation (38), η_{mech} and η_{el} are mechanical efficiency and electrical efficiency. For the VCR model used in MATLAB, they can be found in reference [33]. Because all the models are the same, they are not listed in this paper.

$$W_{\text{Comp}} = \frac{Q_{\text{CR-EVAP}}}{COP_{\text{CR}}} = \sum_{j=5} \frac{Q(i,j)}{COP_{\text{CR}}} \quad (35)$$

$$OC_{\text{Comp}} = \frac{W_{\text{Comp}}}{\eta_{\text{mech}} \cdot \eta_{\text{el}}} \cdot Pe \cdot Hy \quad (36)$$

Equation (37) is used to express the capital cost of the compressor, where M_{ref} is the mass flowrate of the refrigerant. P_{cond} and P_{evap} are the operating pressure of the condenser and evaporator, respectively [36]. M_{ref} can be calculated by Equation (38), ΔH_{EVAP} is the enthalpy change of the evaporator when M_{ref} is 1 kg/s. According to the heat balance of the compression refrigeration cycle, the heat load of the condenser is given by Equation (39).

$$CC_{\text{Comp}} = \left(\frac{573 \cdot M_{\text{ref}}}{0.8996 - \eta_{\text{isen}}}\right)\left(\frac{P_{\text{cond}}}{P_{\text{evap}}}\right) \ln\left(\frac{P_{\text{cond}}}{P_{\text{evap}}}\right) \quad (37)$$

$$M_{\text{ref}} = \frac{Q_{\text{CR-EVAP}}}{\Delta H_{\text{EVAP}}} \quad (38)$$

$$Q_{\text{CR-COND}} = Q_{\text{CR-EVAP}} + W_{\text{Comp}} \quad (39)$$

Then the capital and operation cost of the compression refrigeration cycle are expressed by Equations (40) and (41).

$$CC_{\text{CR}} = CC_{\text{CR-EVAP}} + CC_{\text{CR-COND}} + CC_{\text{Comp}} \quad (40)$$

$$OC_{\text{CR}} = OC_{\text{Comp}} \quad (41)$$

3.6. Cooling Tower Formulation

Cooling towers are usually present wherever water is used as a cooling medium, there are many factors that influence the performance and cost of a cooling tower, such as atmospheric pressure, local air wet bulb temperature and air humidity, cooling water flowrate, and inlet/outlet temperature, etc. To formulate the cooling tower, the following definitions are necessary.

First, the operation cost of the cooling tower includes the cooling tower fan cost, water treatment cost, make-up water cost, and blowdown treatment cost. The constants in Equation (42) are the cost parameters for the calculation.

$$OC_{\text{CT}} = OC_{\text{fan-tower}} + 110 \cdot ft + Pw \cdot Hy \cdot M_{\text{makeup}} + 1138 \cdot B_{\text{blowdown}} \quad (42)$$

In Equation (42), $OC_{\text{fan-tower}}$ is the operation cost of the tower fan, *ft* is the total flowrate of the cooling water, M_{makeup} is the flowrate of make-up water, B_{blowdown} is the flowrate of water blowdown, and Pw is the price of fresh water [37]. $OC_{\text{fan-tower}}$ can be calculated by Equation (43).

$$OC_{\text{fan-tower}} = \frac{Pe \cdot Hy \cdot C_{\text{factor}} \cdot F_{\text{air}}}{\eta_{\text{fan-tower}}} \tag{43}$$

In Equation (43), C_{factor} is the fan factor. To draft each 18,216.44 m^3/h of air, 1 kW is required. F_{air} is the air mass flowrate of the cooling tower. $\eta_{\text{fan-tower}}$ is the fan efficiency.

Mass flowrate of air, make-up water and blowdown water are shown in Equations (44)–(46). E_{vop} is the amount of water evaporation, w_{in} and w_{out} are the inlet and outlet humidity of air, and π_c is the cycle of concentration.

$$F_{\text{air}} = \frac{E_{\text{vop}}}{w_{\text{out}} - w_{\text{in}}} \tag{44}$$

$$M_{\text{makeup}} = E_{\text{vop}} \cdot \frac{E_{\text{vop}}}{\pi_c - 1} \tag{45}$$

$$B_{\text{blowdown}} = \frac{E_{\text{vop}}}{\pi_c - 1} \tag{46}$$

The amount of water evaporation is related to the cooling tower range and water flowrate. *Range* is the difference between the cooling tower inlet temperature T_{cin} and outlet temperature T_{cout}.

$$E_{\text{vop}} = 0.00153 \cdot ft \cdot Range \tag{47}$$

$$Range = T_{\text{cin}} - T_{\text{cout}} \tag{48}$$

The cooling tower inlet air humidity is the local air humidity. The outlet air humidity is calculated by Equations (49)–(51). MW_w and MW_{air} are water and air molecular weight, P_s and P_a are vapor pressure and local atmospheric pressure, and T_{mean} is the mean temperature of the cooling tower [38].

$$w_{\text{out}} = \frac{MW_w}{MW_{\text{air}}} \cdot \frac{P_s}{P_a - P_s} \tag{49}$$

$$LnP_s = 23.1 - \frac{4111}{T_{\text{mean}} + 237.7} \tag{50}$$

$$T_{\text{mean}} = \frac{T_{\text{cout}} + T_{\text{cin}}}{2} \tag{51}$$

Besides, the capital cost of the cooling tower is determined by many factors. Equation (52) is the capital cost of the cooling tower, in which *Approach* is temperature difference between the cooling tower outlet temperature and the air bulb temperature T_{wb}.

$$CC_{\text{CT}} = 746.74 \cdot ft^{0.79} \cdot Range^{0.57} \cdot Approach^{-0.9924} + (0.022 \cdot T_{\text{wb}} + 0.39)^{2.447} \tag{52}$$

$$Approach = T_{\text{cout}} - T_{\text{wb}} \tag{53}$$

In this work, there should be three cooling towers because of the different cooling waters for different cooling requirements, one for the cooling of the hot streams, one for the cooling of the absorber and condenser of the ARC, and one for the cooling of the condenser of the CRC.

3.7. Objective Function

The objective of this work is to determine the optimal heat load distribution of the cascade cooling system with minimizing TAC. TAC includes the heat exchanger capital cost, capital and operation costs of pumps, operation costs of air coolers, capital and operation costs of the absorption refrigeration cycle, capital and operation costs of the compression refrigeration cycle, and capital and operation costs of the cooling towers.

$$\begin{aligned} TAC &= Af \cdot TCC + TOC \\ &= Af \cdot (CC_{HEX} + CC_{Pump} + CC_{PIPE} + CC_{AR} + CC_{CR} + CC_{CT1} + CC_{CT2} + CC_{CT3}) \\ &+ (OC_{AC} + OC_{Pump} + OC_{AR} + OC_{CR} + OC_{CT1} + OC_{CT2} + OC_{CT3}) \end{aligned} \tag{54}$$

In the objective function above, *TCC*, *TOC*, and *TAC* are the total capital cost, total operation cost, and total annual cost of the cooling system, respectively. *Af* is the annualized factor of capital cost which is calculated as Equation (55), where *I* is the annual interest rate and *n* is the lifetime of the equipment in terms of years.

$$Af = \frac{I \cdot (1+I)^n}{(1+I)^n - 1} \tag{55}$$

4. Solution Technique Employed

In this work, Aspen Plus was combined with MATLAB to solve the problem. Aspen Plus was used to simulate the cooling process of hot streams, the LiBr/H_2O absorption refrigeration cycle, and the compression refrigeration cycle, and it provided data for MATLAB calculations. MATLAB was used to assign and read data from Aspen, as well as to control Aspen to open, run, and close [39]. They were combined through an interface program based on ActiveX technology. The interface process between Aspen and MATLAB is shown in Figure 4.

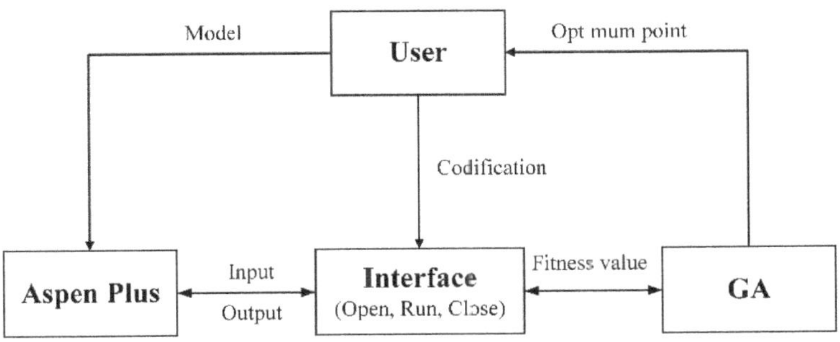

Figure 4. Interface diagram between Aspen and MATLAB.

Because the model contains large amounts of nonlinear terms, a heuristic algorithm such as a genetic algorithm (GA) can solve this problem more effectively. GA simulates natural selection and a genetic mechanism for optimization. It is an effective parallel, random, adaptive search algorithm, which has less mathematical requirements for the problem to be solved. It is a black-box solution. The remarkable characteristics of the genetic algorithms are their inner parallelism and their ability to search for global optimization. The combination of GA and process simulation software not only eliminates the need to establish rigorous mathematic models for complex chemical processes, but also enables global optimization for nonlinear, multivariate and multi-objective problems [40].

Figure 5 is used to show the block diagram of the optimization program which combines both Aspen Plus and GA. The evaporation and condensation temperatures of ARC and CRC determine the operating pressure of the cycles when simulating. In addition, the cascade cooling system determines the amount of waste heat recovered according to the actual refrigeration requirements. Therefore, the Aspen program needs to be run three times in the optimization. In the first run, evaporation and condensation pressure are calculated according to the input evaporation and condensation temperature. In the second run, the COP of ARC and CRC are calculated after the evaporation and condensation pressure are re-assigned. In the third run, according to the COP calculated and the heat load of the absorption refrigeration heat exchanger, the heat load of the waste heat recovery heat exchanger is recalculated and re-assigned, so as to determine the hot stream temperature out of the waste heat recovery heat exchanger and the heat load of the air cooler.

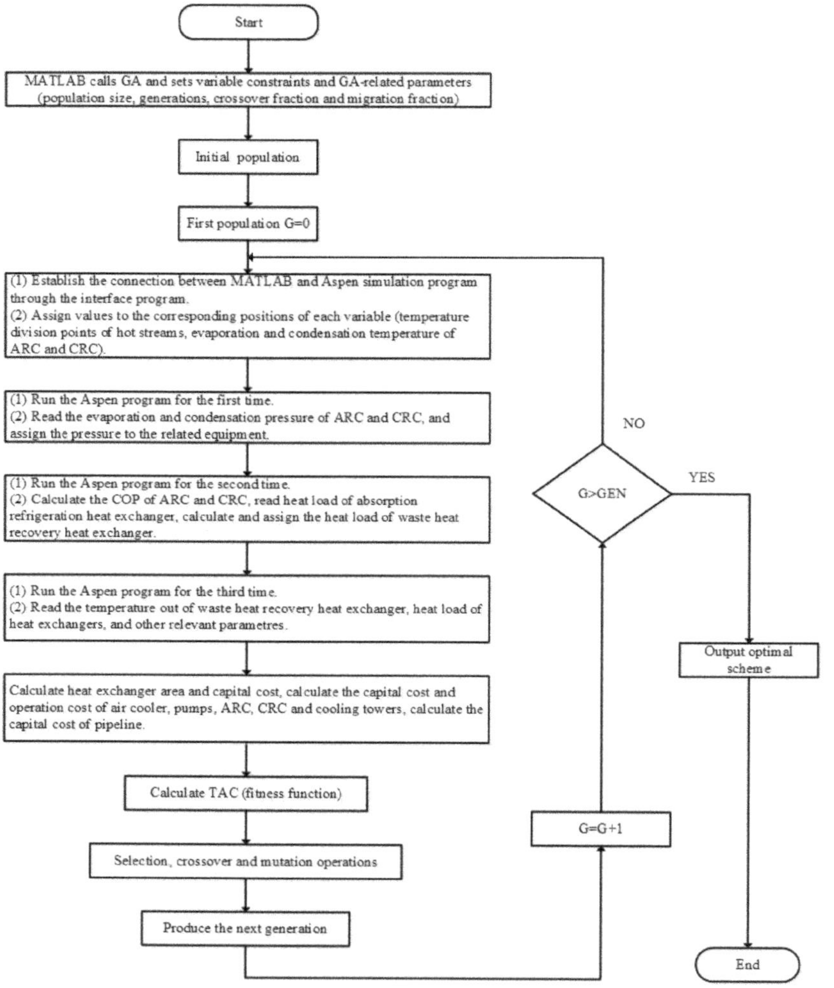

Figure 5. Block diagram of the optimization program.

The variable of the optimization procedure is the temperature in the cascade cooling system, including the breakpoint temperatures of the hot stream cooling process, the evaporation and condensation temperature of refrigeration cycle, the inlet and outlet temperature of cooling medium (e.g., hot water, air, cooling water, chilled water, and ethylene glycol solution). The objective function of optimization is the total annual cost of the system, and the termination condition of the GA is to reach the maximum genetic algebra (GEN).

5. Case Study

To verify the effectiveness of the cascade cooling system, a case study taken from a polysilicon enterprise is employed. The hot streams are the reaction tail gas of a hydrochlorination reaction unit of silicon tetrachloride in the polysilicon enterprise, which are composed of HCl (hydrogen chloride), DCS (dichlorosilane), TCS (trichlorosilane), STC (silicon tetrachloride), and H_2 (hydrogen). In the reactive tail gas treatment process, the tail gas is first used to preheat the raw material STC of the reaction, then it is cooled by series cooling and refrigeration methods. The chlorosilane is condensed down to realize the separation and recycling of H_2 and HCl. In this paper, only the cooling process of hot streams is taken into consideration. The hot streams data is shown in Table 1, *Ts* and *Tt* are the supply temperature after heat exchange with STC and target temperature of hot streams respectively. h_H is the film transfer coefficient of hot streams, but the specific heat capacity of the hot stream is not a constant because it varies with temperature.

Table 1. Hot stream data.

Hot Stream (*i*)	*Ts* (°C)	*Tt* (°C)	*M* ($kg{\cdot}s^{-1}$)	h_H ($W{\cdot}(m^2{\cdot}°C)^{-1}$)	Components (Mass Fraction)
H1	132.19	−5.8	34.0	500	HCl: 0.00036, DCS: 0.00615, TCS: 0.19751, STC: 0.78222, H_2: 0.01376
H2	131.82	−6.7	33.5	500	HCl: 0.00036, DCS: 0.00619, TCS: 0.19828, STC: 0.78122, H_2: 0.01394
H3	131.82	−7.1	33.4	500	HCl: 0.00037, DCS: 0.00619, TCS: 0.19828, STC: 0.78122, H_2: 0.01394

The hot streams have relatively high supply temperature (more than 130 °C) and low target temperature (less than 0 °C). The cooling mediums used in the cooling process are hot water (HW), air (AIR), cooling water (CW), chilled water (CHW), and ethylene glycol solution (EG), respectively. The cooling medium data is shown in Table 2, *tin* and *tout* are the inlet and outlet temperature of the cooling medium in the heat exchangers, ρ is the density of cooling medium, *Cp* is the specific heat capacity, μ is viscosity, κ is thermal conductivity, and h_C is the film transfer coefficient. Since the inlet and outlet temperature of the cooling medium are also the optimization variables, their value ranges are listed in the table. In addition, *hc* is selected through industrial experience.

Table 2. Cooling medium data.

Cooling Medium (*j*)	*tin* (°C)	*tout* (°C)	ρ ($kg{\cdot}m^{-3}$)	*Cp* ($kJ{\cdot}(kg{\cdot}°C)^{-1}$)	μ (Pa·s)	κ ($W{\cdot}(m{\cdot}K)^{-1}$)	h_C ($W{\cdot}(m^2{\cdot}°C)^{-1}$)
1 (HW)	90~110	100~120	950	4.200	0.0003	0.68	2500
2 (AIR)	35~35	60~65	1.169	1.004	0.0011	0.024	475.93
3 (CW)	20~25	45~50	995	4.182	0.001	0.62	2500
4 (CHW)	7~12	12~17	1000	4.182	0.0015	0.57	2500
5 (EG)	−15~−10	−5~0	1082.45	2.582	0.007	0.39	2500

Some other economic parameters and physical parameters of the case study are presented in Tables 3 and 4, respectively [41,42].

Table 3. Economic parameters of case study.

Items	Data
Air coolers capital cost (USD) [1]	$4778 \cdot A^{0.525}$
Other heat exchangers capital cost (USD) [43]	$8500 + 409 \cdot A^{0.85}$
Pump capital cost (USD)	$8600 + 7310(M \cdot \Delta P/\rho)^{0.2}$
Air cooler fan efficiency	70%
Cooling tower fan efficiency	70%
Pump efficiency	70%
Compressor isentropic efficiency	70%
Compressor mechanical efficiency	80%
Compressor electrical efficiency	90%
Price of electricity (USD·kWh^{-1})	0.15
Price of fresh water (USD·t^{-1})	0.5
Plant operation time (h·y^{-1})	8000
Plan lifetime (y)	5
Interest rate	15%
Annualized factor	0.298

Table 4. Other physical parameters and constrains of case study.

Items		Data
Minimum temperature approach difference of heat exchanger ΔT_{min}	Waste heat recovery cooler	10 °C
	Water cooler/air cooler	10 °C
	Absorption refrigeration cooler	3 °C
Allowable outlet temperature of air cooler		50~65 °C
Allowable outlet temperature of water cooler		25~35 °C
Allowable outlet temperature of absorption refrigeration heat exchanger		10~15 °C
Ambient temperature $T_{ambient}$		25 °C
Wet bulb temperature T_{wb}		12 °C
Air saturated humidity at wet bulb temperature		0.011 kgw·kga^{-1}
Atmospheric pressure		101,325 Pa
Air cooler face velocity		3 m·s^{-1}
Air cooler friction factor		0.95
Air cooler number of bundles		4
Cycle of concentration		4
ARC condensation temperature		35~45 °C
ARC evaporation temperature		5~10 °C
ARC cooling water initial temperature		30~32 °C
ARC cooling water final temperature		32~42 °C
CRC condensation temperature		35~45 °C
CRC evaporation temperature		−20~−15 °C
CRC cooling water initial temperature		27~37 °C
CRC cooling water final temperature		32~42 °C

5.1. The Base-Case Cooling System of Hot Streams

In this work, the cooling system containing air cooling, water cooling, and compression refrigeration were optimized as a base case. The NRTL-RK property method was selected for the simulation of the hot streams cooling process in Aspen Plus, because the hot stream is a weakly polarized system. The base-case cooling system of hot streams is shown in Figure 6. The inlet and outlet air temperature in the air cooler are 20 °C and 60 °C, and the inlet and outlet air temperatures in the cooling tower are 20 °C and 35 °C, respectively.

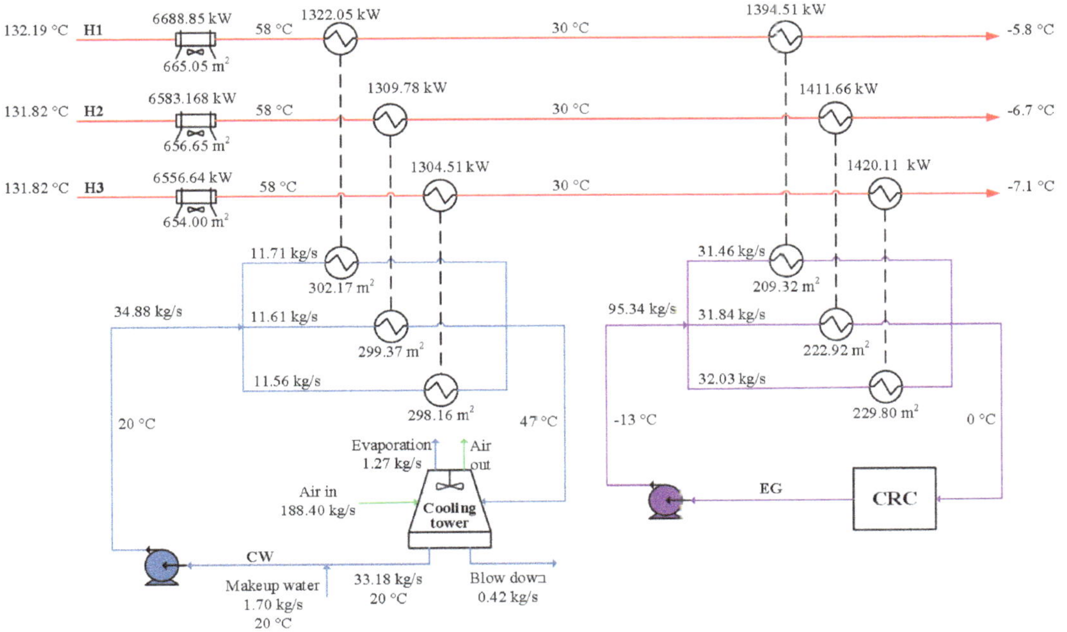

Figure 6. The base-case cooling system of hot streams.

It can be seen from Figure 6 that the inlet and outlet temperature of the cooling water are 20 °C and 47 °C, and the temperature breakpoints of air cooler, water cooler, and compression refrigeration cooler are 58 °C and 30 °C, which are the minimum allowable outlet temperatures of air cooler and water cooler. The purpose is to reduce the heat load of compression refrigeration as much as possible, so as to reduce the operation cost of the compression refrigeration cycle. At the same time, it can reduce the flow rate of the cooling water as well as the costs of related pumps and pipelines

The operation cost of the compression refrigeration cycle in this case is USD 2,514,212·y^{-1}, representing 55.80% of the total annual cost and 81.55% of the total operation cost of the cooling system. In addition, the total heat load of the air cooler is 19,829 kW, in which a large amount of low-grade waste heat available for recovery is directly released into the environment. The total heat load of the compression refrigeration heat exchanger is 4226 kW, and the power of the compressor is 1509 kW. This process will consume a lot of electric energy, resulting in the large operation cost of the compression refrigeration cycle. In the base-case cooling system, there are problems such as the loss of low-grade waste heat resources and large energy consumption of refrigerating. How to recover and utilize low-grade waste heat and reduce cooling energy consumption of the cooling system are important questions for improving the economy of the cooling system.

5.2. The Cascade Cooling System of Hot Streams

The cascade cooling system of hot streams consists of waste heat recovery, air cooling, water cooling, absorption refrigeration, and compression refrigeration. In this system, hot water is used as an intermediate medium to recover waste heat from hot streams and drives the LiBr/H_2O absorption refrigeration cycle, providing sub-ambient cooling capacity for the hot streams cooling process. In this way, the heat load of air cooling, water cooling, and compression refrigeration can be reduced, as well as the operation cost of the compression refrigeration cycle.

Figure 7a,b is used to indicate the heat load distribution of the base-case and cascade cooling systems, taking hot stream H1 as an example. It can be seen from Figure 7a that large amounts of low-grade waste heat are released into the environment form air coolers, and the heat load of compression refrigeration occupies about 15% of the total heat load, resulting in a large energy consumption of the compressor, i.e., the operation cost of the compressor is large. Compared to the base-case cooling system, the cascade cooling system can recover 11.22% of waste heat and provides 8.59% of sub-ambient cooling capacity through the absorption refrigeration cycle. The heat load of compression refrigeration is reduced by 8.59%, which means a lower operation cost of the compressor.

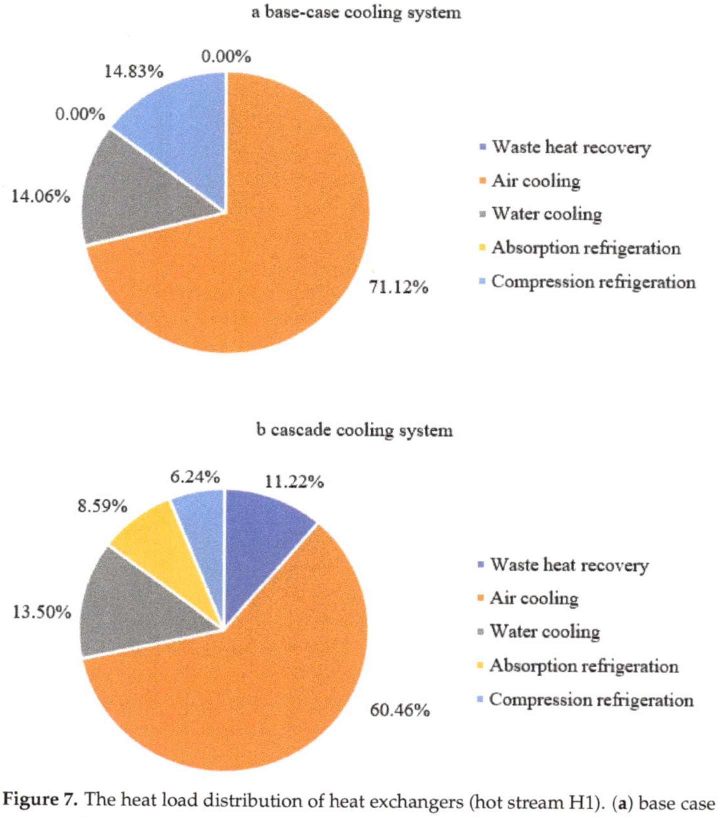

Figure 7. The heat load distribution of heat exchangers (hot stream H1). (**a**) base case cooling system; (**b**) cascade cooling system.

The optimal cascade cooling system of hot streams after optimization is shown in Figure 8. The heat load distribution and temperature breakpoints, the temperature of the cooling medium, and the area of the heat exchangers are also indicated.

As can be seen from Figures 6 and 8, the temperature break point between air cooling and water cooling in the base-case and cascade cooling systems has a small change of only 1 °C. In the cascade cooling system, the temperature break points between air cooling, absorption refrigeration, and compression refrigeration are 30 °C and 10 °C, respectively. The aim was to reduce the heat load of compression refrigeration as much as possible, thus reducing the energy consumption and cost associated with compression refrigeration. In the hot stream cooling process, the cascade cooling system recovered 3139 kW of low-grade waste heat and then generated a refrigerating capacity of 2404 kW through the absorption refrigeration cycle, accounting for 56.87% of the total refrigerating capacity, which effectively reduced the refrigeration load of the compression refrigeration cycle.

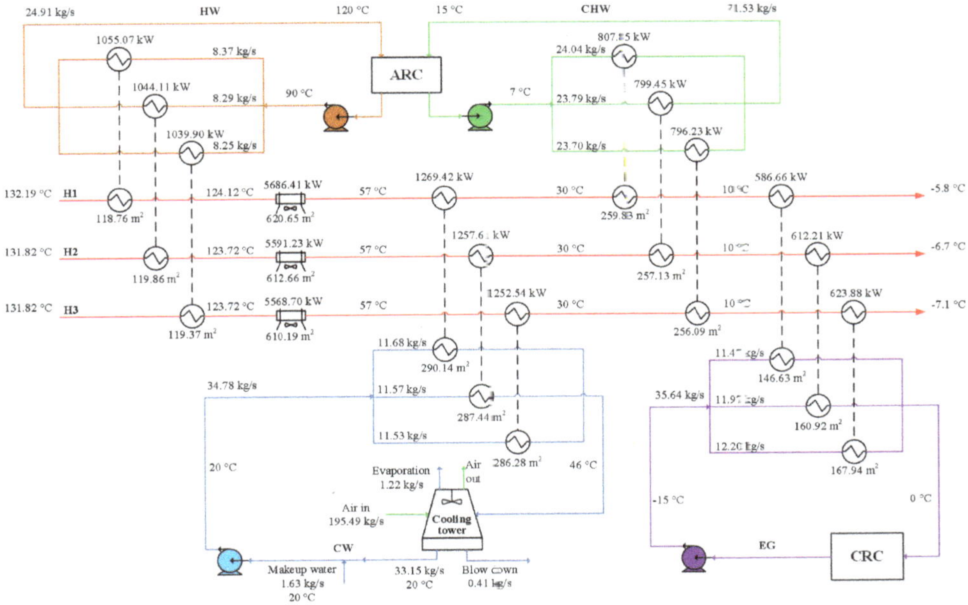

Figure 8. The optimal design of the cascade cooling system for hot streams.

A comparison of the system costs between the base case and the cascade cooling system is shown in Table 5. Compared to the base case, the TAC of the cascade cooling system was reduced by USD 931,025·y^{-1}, primarily due to the reduction of USD 1,305,110·y^{-1} in the TOC of the system. The addition of waste heat recovery and an absorption refrigeration heat exchanger reduced the heat load of air cooling and the compression refrigeration greatly, reducing the operation cost of the air cooler by USD 11,610·y^{-1} and of the compression refrigeration cycle by USD 1,335,272·y^{-1}, which also greatly reduced the total annual cost of the system. However, the operation cost of the compression refrigeration cycle still accounted for 32.98% of the TAC and 66.30% of the TOC, indicating that the compression refrigeration cycle was still the main energy consumption part of the cascade cooling system. The TCC of the system increased by USD 1,255,316, mainly because of the additional investment in an absorption refrigeration cycle, waste heat recovery heat exchangers, absorption refrigeration heat exchangers, and relevant pumps and pipelines. Compared to the base case, the capital cost of the heat exchanger in a cascade cooling system increased by USD 209,630, and the total area of heat exchangers increased from 3537 m^2 to 4314 m^2.

The parameters related to the refrigeration cycle in the base case and the cascade cooling system are shown in Table 6. Compared with the base case, the COP of the compression refrigeration cycle decreased. As a result of the reduction of the compression refrigeration heat load, the flow rates of the refrigerant, secondary refrigerant, and cooling water required were greatly reduced. The capital cost of CRC was reduced by USD 412,115, the operation cost of CRC was reduced by USD 1,335,272·y^{-1}, and the corresponding pump and cooling tower costs were reduced. The cascade cooling system used both absorption refrigeration and compression refrigeration to achieve the refrigerating purpose, the cooling water used was 213.1 kg·s^{-1}, which is 61.3 kg·s^{-1} less than that of the base case only using compression refrigeration.

Table 5. Comparison of results between base-case and cascade cooling systems.

Items	Base-Case Cooling System	Cascade Cooling System
TAC (USD·y^{-1})	4,505,811	3,574,786
TCC (USD)	4,773,854	6,029,170
TOC (USD·y^{-1})	3,083,203	1,778,093
CC_{HEX} (USD)	760,560	970,190
CC_{PUMP} (USD)	203,819	384,554
CC_{PIPE} (USD)	2,894,782	3,961,743
CC_{AR} (USD)	-	208,396
CC_{CR} (USD)	893,692	481,577
CC_{CT1} (USD)	10,267	10,026
CC_{CT2} (USD)	-	7282
CC_{CT3} (USD)	10,734	5403
OC_{PUMP} (USD·y^{-1})	188,834	267,727
OC_{AC} (USD·y^{-1})	77,193	65,583
OC_{AR} (USD·y^{-1})	-	68
OC_{CR} (USD·y^{-1})	2,514,212	1,178,940
OC_{CT1} (USD·y^{-1})	78,092	78,946
OC_{CT2} (USD·y^{-1})	-	98,258
OC_{CT3} (USD·y^{-1})	224,873	88,572
Fan power (kW)	316	276
Compressor power (kW)	1509	982

Table 6. Parameters of the refrigeration cycle.

Items	CRC in Base Case	ARC in Cascade System	CRC in Cascade System
COP	2.8016	0.7657	2.5768
Generating temperature (°C)	-	110	-
Evaporating temperature (°C)	−15	5	−17
Condensing temperature (°C)	35	45	36
Evaporating pressure (Pa)	163,940	873	150,837
Condensing pressure (Pa)	886,981	9590	911,849
Refrigerant flow rate (kg·s^{-1})	30.05	7.18	13.22
Secondary refrigerant flow rate (kg·s^{-1})	95.34	71.53	35.64
Cooling water inlet temperature (°C)	27	30	27
Cooling water outlet temperature (°C)	32	42	33
Cooling water flow rate (kg·s^{-1})	274.39	112.24	100.88

The base case consumed 18,969,640 kWh of electricity per year, and the cascade cooling system consumed 11,168,820 kWh·y^{-1}. Using the cascade cooling system saved 7,800,820 kWh of electricity per year and about 3120 t of standard coal, and it reduced 7778 t of carbon dioxide emissions.

In order to explore the optimal structure of cascade cooling system under different industrial conditions, the cascade cooling system of the same case was optimized in different areas; Nanchang and Xi'an were selected respectively. Unlike Xi'an, where water is extremely scarce, Nanchang is rich in water; hence, the price of fresh water in Nanchang is much lower than in Xi'an, but the electric charge is slightly higher. The costs of the cascade cooling system in two places after optimization are shown in Table 7.

Table 7. Comparison of cascade cooling system results in different regions.

Items	Nanchang	Xi'an
Pe (USD·kWh^{-1})	0.096	0.083
Pw (USD·t^{-1})	0.344	0.843
TAC (USD·y^{-1})	2,941,456	2,836,840
TCC (USD)	6,063,125	5,976,545
TOC (USD·y^{-1})	1,134,644	1,055,830
CC_{HEX} (USD)	968,291	981,735
CC_{PUMP} (USD)	385,700	384,757
CC_{PIPE} (USD)	213,689	208,396
CC_{AR} (USD)	447,170	447,170
CC_{CR} (USD)	10,135	9783
CC_{CT1} (USD)	6827	7282
CC_{CT2} (USD)	5189	5189
CC_{CT3} (USD)	4,026,123	3,932,234
OC_{PUMP} (USD·y^{-1})	174,547	148,380
OC_{AC} (USD·y^{-1})	50,212	36,624
OC_{AR} (USD·y^{-1})	43	37
OC_{CR} (USD·y^{-1})	732,608	633,401
OC_{CT1} (USD·y^{-1})	53,929	74,645
OC_{CT2} (USD·y^{-1})	68,128	99,737
OC_{CT3} (USD·y^{-1})	55,178	66,006

According to the results, the TAC of Nanchang system is USD 168,142·y^{-1} higher than that of Xi'an system, the TOC is USD 156,558·y^{-1} higher than that of the Xi'an system, and the TCC is only USD 38,872·y^{-1} higher. The main reason is that the electricity price in Nanchang is higher than that in Xi'an; hence, the operation cost of compression refrigeration cycle is USD 186,324·y^{-1} higher, and the operation cost of the pumps is 24,096 USD·y^{-1} higher than that in Xi'an. Because the fresh water price in Xi'an is much higher than that in Nanchang, the operation costs of cooling towers in the Xi'an system are higher than those in the Nanchang system. It also can be seen from the results that the fresh water price has a small impact on the system cost, while the electricity price has a greater influence on the system cost.

5.3. The Sensitivity Analysis for the Cascade Cooling System

In this section, sensitivity analyses of the electricity charge and fresh water charge were carried out to study their influences on the cascade cooling system. The influences of the electricity charge and fresh water charge on the TCC, TOC, and TAC are shown in Figure 9a,b.

Both the electricity charge and the fresh water charge had little influence on TCC. TOC and TAC increased with the electricity charge, while they increased a little with the fresh water charge. That is because the electricity charge is related to all the operation costs, especially the operation cost of the compressor, but the fresh water charge is only related to the operation cost of the cooling tower. Fresh water was only used as the make-up water in the cooling tower, and its mass flowrate is small in this system; hence, it had little influence on the cost of the system. Figure 9 also shows the change of the operation cost of the compression refrigeration cycle in the cascade cooling system with electricity and fresh water prices. It increased with the increase of the electric charge, while it changed very little with the fresh water charge.

Taking the impact of electric charge and fresh water charge on the total cost of the system into account, as well as the comparative analysis of cascade cooling systems in different regions above, the cascade cooling system proposed in this paper is less dependent on water resources, and can be applied in areas with abundant and scarce water resources. However, it has more advantages in areas with a low electricity price. It is noted that the environment temperature was assumed to be the same in all situations.

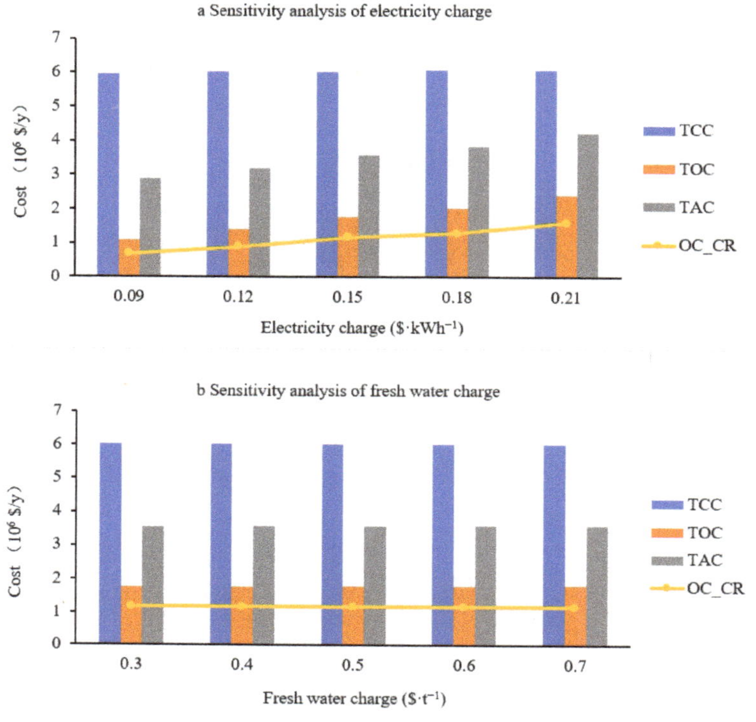

Figure 9. Sensitivity analysis of the cascade cooing system. (**a**) sensitivity analysis of electricity charge; (**b**) sensitivity analysis of fresh water charge.

5.4. The Comparison of Optimization Method with Using MATLAB Only

In this paper, Aspen and MATLAB were combined to optimize the cascade cooling system. The shortcoming of this optimization method is its low computational efficiency mainly because the interaction process between Aspen and MATLAB consumes too much time. There is another optimization method to solve the problem, which uses MATLAB only. In this method, the process simulated in Aspen, such as the cooling process of hot streams, the absorption refrigeration cycle, and the compression refrigeration cycle are modeled and calculated in MATLAB. The method using MATLAB can greatly save computation time and improve optimization efficiency.

It can be seen from Table 8 that the population size and maximum generation in the method using MATLAB is 10 times higher than that in the method using Aspen and MATLAB. However, the computation time of the method using Aspen and MATLAB is 94.57 h, which is 180 times the computation time of the method using MATLAB. Meanwhile, in the base-case cooling system, the number of variables is 16, but the computation time using Aspen and MATLAB is 72.39 h. Thus, the reduction of the number of variables is helpful to improve the computational efficiency of the method using Aspen and MATLAB, but the main reason for the low efficiency of this method is still that the interaction between Aspen and MATLAB consumes too much time.

Table 8. Parameter comparison of the optimization methods.

	Aspen and MATLAB	MATLAB
Population size	30	300
Maximum generation	500	5000
Computation time (s)	340,471	1884

The method combining Aspen and MATLAB directly models the processes. By selecting the appropriate physical property method, the required data, such as temperature, pressure, and heat load, can be read directly after simulation and calculation. This method is intuitive and convenient. The method using MATLAB only requires not only the establishment of the thermodynamic model of the processes, but also the modeling and solving of the related physical properties of the hot streams and refrigeration working medium, which is a complicated and cumbersome process.

The modeling of the hot streams cooling process, absorption refrigeration cycle, and compression refrigeration cycle is different between the two optimization methods, which results in different calculation results. Aspen and MATLAB combined use a series heat exchanger model to model the cooling process of the heat streams; the heat load of heat exchangers can be obtained directly through the simulation and calculation of Aspen after assigning the outlet temperature of the heat exchanger. The method using only MATLAB calculates the heat load through the enthalpy difference between the inlets and outlets of heat exchangers. Although the enthalpy calculation equation of the hot streams is obtained through the data fitting from Aspen, there is a certain error between the fitting equation and the simulation data; hence, the calculation result is not as accurate as that in Aspen.

Table 9 shows the heat load of the heat exchangers obtained by the two optimization methods which takes H1 as an example. The total heat load of Aspen and MATALB is 9,405.41 kW, and that of MATLAB is 9,498.69 kW. This constitutes an increase of 93.28 kW, and the margin of error is 0.99%. Meanwhile, compared to Aspen and MATLAB, although the temperature breakpoints between water cooling, absorption refrigeration, and compression refrigeration are the same, the heat load of the heat exchangers in the method using MATLAB is smaller than that in Aspen and MATLAB. The heat load of the absorption refrigeration heat exchanger is reduced by 27.36 kW, and the error is 3.39%. The heat load of the compression refrigeration heat exchanger is reduced by 26.08 kW with an error of 4.45%.

Table 9. Heat exchanger load of cascade cooling system (H1).

Heat Exchanger Load (kW)	Waste Heat Recovery	Air Cooling	Water Cooling	Absorption Refrigeration	Compression Refrigeration	Total Heat Load
Aspen and MATLAB	1055.07	5686.41	1269.42	807.85	586.67	9405.41
MATLAB	1062.51	5692.98	1402.14	780.49	560.58	9498.69

In addition, the calculations of the ARC and CRC, as well as the physical properties of the refrigeration working medium in the two optimization methods are different. Therefore, the calculation results of the two optimization methods are different. Table 10 shows the costs comparison of the cascade cooling system optimized by different methods.

As can be seen from Table 10, the difference of cascade cooling system costs obtained by the two optimization methods is mainly the investment in and operation cost of the absorption refrigeration cycle and compression refrigeration cycle. Compared to the method using MATLAB only, the capital cost of ARC in Aspen and MATLAB is USD 29,946 higher, the capital cost of CRC is USD 96,195 higher, and the operation cost is USD 27,808$\cdot$y^{-1} higher. Costs of cooling water, pipeline, and heat exchangers related to the two refrigeration cycles increase, but other costs have a small change.

Table 10. Comparison of optimization results of cascade cooling system with different methods.

Items	Aspen and MATLAB	MATLAB
TAC ($USD \cdot y^{-1}$)	3,574,786	3,475,097
TCC (USD)	6,029,170	5,842,860
TOC ($USD \cdot y^{-1}$)	1,778,093	1,733,924
CC_{HEX} (USD)	970,190	925,009
CC_{PUMP} (USD)	384,554	384,068
CC_{PIPE} (USD)	3,961,743	3,947,379
CC_{AR} (USD)	208,396	178,450
CC_{CR} (USD)	481,577	385,382
CC_{CT1} (USD)	10,026	10,483
CC_{CT2} (USD)	7282	6737
CC_{CT3} (USD)	5403	5352
OC_{PUMP} ($USD \cdot y^{-1}$)	267,727	265,594
OC_{AC} ($USD \cdot y^{-1}$)	65,583	65,304
OC_{AR} ($USD \cdot y^{-1}$)	68	55
OC_{CR} ($USD \cdot y^{-1}$)	1,178,940	1,151,132
OC_{CT1} ($USD \cdot y^{-1}$)	78,946	72,229
OC_{CT2} ($USD \cdot y^{-1}$)	98,258	89,065
OC_{CT3} ($USD \cdot y^{-1}$)	88,572	90,545

6. Conclusions

In this paper, the cascade cooling system containing waste heat recovery, air cooling, water cooling, absorption refrigeration, and compression refrigeration is proposed and optimized. An $LiBr/H_2O$ absorption refrigeration cycle was used to recycle low-grade waste heat and provide sub-ambient cooling capacity for the hot streams cooling process.

The optimal design of the cascade cooling system was determined based on minimum TAC. The optimal cascade cooling system has a TAC of USD $3{,}574{,}786 \cdot y^{-1}$ and saves USD $931{,}025 \cdot y^{-1}$ compared to the base-case cooling system, which is a 20.66% reduction. As can be seen from the results of both the base-case and the cascade cooling system, the compressor is the main energy consumption of the cooling system, and the operation cost of the compressor accounts for more than 50% and 30% of the TAC in the base-case and cascade cooling systems, respectively. In the cascade cooling system, the absorption refrigeration cycle can recycle 3139.08 kW of waste heat and provide 2403.53 kW of sub-ambient cooling capacity. The COP of the absorption refrigeration cycle and compression refrigeration cycle are 0.7657 and 2.5768. Using a cascade cooling system can save 7,800,820 kWh of electricity per year. The cascade cooling system proposed can achieve good energy savings and economic benefits.

A sensitivity analysis of electricity prices and fresh water prices was also carried out for the cascade cooling system. The price of electricity mainly influences the operation costs of the cascade cooling system, especially the operation cost of the compressor, while the price of fresh water only influences the operation costs of the cooling towers. The electricity charge has a large influence on the TAC of the system, but the fresh water charge has little influence on the TAC. The comparative analysis of the cascade cooling system in different regions showed that the cascade cooling system is less dependent on water resources and has more advantages in areas with low electric charge.

Compared to the method using MATLAB only, the optimization method combining Aspen and MATLAB has the advantages of convenient modeling and accurate calculation results. However, this method has some disadvantages. Firstly, the optimization calculation efficiency using Aspen and MATLAB is low, mainly because the interaction between Aspen and MATLAB is too time-consuming, and Aspen runs the calculation slowly. In addition, it is difficult to optimize convergence and obtain ideal results.

This work considers more cooling methods in a traditional cooling system. To achieve the cooling duty distribution obtained by the proposed method, the control system must be enhanced. This part will be studied in future.

Author Contributions: Conceptualization, Y.W. (Yufei Wang); methodology, S.Y. and Y.W. (Youlei Wang); software, S.Y.; validation, Y.W. (Youlei Wang); formal analysis, S.Y.; investigation, Y.W. (Youlei Wang); resources, Y.W. (Yufei Wang); writing—original draft preparation, S.Y. and Y.W. (Youlei Wang); writing—review and editing, Y.W. (Yufei Wang); supervision, Y.W. (Yufei Wang); funding acquisition, Y.W. (Yufei Wang). All authors have read and agreed to the published version of the manuscript.

Funding: Financial support from the National Natural Science Foundation of China under Grant (No. 22022816 and 22078358) are gratefully acknowledged.

Institutional Review Board Statement: Not applicable.

Informed Consent Statement: Not applicable.

Conflicts of Interest: The authors declare no conflict of interest.

Nomenclature

a, b, c	Constants of heat exchanger capital cost
A	Area of heat exchanger, m^2
A_1, A_2, A_3, A_4	Parameters of pipe capital cost
Af	Annualized factor
$Approach$	Temperature difference between cooling tower outlet temperature and air bulb temperature, °C
$B_{blowdown}$	Mass flowrate of blowdown water in cooling tower, $kg \cdot s^{-1}$
CC	Capital cost, USD
C_{factor}	Tower fan factor
COP	Coefficient of performance for refrigeration cycle
Cp_{CW}	Specific heat capacity of cooling water, $kJ \cdot (kg \cdot °C)^{-1}$
Cp_m	Specific heat capacity of cooling medium, $kJ \cdot (kg \cdot °C)^{-1}$
dt_{in}	Inlet temperature difference of heat exchanger, °C
dt_{out}	Outlet temperature difference of heat exchanger, °C
D_{in}	Pipe inner diameter, m
D_{out}	Pipe outer diameter, m
Dt_{in}	Inner diameter of heat exchanger, m
Dt_{out}	Outer diameter of heat exchanger, m
E_{vop}	The amount of water evaporation in cooling tower, $kg \cdot s^{-1}$
F_{air}	Mass flowrate of air in cooing tower, $kg \cdot s^{-1}$
$f_{friction}$	Factor of friction
ft	Total mass flowrate of cooling water, $kg \cdot s^{-1}$
g	Gravitational constant
G	Mass velocity rate of air, $kg \cdot (m^2 \cdot s)^{-1}$
G_{max}	Maximum mass velocity rate of air, $kg \cdot (m^2 \cdot s)^{-1}$
h_a	Film transfer coefficient of air, $W \cdot (m^2 \cdot °C)^{-1}$
hin	Film transfer coefficient inside of the heat exchanger, $W \cdot (m^2 \cdot °C)^{-1}$
$hout$	Film transfer coefficient outside of the heat exchanger, $W \cdot (m^2 \cdot °C)^{-1}$
h_t	Film transfer coefficient, $W \cdot (m^2 \cdot °C)^{-1}$
Hy	Plant operation time, $s \cdot y^{-1}$
ΔH_{EVAP}	Enthalpy change of evaporator in compression refrigeration cycle, $kJ \cdot kg^{-1}$
K	Total heat transfer coefficient, $W \cdot (m^2 \cdot °C)^{-1}$
K_t	Constant for heat exchanger tube side pressure drop
L	Transmission distance, m
i	Hot streams
I	Annual interest rate
Iin	Inlet specific enthalpy of hot stream, $kJ \cdot kg^{-1}$
$Iout$	Outlet specific enthalpy of hot stream, $kJ \cdot kg^{-1}$
j	Cooling medium
m	Mass flow rate of cooling medium, $kg \cdot s^{-1}$
M	Mass flow rate of hot stream, $kg \cdot s^{-1}$
$M_{AR\text{-}CW}$	Mass flow rate of cooling water in absorption refrigeration, $kg \cdot s^{-1}$

M_{LiBr}	Mass flow rate of working fluid in absorption refrigeration, kg·s^{-1}
M_{makeup}	Mass flowrate of make-up water in cooling tower, kg·s^{-1}
M_{ref}	Mass flowrate of refrigerant in compression refrigeration cycle, kg·s^{-1}
MW_{air}	Molecular weight of air
MW_w	Molecular weight of water
n	Lifetime of equipment, y
N_b	Number of bundles
OC	Operation cost, USD·y^{-1}
P_a	Local atmospheric pressure, Pa
P_{cond}	Pressure of condenser in compression refrigeration cycle, MPa
P_{evap}	Pressure of evaporator in compression refrigeration cycle, MPa
P_{fan}	Air cooler fan power consumption, kW
P_s	Vapor pressure, Pa
$Pcul$	Capital cost of pope per unit length, USD·m^{-1}
Pe	Unit cost of electricity, USD·(kWh)$^{-1}$
Pw	Unit cost of fresh water, USD·t^{-1}
Δp_{air}	Air cooler fan pressure drop, Pa
ΔP	Pump pressure drop, Pa
ΔP_t	Tube side pressure drop, Pa
q_{GEN}	Heat load of generator when working fluid flowrate is 1 kg·s^{-1}, kW
Q	Heat load of heat exchanger, kW
$Range$	Difference between cooling tower inlet and outlet temperature, °C
Re	Reynolds number
tin	Inlet temperature of cooling medium, °C
$tout$	Outlet temperature of cooling medium, °C
TAC	Total annual cost, USD·y^{-1}
TCC	Total capital cost, USD
TOC	Total operation cost, USD·y^{-1}
$T_{ambient}$	Ambient temperature, °C
T_{cin}	Cooling water inlet temperature of cooling tower, °C
T_{cout}	Cooling water outlet temperature of cooling tower, °C
T_{CW1}	Inlet temperature of cooling water of absorber, °C
T_{CW2}	Outlet temperature of cooling water of absorber, °C
T_{CW3}	Outlet temperature of cooling water of condenser, °C
T_{mean}	Mean temperature of cooling tower, °C
T_{wb}	Air bulb temperature, °C
T_{in}	Inlet temperature of hot stream, °C
T_{out}	Outlet temperature of hot stream, °C
ΔT_m	Logarithmic mean temperature difference of heat exchanger, °C
ΔT_{min}	Minimum temperature approach difference of heat exchanger, °C
u	The flow velocity, m·s^{-1}
U	Total heat transfer coefficient of heat exchanger, kW·(m^2·°C)$^{-1}$
V_{air}	Volumetric flow rate of air, m^3·s^{-1}
V_F	Face velocity of air cooler, m·s^{-1}
V_{NF}	Actual face velocity of air cooler, m·s^{-1}
w_{in}	Air inlet humidity of cooling tower
w_{out}	Air outlet humidity of cooling tower
W_{Comp}	Non-isentropic work of compressor, kW
Wt	Pipe weight per unit length, kg·m^{-1}
Greek letter	
α, β, γ	Parameters of pump capital cost
η_{el}	Electrical efficiency
η_{fan}	Air cooler's fan efficiency
$\eta_{fan\text{-}tower}$	Cooling tower fan efficiency
η_{isen}	Isentropic efficiency for compressor
η_{mech}	Mechanical efficiency
η_{Pump}	Pump efficiency
κ	Heat conductivity, W·(m·K)$^{-1}$

λ	Friction factor
μ	Viscosity, Pa·s
ρ	Density of cooling medium, kg m^{-3}
ρ_{air}	Density of air, kg·m^{-3}
π_c	Cycle of concentration for cooling tower
Subscripts	
ABS	Absorber
AC	Air cooler
AR	Absorption refrigeration cycle
Comp	Compressor
COND	Condenser
CR	Compression refrigeration cycle
CT	Cooling tower
EVAP	Evaporator
fan-tower	Fan for cooling tower
GEN	Generator
HEX	Heat exchanger
Pipe	Pipeline
Pump	Pumps for cooling mediums
Pump-CW	Pump for cooling water used in absorption refrigeration cycle
Pump-LiBr	Pump for LiBr/H_2O solution in absorption refrigeration cycle

References

1. Doodman, A.; Fesanghary, M.; Hosseini, R. A robust stochastic approach for design optimization of air cooled heat exchangers. *Appl. Energy* **2009**, *86*, 1240–1245. [CrossRef]
2. Manassaldi, J.; Scenna, N.; Mussati, S. Optimization mathematical model for the detailed design of air cooled heat exchangers. *Energy* **2014**, *64*, 734–746. [CrossRef]
3. Fahmy, M.; Nabih, H. Impact of ambient air temperature and heat load variation on the performance of air-cooled heat exchangers in propane cycles in LNG plants—Analytical approach. *Energy Convers. Manag.* **2016**, *121*, 22–35. [CrossRef]
4. Chen, L.; Yang, L.; Du, X. Anti-freezing of air-cooled heat exchanger by air flow control of louvers in power plants. *Appl. Therm. Eng.* **2016**, *106*, 537–550. [CrossRef]
5. Kuruneru, S.; Sauret, E.; Saha, S. Numerical investigation of the temporal evolution of particulate fouling in metal foams for air-cooled heat exchangers. *Appl. Energy* **2016**, *184*, 531–547. [CrossRef]
6. Kim, J.; Smith, R. Cooling water system design. *Chem. Eng. Sci.* **2001**, *56*, 3641–3658. [CrossRef]
7. Panjeshahi, M.; Ataei, A. Application of an environmentally optimum cooling water system design to water and energy conservation. *Int. J. Environ. Sci. Technol.* **2008**, *5*, 251–262. [CrossRef]
8. Castro, M.; Song, T.; Pinto, J. Minimization of Operational Costs in Cooling Water Systems. *Chem. Eng. Res. Des.* **2000**, *78*, 192–201. [CrossRef]
9. Ponce-Ortega, J.; Serna-Gonzalez, M.; Jimenez-Gutierrez, A. MINLP synthesis of optimal cooling networks. *Chem. Eng. Sci.* **2007**, *62*, 5728–5735. [CrossRef]
10. Srikhirin, P.; Aphornratana, S.; Chungpaibulpatana, S. A review of absorption refrigeration technologies. *Renew. Sustain. Energy Rev.* **2001**, *5*, 343–372. [CrossRef]
11. Sun, J.; Fu, L.; Zhang, S. A review of working fluids of absorption cycles. *Renew. Sustain. Energy Rev.* **2012**, *16*, 1899–1906. [CrossRef]
12. Karamangil, M.; Coskun, S.; Kaynakli, O. A simulation study of performance evaluation of single-stage absorption refrigeration system using conventional working fluids and alternatives. *Renew. Sustain. Energy Rev.* **2010**, *14*, 1969–1978. [CrossRef]
13. Kaynakli, O.; Kilic, M. Theoretical study on the effect of operating conditions on performance of absorption refrigeration system. *Energ. Convers. Manag.* **2007**, *48*, 599–607. [CrossRef]
14. Ebrahimi, K.; Jones, G.; Fleischer, S. Thermo-economic analysis of steady state waste heat recovery in data centers using absorption refrigeration. *Appl. Energy* **2015**, *139*, 384–397. [CrossRef]
15. Kalinowski, P.; Hwang, Y.; Radermacher, R.; Al Hashimi, S.; Rodgers, P. Application of waste heat powered absorption refrigeration system to the LNG recovery process. *Int. J. Refrig.* **2009**, *32*, 687–694. [CrossRef]
16. Zhang, B.; Zhang, Z.; Liu, K. Network Modeling and Design for Low Grade Heat Recovery, Refrigeration, and Utilization in Industrial Parks. *Ind. Eng. Chem. Res.* **2016**, *55*, 9725–9737. [CrossRef]
17. Salmi, W.; Vanttola, J.; Elg, M. Using waste heat of ship as energy source for an absorption refrigeration system. *Appl. Therm. Eng.* **2017**, *115*, 501–516. [CrossRef]
18. Yang, S.; Deng, C.; Liu, Z. Optimal design and analysis of a cascade LiBr/H2O absorption refrigeration/transcritical CO2 process for low-grade waste heat recovery. *Energy Convers. Manag.* **2019**, *192*, 232–242. [CrossRef]

19. Dalkilic, A.; Wongwises, S. A performance comparison of vapour-compression refrigeration system using various alternative refrigerants. *Int. Commun. Heat Mass* **2010**, *37*, 1340–1349. [CrossRef]
20. Ahamed, J.; Saidur, R.; Masjuki, H. A review on exergy analysis of vapor compression refrigeration system. *Renew. Sustain. Energy Rev.* **2011**, *15*, 1593–1600. [CrossRef]
21. Arora, A.; Kaushik, S. Theoretical analysis of a vapour compression refrigeration system with R502, R404A and R507A. *Int. J. Refrig.* **2008**, *31*, 998–1005. [CrossRef]
22. Selbas, R.; Kizilkan, O.; Sencan, A. Thermoeconomic optimization of subcooled and superheated vapor compression refrigeration cycle. *Energy* **2006**, *31*, 2108–2128. [CrossRef]
23. Kaushik, S.; Arora, A. Energy and exergy analysis of single effect and series flow double effect water-lithium bromide absorption refrigeration systems. *Int. J. Refrig.* **2009**, *32*, 1247–1258. [CrossRef]
24. Yerramsetty, K.; Murty, C. Synthesis of cost-optimal heat exchanger networks using differential evolution. *Comput. Chem. Eng.* **2008**, *32*, 1861–1876. [CrossRef]
25. Chang, C.; Wang, Y.; Feng, X. Indirect heat integration across plants using hot water circles. *Chin. J. Chem. Eng.* **2015**, *23*, 992–997. [CrossRef]
26. Gu, Y. Friction factor of fluids in pipes. *Chem. Eng.* **1936**, *3*, 3–14.
27. Soltani, H.; Shafiei, S. Heat exchanger networks retrofit with considering pressure drop by coupling genetic algorithm with LP (linear programming) and ILP (integer linear programming) methods. *Energy* **2011**, *36*, 2381–2391. [CrossRef]
28. Zhu, X.; Nie, X. Pressure drop considerations for heat exchanger network grassroots design. *Comput. Chem. Eng.* **2002**, *26*, 1661–1676. [CrossRef]
29. Stijepovic, M.; Linke, P. Optimal waste heat recovery and reuse in industrial zones. *Energy* **2011**, *36*, 4019–4031.
30. Briggs, D.; Young, E. Convection Heat Transfer and Pressure Drop of Air Flowing Across Triabgular Pitch Banks of Finned Tubes. *Chem. Eng. Prog. Symp. Ser.* **1963**, *59*, 1–10.
31. Wang, M.; Wang, Y.; Feng, X. Energy Performance Comparison between Power and Absorption Refrigeration Cycles for Low Grade Waste Heat Recovery. *ACS Sustain. Chem. Eng.* **2018**, *6*, 4614–4624. [CrossRef]
32. Reid, R.; Prausnitz, J.; Poling, B. *The Properties of Gases & Liquids*; McGraw-Hill: New York, NY, USA, 1988.
33. Shi, S.; Wang, Y.; Wang, Y.; Feng, X. A new optimization method for cooling systems considering low-temperature waste heat utilization in a polysilicon industry. *Energy* **2021**, *238*, 121800. [CrossRef]
34. Kilicarslan, A.; Muller, N. A comparative study of water as a refrigerant with some current refrigerants. *Int. J. Eenrgy Res.* **2005**, *29*, 947–959. [CrossRef]
35. Chen, Q. Simulation of a vapour-compression refrigeration cycles using HFC134A and CFC12. *Int. Commun. Heat Mass* **1999**, *26*, 513–521. [CrossRef]
36. Sayyaadi, H.; Nejatolahi, M. Multi-objective optimization of a cooling tower assisted vapor compression refrigeration system. *Int. J. Refrig.* **2011**, *34*, 243–256. [CrossRef]
37. Panjeshahi, M.; Ataei, A.; Gharaie, M. Optimum design of cooling water systems for energy and water conservation. *Chem. Eng. Res. Des.* **2009**, *87*, 200–209. [CrossRef]
38. Kim, J.; Smith, R. Automated retrofit design of cooling-water systems. *AICHE J.* **2003**, *49*, 1712–1730. [CrossRef]
39. Lee, U.; Mitsos, A.; Han, C. Optimal retrofit of a CO2 capture pilot plant using superstructure and rate-based models. *Int. J. Grennh. Gas Control* **2016**, *50*, 57–69. [CrossRef]
40. Lee, U.; Jeon, J.; Han, C. Superstructure based techno-economic optimization of the organic rankine cycle using LNG cryogenic energy. *Energy* **2017**, *137*, 83–94. [CrossRef]
41. Ma, J.; Wang, Y.; Feng, X. Synthesis cooling water system with air coolers. *Chem. Eng. Res. Des.* **2018**, *131*, 643–655. [CrossRef]
42. Towler, G.; Sinnott, R. *Chemical Engineering Design: Principles, Practice and Economics of Plant and Process Design*; Butterworth-Heinemann: Oxford, UK, 2007.
43. Jabbari, B.; Tahouni, N.; Ataei, A. Design and optimization of CCHP system incorporated into kraft process, using Pinch Analysis with pressure drop consideration. *Appl. Therm. Eng.* **2013**, *61*, 88–97. [CrossRef]

MDPI
St. Alban-Anlage 66
4052 Basel
Switzerland
Tel. +41 61 683 77 34
Fax +41 61 302 89 18
www.mdpi.com

Processes Editorial Office
E-mail: processes@mdpi.com
www.mdpi.com/journal/processes

www.ingramcontent.com/pod-product-compliance
Lightning Source LLC
Chambersburg PA
CBHW040800250926
51850CB00007BA/1044